Minerals and Rocks   9

Editor in Chief
P. J. Wyllie, Chicago, IL

Editors
A. El Goresy, Heidelberg
W. von Engelhardt, Tübingen · T. Hahn, Aachen

Jochen Hoefs

Stable Isotope Geochemistry

Third, Completely Revised and Enlarged Edition

With 62 Figures

Springer-Verlag
Berlin Heidelberg NewYork
London Paris Tokyo

Professor Dr. JOCHEN HOEFS
Geochemisches Institut der Universität
Goldschmidtstraße 1
D-3400 Göttingen, FRG

Volumes 1 to 9 in this series appeared under the title
Minerals, Rocks and Inorganic Materials

ISBN 3-540-17341-2 3. Aufl. Springer-Verlag Berlin Heidelberg New York
ISBN 0-387-17341-2 3rd ed. Springer-Verlag New York Berlin Heidelberg

ISBN 3-540-09917-4 2. Aufl. Springer-Verlag Berlin Heidelberg New York
ISBN 0-387-09917-4 2nd ed. Springer-Verlag New York Berlin Heidelberg

Library of Congress Cataloging-in-Publication Data. Hoefs, Jochen. Stable
isotope geochemistry. (Minerals and rocks; 9). Bibliography: p. 1. Geochemistry.
2. Isotope geology. I. Title. II. Series. QE515.H54 1987 551.9 86-33876.

Typesetting, printing and binding: Brühlsche Universitätsdruckerei, Giessen
2132/3130-543210

Preface to the Third Edition

The first edition of this book was published in 1973, the second, totally rewritten, followed 7 years later in 1980. Because the field of stable isotopes is still growing and exerting an increasing influence on geosciences in general, it seems to be necessary, after a further 7 years, to revise the edition again accordingly.

Not only has the previous edition been updated, but two completely new chapters on the isotopic composition of mantle-derived material and on the isotopic composition of the ocean during the geologic past, have been added.

The references concentrate on recent literature. In some cases, older references have been omitted to save space. I do not intend to underrate the value of older publications, but only to keep the reference list — already very voluminous in relation to the total length — from becoming even larger.

An early draft has been reviewed by Russell Harmon and Alan Matthews. John Valley has sent me a preprint of an article on metamorphic rocks. To all three of them I owe my deepest thanks.

Göttingen, January 1987   Jochen Hoefs

Preface to the Second Edition

Since the first edition of this book appeared in 1973 knowledge in the field of isotope geochemistry has grown so fast that it appeared necessary to revise it accordingly. Although the main subdivisions have remained the same, the book has been totally revised and rewritten. Some reviewers of the first edition have criticized the subdivisions and proposed a more appropriate subdivision of the book along the line of different chemical elements. Since this book is mainly written for earth scientists and not for chemists, I believe the present subdivision to be more appropriate. I am fully aware that any subdivision is problematical and debatable (nature is indivisible), however, for practical purposes, geochemists have, for a long time, tried to subdivide the earth into certain "spheres". This book follows the classical scheme of subdivision with all its disadvantages, because I have no better one.

I am especially grateful to my colleagues who, during the various stages of the preparation of the manuscripts have read and criticized parts or whole drafts of the manuscript. I owe my deepest thanks to the following persons (in alphabetical order): W. Deuser (Woods Hole, Mass.), R. Harmon (East Kilbride, Scotland), T. Hoering (Washington, D.C.), H. Hubberten (Karlsruhe, FRG), Y. Kolodny (Jerusalem), J. O'Neil (Menlo Park, Calif.), B. Robinson (Wellington, N.Z.), W. Sackett (College Station, Texas), H. Sakai (Misasa, Japan), M. Schoell (Hannover, FRG), E. Usdowski (Göttingen, FRG).

However, I take, of course, full responsibility for any shortcomings.

Göttingen, January 1980   Jochen Hoefs

Contents

Chapter 1 Theoretical and Experimental Principles

1.1 General Characteristics of Isotopes

Isotopes are defined as atoms whose nuclei contain the same number of protons but a different number of neutrons. The term "isotopes" is derived from Greek (meaning equal places) and indicates that isotopes occupy the same position in the Periodic Table.

It is convenient to denote isotopes in the form ^m_nE, where the superscript m represents the mass number and the subscript n represents the atomic number of an element E. For example, $^{12}_6C$ is the isotope of carbon which has six protons and six neutrons in its nucleus. This isotope has been assigned an atomic weight of exactly 12. All atomic weights are referred to this standard carbon isotope. The atomic weights of naturally occurring elements are averages of the weights contributed by the various kinds of isotopic nuclei.

Isotopes can be divided into stable and unstable (radioactive) species. The number of stable isotopes is about 300; whilst over 1200 unstable ones have been discovered so far. The term "stable" is relative, depending on the detection limits of radioactive decay times. In the range of atomic numbers from 1 (H) to 83 (Bi), stable nuclides of all masses except 5 and 8 are known. Only 21 elements are pure elements, in the sense that they have only one stable isotope. All other elements are mixtures of at least two isotopes. In some elements, the different isotope may be present in substantial proportions. In copper, for example, $^{63}_{29}Cu$ accounts for 69% and $^{65}_{29}Cu$ accounts for 31%. In most cases one isotope is predominant, the others being present only in trace amounts.

The stability of nuclides is characterized by several important rules, two of which are briefly discussed here. The first is the so-called symmetry rule, which states that in a stable nuclide with low atomic number, the number of protons is approximately equal to the number of neutrons, or the neutron-to-proton ratio, N/Z, is approximately equal to unity. In stable nuclei with more than 20 protons or neutrons, the N/Z ratio is always greater than unity, with a maximum value of about

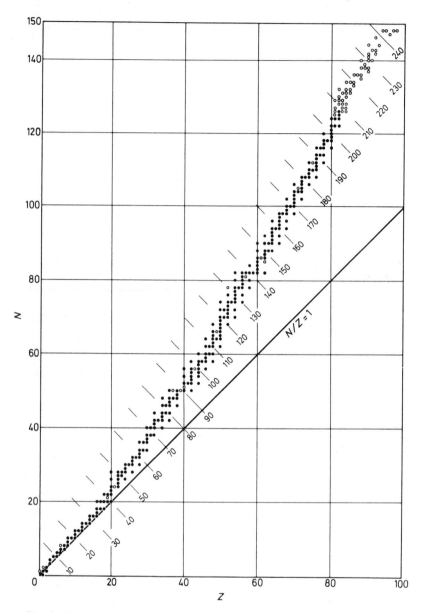

Fig. 1. Plot of number of protons (Z) and number of neutrons (N) in stable (●) and unstable (○) nuclides

Table 1. Types of atomic nuclei and their frequency of occurrence

Z–N combination	Number of stable nuclides
Even–even	160
Even–odd	56
Odd–even	50
Odd–odd	5

1.5 for the heaviest stable nuclei. The electrostatic Coulomb repulsion of the positively charged protons grows rapidly with increasing Z. To maintain the stability in the nuclei, electrically more neutral neutrons than protons are incorporated into the nucleus (see Fig. 1).

The second rule is the so-called Oddo-Harkins rule, which states that nuclides of even atomic numbers are more abundant than those with odd numbers. As shown in Table 1, the most common of the four possible combinations is even–even, the least common odd–odd.

The same relationship is demonstrated in Fig. 2, which shows that there are more stable isotopes with even than with odd proton numbers.

Radioactive isotopes can be classified into artificial and natural. Only the latter are of interest in geology, because they are the basis for radiometric age-dating methods. Radioactive decay processes are spontaneous nuclear reactions, characterized by the radiation emitted. This may be classified into α-β-γ-radiation and electron capture.

Radioactive decay is one process that produces isotope abundance variations. The second process is that of isotopic fractionation caused by small chemical and physical differences between the isotopes of an element. It is exclusively this process that we are discussing in the chapters which follow.

1.2 Isotope Effects

Differences in chemical and physical properties arising from differences in atomic mass of an element are called isotope effects. It is well known that the extranuclear structure of an element essentially determines its chemical behavior, whereas the nucleus is more or less responsible for its physical properties. Because all isotopes of a given element contain the same number and arrangement of electrons, a far-reaching similarity in chemical behavior is the logical consequence. But this similarity is

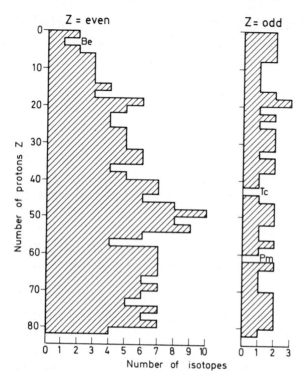

Fig. 2. Number of stable isotopes of elements with even and odd numbers of protons (radioactive isotopes with half-lifes greater than 10^9 years are included)

not unlimited; certain differences exist in physicochemical properties due to the mass differences of different isotopes. The replacement of any atom in a molecule by one of its isotopes is one of the smallest of all the perturbations in chemical behavior. However, the addition of one neutron can, for instance, depress the rate of chemical reaction considerably. Furthermore, it leads, for example, to a shift of the lines in the Raman and IR spectra. These mass differences are most pronounced among the lightest elements. For example, in Table 2, some

Table 2. Characteristic constants of H_2O, D_2O, and $H_2{}^{18}O$

Constants	$H_2{}^{16}O$	$D_2{}^{16}O$	$H_2{}^{18}O$
Density (20 °C, in g cm^{-3})	0.9979	1.1051	1.1106
Temperature of greatest density (°C)	3.98	11.24	4.30
Melting point (760 Torr, in °C)	0.00	3.81	0.28
Boiling point (760 Torr, in °C)	100.00	101.42	100.14
Vapor pressure (at 100 °C, in Torr)	760.00	721.60	
Viscosity (at 20 °C, in centipoise)	1.002	1.247	1.056

differences in physicochemical properties of $H_2{}^{16}O$, $H_2{}^{18}O$, and $D_2{}^{16}O$ are listed. To summarize, the properties of molecules differing only in isotopic substitution are qualitatively the same, but quantitatively different.

Since the discovery of the isotopes of hydrogen by Urey et al. (1932a,b), differences in the chemical properties of the isotopes of the elements H, C, N, O, S, and other elements have been calculated by the methods of statistical mechanics and also determined experimentally. These differences in the chemical properties can lead to considerable isotope effects in chemical reactions.

The theory of isotope effects and a related isotope fractionation mechanism will be discussed very briefly. For a more detailed introduction to the theoretical background see Bigeleisen and Mayer (1947), Urey (1947), Melander (1960), Roginsky (1962), Bigeleisen (1965), Bottinga and Javoy (1973), Javoy (1977), Richet et al. (1977), Hulston (1978), and others.

Differences in the physicochemical properties of isotopes arise as a result of quantum mechanical effects. Figure 3 shows schematically the energy of a diatomic molecule as a function of the distance between the two atoms. According to the quantum theory, the molecule cannot retain any energy on the continuous curve shown in Fig. 3, but is restricted to certain discrete energy levels. The lowest level is not at the minimum of the energy curve, but above it by an amount of 1/2 hv, where h is Planck's constant and v is the frequency with which the atoms in the molecule vibrate with respect to one another. Thus, even in the ground state at absolute zero temperature the vibrating molecule posesses a certain energy above the minimum of the potential energy curve of the molecule. It vibrates with its fundamental frequency which

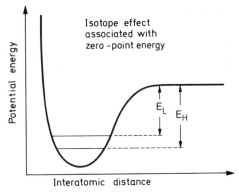

Fig. 3. Schematic potential-energy curve for the interaction of two atoms in a stable molecule or between two molecules in a liquid or solid. (After Bigeleisen 1965)

depends on the mass of the isotopes. Furthermore, it is worth mentioning that only vibrational motions cause isotope effects, rotational and translational motions have no effect on isotope separations. Therefore, different isotopic species will have different zero-point energies in molecules with the same chemical formula: the molecule of the heavy isotope will have a lower zero-point energy than the molecule of the light isotope. This is shown schematically in Fig. 3, where the upper horizontal line (E_L) represents the dissociation energy of the light molecule and the lower line (E_H), that of the heavy one. E_L is actually not a line, but an energy interval between the zero-point energy level and the "continuous" level. This means that the bonds formed by the light isotope are weaker than bonds involving the heavy isotope. Thus, during a chemical reaction, molecules bearing the light isotope will, in general, react slightly more readily than those with the heavy isotope.

1.3 Isotope Fractionation Processes

The largest isotope effect will not cause any fractionation if the reaction with which it is associated occurs quantitatively. Thus, an isotope fractionation will be observed when a reaction has an isotope effect and the formation of product is not quantitative. The partitioning of isotopes between two substances with different isotope ratios is called isotope fractionation. The main phenomena producing isotope fractionations are:

1. isotope exchange reactions,
2. kinetic processes, mainly depending on differences in reaction rates of isotopic molecules.

1.3.1 Isotope Exchange

This includes processes with very different mechanisms. In the following, the term "isotope exchange" is used for all processes in which ordinary changes in the chemical system do not occur, but in which the isotope distribution changes between different chemical substances, between different phases, or between individual molecules.

Isotope exchange reactions are a special case of general chemical equilibrium and can be written

$$aA_1 + bB_2 \rightleftharpoons aA_2 + bB_1 \, ,$$

where the subscripts indicate that species A and B contain either the light or heavy molecule 1 or 2. For this reaction the equilibrium constant will be equal to

$$K = \frac{\left(\dfrac{A_2}{A_1}\right)^a}{\left(\dfrac{B_2}{B_1}\right)^b} , \tag{1}$$

where the terms in parentheses may be, for example, the molar ratios of any species. Using the methods of statistical mechanics, the isotopic equilibrium constant may be expressed in terms of the partition functions Q of the various species

$$K = \frac{Q_{A_2}}{Q_{A_1}} \bigg/ \frac{Q_{B_2}}{Q_{B_1}} . \tag{2}$$

The equilibrium constant then is simply the product or quotient of two partition function ratios, one for the two isotopic species of A, the other for B.

The partition function Q is defined by

$$Q = \Sigma_i \left(g_i \exp(- E_i/kT)\right) , \tag{3}$$

where the summation is over all the allowed energy levels E_i of the molecules and g_i is the degeneracy or statistical weight of the ith level [of E_i] and T is the temperature. Urey (1947) has shown that for the purpose of calculating partition function ratios of isotopic molecules, it is very convenient to introduce, for any chemical species, the ratio of its partition function to that of the corresponding isolated atom, which is called the reduced partition function. This reduced partition function ratio can be used in exactly the same way as the normal partition function ratio. The partition function of a molecule can be separated into factors corresponding to each type of energy: translation, rotation, and vibration

$$Q_2/Q_1 = (Q_2/Q_1)_{trans} \times (Q_2/Q_1)_{rot} \times (Q_2/Q_1)_{vib} . \tag{4}$$

The difference of the translation and rotation is more or less the same among the compounds appearing at the left- and right-hand side of the exchange reaction equation, except for hydrogen, where rotation must be taken into account. This leaves differences in vibrational energy as the source of "isotope effects". The vibrational energy term can be separated into two factors, the first is related to the zero-point energy difference and accounts for most of the variation with temperature. The second term represents the contributions of all the other bound states and is not very different from unity. The complications which may occur relative to this simple model are mainly that the oscillator is not perfectly harmonic, so an "anharmonic" correction has to be added.

For geologic purposes the dependence of the equilibrium constant K on temperature is the most important property [Eq. (3)]. In principle, isotope fractionation factors for isotope exchange reactions are also slightly pressure-dependent. Experimental studies up to 20 kbar by Clayton et al. (1975) have shown that the pressure dependence is, however, less than the limit of detection. Thus, the pressure dependence seems to be of no importance for crustal and upper mantle environments. Isotope fractionations are equal to 1 at very high temperatures. However, the per mil fractionations do not decrease to zero monotonically with increasing temperatures. At higher temperatures, the fractionations may change sign and may increase in magnitude, but they must return a priori to zero at very high temperatures. Such crossover phenomena are due to the complex manner by which thermal excitation of the vibration of atoms contributes to an isotope effect (Stern et al. 1968). Some consequences of the crossover phenomenon for geologic material have been discussed by Muehlenbachs and Kushiro (1974).

Approaching $0°$ Kelvin, the equilibrium constant K tends towards zero, corresponding to complete isotope separation. Isotope fractionation, in general, disappears at high temperatures where the energy of an oscillator is given by the product kT, regardless of the mass of the vibrating atom or of the strength of the bond it forms with neighboring atoms.

For ideal gas reactions, there are two temperature regions where the behavior of the equilibrium constant K is simple: at low temperatures (generally much below room temperature) K follows in K $\sim 1/T$ where T is the absolute temperature. At high temperatures the approximation becomes ln K $\sim 1/T^2$.

The definition of high and low temperature depends on the vibrational frequencies of the molecules involved in the reaction. We have seen that for the calculation of a partition function ratio for a pair of isotopic molecules, we have to know the vibrational frequencies of each. When solid materials are considered, the evaluation of partition function ratios becomes even more complicated, because it is necessary not only to take into account the independent internal vibrations of each molecule, but also to consider the lattice vibrations.

1.3.1.1 Fractionation Factor (α)

Usually, we are interested in the fractionation factor, rather than in the equilibrium constant. The *fractionation factor* α is defined as the ratio of the numbers of any two isotopes in one chemical compound A divided by the corresponding ratio for another chemical compound B:

$$\alpha_{A-B} = \frac{R_A}{R_B} . \tag{5}$$

If the isotopes are randomly distributed over all possible positions in the compounds A and B, α is related to the equilibrium constant K by

$$\alpha = K^{1/n} , \tag{6}$$

where n is the number of atoms exchanged. For simplicity, isotope exchange reactions are written such that only one atom is exchanged [Eq. (7)]. In these cases, the equilibrium constant is identical with the fractionation factor, $K = \alpha$.

For example, the fractionation factor for the exchange of ^{18}O and ^{16}O between water and $CaCO_3$ according to

$$H_2 {}^{18}O + \frac{1}{3} CaC^{16}O_3 \rightleftharpoons H_2 {}^{16}O + \frac{1}{3} CaC^{18}O_3 \tag{7}$$

is given by

$$\alpha_{CaCO_3 - H_2O} = \frac{(^{18}O/^{16}O)_{CaCO_3}}{(^{18}O/^{16}O)_{H_2O}} = 1.031 \text{ at } 25\,^{\circ}C . \tag{8}$$

1.3.1.2 The Delta Value (δ)

The isotopic composition of two compounds A and B actually measured in the laboratory are expressed by δ-values:

$$\delta_A = \left(\frac{R_A}{R_{St}} - 1 \right) \cdot 10^3 \ (\%_0) \tag{9}$$

and

$$\delta_B = \left(\frac{R_B}{R_{St}} - 1 \right) \cdot 10^3 \ (\%_0) , \tag{10}$$

where R_{St} is the defined isotope ratio of a standard sample.

Table 3. Comparison between Δ, α, and $10^3 \ln \alpha$

δ_A	δ_B	Δ_{A-B}	α_{A-B}	$10^3 \ln \alpha_{A-B}$
1.00	0	1	1.001	0.9995
10.00	0	10	1.01	9.95
20.00	0	20	1.02	19.80
10.00	5.00	4.98	1.00498	4.96
20.00	15.00	4.93	1.00493	4.91
30.00	20.00	9.80	1.00980	9.76
30.00	10.00	19.80	1.01980	19.61

The δ-values and fractionation factor α are related by

$$\delta_A - \delta_B \cong \Delta_{A-B} \cong 10^3 \ln \alpha_{A-B} \ . \tag{11}$$

Table 3 illustrates the closeness of the approximation. Considering experimental errors, approximations are excellent for differences in δ-values of less than about 10.

1.3.1.3 Evaporation-Condensation Processes

Of special interest in stable isotope geochemistry are evaporation-condensation processes, because differences in the vapor pressures of isotopic compounds lead to fractionations. For example, from the vapor pressure data for water it is evident that the lighter molecular species are preferentially enriched in the vapor phase, the extent depending upon the temperature. Such an isotopic separation process can be treated theoretically in terms of fractional distillation or condensation under equilibrium conditions and is expressed by a Rayleigh (1896) equation. For a condensation process this equation is

$$\frac{R_v}{R_{v_0}} = f^{(\alpha-1)} \ , \tag{12}$$

where R_{v_0} is the isotope ratio of the initial bulk composition and R_v is the instantaneous ratio of the remaining vapor (v); f is the fraction of the residual vapor, and the fractionation factor is $\alpha = R_l/R_v$ (l = liquid). The instantaneous isotope ratio of the condensate leaving the vapor (R_l) is given by

$$\frac{R_l}{R_{v_0}} = \alpha \ f^{(\alpha-1)} \tag{13}$$

and the average isotope ratio of the separated and accumulated condensate ($\bar{R}_l$) at any time of condensation is expressed by

$$\frac{\overline{R}_l}{R_{v_0}} = \frac{1 - f^\alpha}{1 - f} \; . \tag{14}$$

For a distillation process the instantaneous isotope ratios of the remaining liquid and the vapor leaving the liquid are given by

$$\frac{R_l}{R_{l_0}} = f^{\left(\frac{1}{\alpha} - 1\right)} \tag{15}$$

and

$$\frac{R_v}{R_{l_0}} = \frac{1}{a} f^{\left(\frac{1}{\alpha} - 1\right)} \; . \tag{16}$$

The average isotope ratio of the separated and accumulated vapor is expressed by

$$\frac{\overline{R}_v}{R_{l_0}} = \frac{1 - f^{\frac{1}{\alpha}}}{1 - f} \quad (f = \text{fraction of residual liquid}) \; . \tag{17}$$

Any isotope fractionation carried out in such a way that the products are isolated from the reactants immediately after formation will show a characteristic trend in isotopic composition. As condensation or distillation proceed the residual vapor or liquid become progressively depleted or enriched with respect to the heavy isotope. A natural example is the fractionation between oxygen isotopes in the water vapor of a cloud and the raindrops released from that cloud. The resulting depletion of the $^{18}O/^{16}O$ ratio in the residual vapor and the instantaneous isotopic composition of the raindrops released from that cloud are given as a function of the fraction of vapor remaining in the cloud (Fig. 4).

1.3.2 Kinetic Effects

These are the second main phenomena producing fractionations. The theory of kinetic isotope effects has been reviewed by Bigeleisen and Wolfsberg (1958), Melander (1960), and Melander and Saunders (1980). Knowledge of kinetic isotope effects is very important, because it can provide information about details of reaction pathways. A kinetic isotope effect occurs when the rate of a chemical reaction is sensitive to atomic mass at a particular position in one of the reacting species.

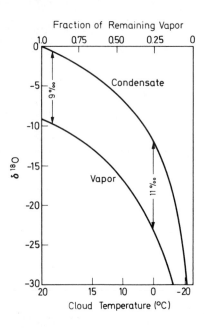

Fig. 4. $\delta^{18}O$ in a cloud vapor and condensate plotted as a function of the fraction of remaining vapor in the cloud for a Rayleigh process. The temperature of the cloud is shown on the lower axis. The increase in fractionation with decreasing temperature is taken into account. (After Dansgaard 1964)

Quantitatively, many observed deviations from the simple equilibrium processes can be interpreted as consequences of the various isotopic components having different rates of reaction.

Isotope fractionation measurements taken during unidirectional chemical reactions always show a preferential enrichment of the lighter isotope in the reaction products. The isotope fractionation introduced in the course of a unidirectional reaction may be considered in terms of the ratio of rate constants for the isotopic substances.

Thus, for two competing isotopic reactions

$$A_1 \overset{k_1}{\to} B_1 \quad \text{and} \quad A_2 \overset{k_2}{\to} B_2 \tag{18}$$

the ratio of rate constants for the reaction of light and heavy isotope species k_1/k_2 as in the case of equilibrium constants, is expressed in terms of two partition function ratios, one for the two isotopic reactant species, and one for the two isotopic species of the activated complex or transition state A^*:

$$\frac{k_1}{k_2} = \left[\frac{Q^*(A_2)}{Q^*(A_1)} \middle/ \frac{Q^*(A_2^*)}{Q^*(A_1^*)} \right] \frac{v_1}{v_2}. \tag{19}$$

The factor v_1/v_2 in the expression is a mass term ratio for the two isotopic species. The determination of the ratio of rate constants is, therefore, principally the same as the determination of an equilibrium constant, although the calculations are not so precise because of the need for detailed knowledge of the transition state. By "transition state" is meant that molecular configuration which is most difficult to attain along the path between the reactants and the products. This theory is based on the idea that a chemical reaction proceeds from some initial state to a final configuration by a continuous change, and that there is some critical intermediate configuration called the activated species or transition state. There are a small number of activated molecules in equilibrium with the reacting species and the rate of reaction is controlled by the rate of decomposition of these activated species.

1.3.3 Diffusion

The process of diffusion can cause significant isotope fractionations. As is well known, gaseous diffusion is used in the nuclear industry to separate ^{235}U from ^{238}U. Generally light isotopes are more mobile than heavy isotopes. The ratio of the diffusion coefficients expresses the isotopic enrichment factors. For gases the ratio is equivalent to the square root of their masses.

In solutions and solids the relationships are much more complicated. The term "solid state diffusion" generally includes volume diffusion and diffusion mechanisms where the atoms move along paths of easy diffusion such as grain boundaries and surfaces. The diffusion coefficient D is usually represented as depending exponentially on an activation energy E and the absolute temperature T according to

$$D = D_0^{(-E/RT)} \text{ ,}$$

where R is the universal gas constant and D_0 is a term which is temperature-independent. Attempts to determine diffusion coefficients have been carried out by Yund and Anderson (1974), Giletti and Anderson (1975), Giletti et al. (1978), Graham (1981), Freer and Dennis (1982), and Elphick et al. (1986). Perhaps the most promising method is the use of the Secondary Ion Mass Spectrometer (SIMS).

With the acquisition of kinetic data for oxygen diffusion in various minerals the effect of diffusion during cooling can be evaluated. Figure 5 shows an Arrhenius plot of diffusion coefficients versus reciprocal temperature for several minerals. At 600 °C, for example, feldspar, quartz, and hornblende have diffusion coefficients that differ by three orders

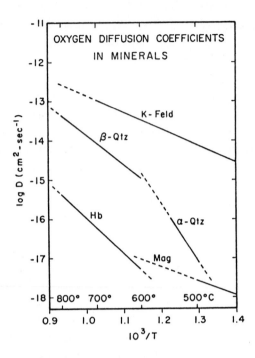

Fig. 5. Arrhenius plots of oxygen self-diffusion coefficients for K-feldspar, quartz, hornblende, and magnetite. Silicate data are for $p_{H_2O} = 1$ kbar, magnetite data are for approximately 1 bar and $(p_{H_2}/p_{H_2O}) = 1.0$. *Solid lines* show range of observed data (Giletti 1986)

of magnitude. That means in a given rock these minerals exchange oxygen at different rates and become closed systems at different temperatures.

Oxygen diffusion in silicate minerals under hydrothermal conditions is characterized by low activation energies (typically 30 kcal mol^{-1}) and comparatively high diffusion rates, whereas under essentially dry conditions several orders of magnitude higher activation energies and lower diffusion rates are observed. The differences are believed to arise from variations in the oxygen exchange and transport mechanisms. The study of Matthews et al. (1983c) on oxygen isotope exchange between quartz or feldspar and water has shown that exchange mechanisms are a complex function of $p_{(H_2O)}$, T, extent of exchange, grain size and morphology, and other variables and that, particularly for quartz, it may be unsafe to generalize on the mechanisms of oxygen exchange from diffusion data. Nevertheless, the determination of diffusion coefficients is of considerable importance in interpreting oxygen isotope temperatures (Giletti 1986), in studying hydrothermal water-rock interactions, and in estimating the rates of diffusion-controlled ductile deformation processes in minerals (Giletti and Yund 1984).

1.3.4 Nonmass-Dependent Isotope Effects

It has been a common belief that chemically produced isotope effects arise solely because of differences in isotopic mass. This means that for an element with more than two isotopes such as oxygen the enrichment of ^{18}O relative to ^{16}O is expected to be approximately twice as large as the enrichment of ^{17}O relative to ^{16}O. This yields a slope of 0.52 (Matsuhisa et al. 1979) on a three-isotope correlation diagram of $\delta^{17}O$ versus $\delta^{18}O$. On this basis, nonmass-dependent isotope fractionations have been ascribed solely to nuclear processes. However, Heidenreich and Thiemens (1983, 1985) have shown experimentally that during the formation of ozone from molecular oxygen and during the dissociation of carbon dioxide in a high frequency discharge, nonmass-dependent isotope fractionations do occur. To explain those unexpected isotope effects Heidenreich and Thiemens (1983) first assumed an isotopic self-shielding process which leads to the preferential dissociation of ^{18}O-containing molecules, but later (1985) demonstrated that during the dissociation of CO_2 the reaction product (O_2) is also depleted in ^{17}O and ^{18}O which would not be the case in a self-shielding process. From a kinetic standpoint it appears that both the ozone formation and the CO_2 dissociation represent the same phenomena. According to Heidenreich and Thiemens (1985), the most likely process in producing this effect is either the formation or relaxation of ozone in an excited electronic state. More experiments are needed to estimate the frequency and the importance of such nonmass-dependent isotope effects. The effects described may be of relevance to cosmochemical studies, whereas it is very likely that terrestrial processes are accompanied by mass-dependent fractionations.

1.3.5 Variation of Isotopic Composition with Chemical Composition and Crystal Structure

This aspect has been reviewed by O'Neil (1977).

Chemical Composition. Qualitatively, the isotopic composition of a mineral depends to a very high degree upon the nature of the chemical bonds within the mineral and to a smaller degree upon the atomic mass. In general, bonds to ions with a high ionic potential and low atomic mass are associated with high vibrational frequencies and have a tendency to incorporate preferentially the heavy isotope. This relationship can be easily demonstrated considering the bonding of oxygen to the small, highly charged Si^{4+} ion compared to the relatively large Fe^{2+} ion: In natural mineral assemblages quartz is always the most ^{18}O-rich mineral and magnetite is always the most ^{18}O-deficient mineral. Furthermore, carbonates are always enriched in ^{18}O relative to most other mineral groups because oxygen is bonded to the small, highly charged C^{4+} ion. The nature of the divalent cation is only of secondary importance to the C–O bonding. However, the mass effect is apparent in ^{34}S distributions among sulfides where, for example, ZnS always concentrates ^{34}S relative to coexisting PbS.

Crystal Structure. Structural effects are secondary in importance to those arising from the primary chemical bonding: the heavy isotope being concentrated in the more closely packed or well-ordered structures. Such effects can be large, for example, between graphite and diamond. The calculated diamond-graphite fractionation ranges from 11.5 ‰ at 0 °C to 0.4 ‰ at 1000 °C (Bottinga 1969b), conversely, the ^{18}O and ^{13}C fractionations between aragonite and calcite at 25 °C are relatively small at 0.6 ‰ and 1.8 ‰, respectively (Rubinson and Clayton 1969; Tarutani et al. 1969).

Stable isotope studies can also provide information on details of crystal structure. For instance, Heinzinger (1969) identified two kinds of water released from $CuSO_4 \times 5 H_2O$ below and above 50 °C, differing in their bonding characteristic in the crystal. The oxygens of four water molecules are bonded to the copper ion, that of the fifth molecule is hydrogen-bonded. Heinzinger (1969) demonstrated that this fifth hydrogen-bonded molecule is enriched in deuterium by 57‰ relative to the water coordinated by the copper ion.

1.3.6 Isotope Geothermometers

Isotopic thermometry has become well established since the classic paper of Urey (1947) on the thermodynamic properties of isotopic substances. The partitioning of two stable isotopes of an element between two mineral phases can be viewed as a special case of element partitioning between two minerals. There are, however, quantitative differences between these two exchange reactions, the most important being that isotope partitioning is more or less pressure independent, which represents the greatest advantage relative to the numerous other geothermometers.

Recently, Rumble (1982), however, argued that changing pressure has a significant influence on isotope fractionations in rocks. The pressure effect arises because changing pressure causes changes in the proportions of volatile species in fluids, which in turn leads to changes in fractionation between bulk fluid and bulk rock.

The necessary condition to apply the different geothermometers is isotope equilibrium. Conclusions concerning the nature and extent of isotope equilibrium are influenced by the criteria used to test for attainment of equilibrium and the spatial scale over which measurements have been made.

In a mineral assemblage of n-phases we can obtain n−1 independent temperatures, one temperature for each mineral pair. If each mineral pair gives concordant temperatures, we can be nearly certain that isotope equilibrium was attained and that equilibrium was frozen in at the same temperature in every mineral. A necessary requirement of the concordant temperature approach is that temperature calibrations must be accurate, which, however, is far from being the case. Theoretical studies show that the fractionation factor α, for isotope exchange between minerals is a linear function of $1/T^2$, where T is in degrees Kelvin at crustal temperatures. Bottinga and Javoy (1973) were the first to show that isotope fractionations between anhydrous mineral pairs at $T > 500\,°C$ can be expressed in terms of the equation:

$$1000 \ln \alpha = A/T^2 \ ,$$

which means that for a temperature determination factor A has to be known. Fractionations between minerals and fluids at $< 500\,°C$ can, on the other hand, be expressed by the equation:

$$1000 \ln \alpha = A/T^2 + B \ .$$

Three different methods have been used to determine the equilibrium constants for isotope exchange reactions:

1. calculation from statistical mechanical theory, which is specially suitable for gas reactions,
2. experimental determination in the laboratory;
3. calibration on an empirical basis.

The latter method is based on the idea that the calculated "formation temperature" of a rock in which other minerals are also present serves as a calibration to the measured fractionations of other minerals, providing that all minerals were at equilibrium. However, because there is evidence that totally equilibrated systems are not very common in nature, such empirical calibrations should be regarded with extreme caution.

The theoretical calculation of isotope fractionation factors for solids is exceedingly difficult, because all vibrational frequencies of the crystalline lattice must be taken into account. Calculation of equilibrium isotope fractionation factors have been particularly successful for gases (Richet et al. 1977) and various quartz-silicate systems (Kieffer 1982). Theoretical methods are likely to play an increasing role in improving the quality of the basis for extrapolating the experimental fractionation data outside of the practical laboratory temperature range.

In a number of cases two or more approaches have been applied. For example, for the system calcite-H_2O, theoretical, experimental, and empirical methods have given coherent results.

The most promising approach seems to be the experimental determination of isotope fractionation factors. In principle, the experimental determinations of isotope exchange equilibrium constants can be carried out by simply holding the phases at a fixed temperature. By a suitable choice of isotopic compositions of the starting materials, it is possible to approach equilibrium from opposite directions, thus satisfying the classical criterion for equilibrium. However, the driving forces for the exchange reactions are small and rates of exchange are often very low. In such cases a variety of techniques have been used to facilitate exchange, and have been summarized by Clayton (1981):

1. recrystallization of a very finely ground powder;
2. crystallization of a gel or glass;
3. crystallization as a result of polymorphic phase transition;
4. synthesis of a new phase by cation exchange;
5. complete mineral synthesis.

All of these techniques depart from an ideal exchange experiment in that there are driving forces for a reaction other than the differences in isotopic composition. These obvious limitations result in various calibration curves for which significant discrepancies exist.

Experimental calibrations of isotopic geothermometers have been typically performed between 250° and 800 °C. The upper temperature limit is generally determined by the stability of the mineral or by limitations of the equipment. The lower temperature limit is determined by the decreasing rate of exchange. Various procedures have been used to establish isotope equilibrium. For direct mineral-fluid exchange experiments isotope equilibrium can be approached from opposite directions. For mineral-fluid experiments where only an approach towards equilibrium is achieved an extrapolation technique is used (Northrop and Clayton 1966), which has been widely applied for the low temperature experiments. However, the predicted "equilibrium" fractionation is often larger than the actual equilibrium value. An important modification of the Northrop and Clayton method has been introduced by Matsuhisa et al. (1979) and Matthews et al. (1983a,b), by using a three-isotope exchange method, which is discussed in more detail in Sect. 2.3.

Many of the experiments designed to calibrate an isotope thermometer yield additional information concerning the kinetics and mechanisms involved in the isotope exchange reaction. Such information is important to understand fluid/rock interaction processes. In the simplest case of direct exchange experiments, where no material is created or destroyed during the exchange process, the grain size remains constant and the volume diffusion of the element is the mechanism for isotope exchange. In many of the other hydrothermal experiments the grain size was observed to change due to chemical breakdown (e.g., feldspars), recrystallization, and/or the solution-reprecipitation process. The rate of isotope exchange in such systems may vary with time and thus kinetic effects cannot be dismissed (Matthews et al. 1983a). For example, in feldspar-water isotope exchange, diffusion appears to be the dominant process rather than dissolution and redeposition (O'Neil and Taylor 1967). On the other hand, in other systems, such as quartz-water, both recrystallization and diffusional exchange mechanisms may be occurring (Matthews et al. 1983c).

1.4 Basic Principles of Mass Spectrometry

Mass spectrometric methods are by far the most effective means of measuring isotope abundances. A mass spectrometer separates charged atoms and molecules on the basis of their masses based on their motions in magnetic and/or electrical fields. The design and the applications of the many types of mass spectrometers are too broad to cover here. Only the principles of mass analysis will be briefly discussed.

In principle, a mass spectrometer may be divided into four different parts: (1) the inlet system, (2) the ion source, (3) the mass analyzer, and (4) the ion detector (see Fig. 6).

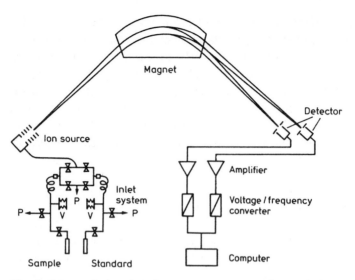

Fig. 6. Schematic drawing of a mass spectrometer for stable isotope measurements. P denotes pumping system, V denotes a variable volume

1. Special arrangements for the *inlet system* are necessary because the instability of the ions produced and the mass separation require a high vacuum. If the mean free path length (flight without collision with other molecules) of molecules is large compared with the dimensions of the tubing through which the gas is flowing, then we refer to this as molecular flow. During molecular flow the gas particles do not influence each other. Therefore, the gas flow velocity of the lighter component is greater than that of the heavier component, and this means that the heavier isotope becomes enriched in the reservoir from which the gas flows into the mass spectrometer. To avoid such a mass discrimination, normally the isotope abundance measurements of gaseous substances are carried out utilizing viscous gas flow. During the viscous gas flow the free path length of molecules is small and no mass separation takes place. The normal gas pressure is around 100 Torr. At the end of the inlet system through which we have viscous gas flow, there is a "leak", a constriction in the flow line.

2. The *ion source* is the part of the mass spectrometer where ions are formed, accelerated, and focused into a narrow beam. In the ion source, the gas flow is always molecular.

In general, ions are produced thermally or by electron impact. Ions of gaseous samples are provided most reliably by electron bombardment. A beam of electrons is emitted by a heated filament, usually tungsten or rhenium, and is directed to pass between two parallel plates. The beam is collimated by means of a weak magnetic field. Positive ions are formed between the two parallel plates as a result of gas molecule-electron collisions. The ions are drawn out of the electron beam by the action of an electric field, and are further accelerated up to several kV. The positive ions entering the magnetic field are essentially monoenergetic, i.e., they will possess the same kinetic energy, according to the equation:

$$1/2 \; mv^2 = eV \; .$$

There is a minimum threshold energy below which ionization does not occur. The energy of electrons used for ionization generally is about 50 to 70 V, because this range of energy maximizes the efficiency of single ionization, but is too low to produce a significant number of multiply charged ions. The principal advantage of an electron-bombardment ion source is the stability of the resulting ion beam, the disadvantage is that the vacuum system must be extremely clean, because the electrons will ionize any gases present in the ionization chamber.

3. The *mass analyzer* separates the ion beams emerging from the ion source according to their M/e (mass/charge) ratios. From the many possible mass analyzer configurations only the first-order direction, focusing mass analyzer is used in stable isotope research. As the ion beam passes through the magnetic field, the ions are deflected into circular paths, the radii of which are proportional to the square root of M/e. Thus, the ions are separated into beams, each with a particular value of M/e.

In 1940 Nier introduced the sector magnetic analyzer. In this type of analyzer, deflection takes place in a wedge-shaped magnetic field. The ion beam enters and leaves the field at right angles to the boundary, so the deflection angle is equal to the wedge angle, for instance, 60°. The sector instrument has the advantage of its source and detector being comparatively free from the mass-discriminating influence of the analyzer field.

4. After passing through the magnetic field, the separated ions are collected in the *ion detector* and converted into an electrical impulse, which is then fed into an amplifier. For relatively large ion currents a simple metal cup (Faraday cage) is used. The cup is grounded through a high ohmic resistor. As the ion current passes to the ground, the potential drop in the resistor acts as a measure of the ion current.

By collecting two ions beams of the isotopes in question simultaneously, and by measuring the ratio of this ion current directly, a much higher precision can be obtained than from a single ion beam collection. With simultaneous collection, the isotope ratios of two samples can be compared quickly under nearly identical conditions. Nier et al. (1947) developed this technique for routine measurements and McKinney et al. (1950) improved this type of mass spectrometer, which became the standard type for isotope ratio analysis for many years. The double collecting mass spectrometer employed a precision voltage divider in a null circuit (Kelvin bridge type). With this technique the isotope ratios could be accurately measured using the chart recorder output of a vibrating reed electrometer.

During the 1960's and early 1970's, instrument makers automated their mass spectrometers, changing the measurement system from the null technique to one employing voltage-to-frequency converters and counters on each electrometer output. Today, the newest mass-spectrometer generation is fully automated and computerized, improving the reproducibility to values better than $\pm 0.02\%o$.

The overall instrumental error of the mass-spectrometric measurement may be increased by nonlinearities within the individual measurement devices. The probable variation between different instruments may reach a level of 1% to 2% of the measured δ-values. This is not critical for small differences in isotopic composition. However, the uncertainty in comparing data from different laboratories increases when samples of very different isotopic compositions are compared. Blattner and Hulston (1978), by distributing a pair of calcite reference samples, showed that the individual differences between $\delta^{18}O$ determinations in more than 10 laboratories range from $23.0\%o$ to $23.6\%o$.

1.5 Standards

The accuracy with which *absolute* isotope abundances can be measured is substantially poorer than the precision with which *relative* differences in isotope abundances between two samples can be determined. Nevertheless, the determination of absolute isotope ratios is very important, because these numbers form the basis for the calculation of the relative differences, the δ-values. Table 4 summarizes absolute isotope ratios of primary standards.

Table 4. Absolute isotope ratios of international standards. (After Hayes 1983)

Standard	Ratio	Accepted value ($\times 10^6$) (with 95% confidence interval)	Source
SMOW	D/H	155.76 ± 0.10	Hagemann et al. (1970)
	$^{18}O/^{16}O$	$2,005.20 \pm 0.43$	Baertschi (1976)
	$^{17}O/^{16}O$	373 ± 15	Nier (1950), corrected by Hayes (1983)
PDB	$^{13}C/^{12}C$	$11,237.2 \pm 2.9$	Craig (1957)
	$^{18}O/^{16}O$	$2,067.1 \pm 2.1$	
	$^{17}O/^{16}O$	379 ± 15	
Air nitrogen	$^{15}N/^{14}N$	$3,676.5 \pm 8.1$	Junk and Svec (1958)
Canyon Diablo Troilite (CDT)	$^{34}S/^{32}S$	$45,004.5 \pm 9.3$	Jensen and Nakai (1962)

Irregularities and problems concerning standards have been evaluated by Friedman and O'Neil (1977), Gonfiantini (1978, 1984), and Coplen et al. (1983). The accepted unit of isotope ratio measurements is the delta value (δ), given in per mil (‰). The δ-value is defined as

$$\delta \text{ in } ‰ = \frac{R_{(sample)} - R_{(standard)}}{R_{(standard)}} \times 1000 ,$$

where R represents the isotope ratio. If $\delta_A > \delta_B$, we speak of A being enriched in the rare isotope or "heavier" than B. Unfortunately, not all of the δ-values cited in the literature are given relative to a single universal standard, so that often several standards of one element are in use. To convert δ-values from one standard to another, the following equation may be used

$$\delta_{X-A} = \left[\left(\frac{\delta_{B-A}}{10^3} + 1 \right) \left(\frac{\delta_{X-B}}{10^3} + 1 \right) - 1 \right] 10^3 ,$$

where X represents the sample, A and B different standards.

For different elements a convenient "working standard" is used in each laboratory. However, all values measured relative to the "working standard" are reported in the literature relative to a universal standard. Unfortunately, there has sometimes been a lack of agreement among researchers in this field as to what standard should be designated as the universal standard. A standard should fulfill the following requirements:

1. be used worldwide as the zero point;
2. be homogeneous in composition;
3. be available in relatively large amounts;
4. be easy to handle for chemical preparation and isotopic measurement;
5. have an isotope ratio near the middle of the natural variation range.

Among the reference samples now used, relatively few meet all of these requirements. For example, the most widely used carbon isotope standard, PDB, a sample of a belemnite guard, is exhausted. The worldwide standards now in use are given in Table 5.

The problems related to standards have been discussed by an advisory group, who met in 1983 for the third time in Vienna. As a result of these meetings (Coplen et al. 1983; Confiantini 1984) several new standards can be cited.

A further advancement comes from interlaboratory comparison of two standards having different isotopic composition, which, for instance, has been carried out by Blattner and Hulston (1978) on two carbonates. Such an interlaboratory calibration can then be used for a normalization procedure, which corrects for all proportional errors due to the mass spectrometer and to the sample preparation. Ideally, the two standard samples should have isotope ratios as different as possible, but still within the range of natural variations. There are, however, some problems connected with the data normalization, which are still under debate. For example, the CO_2 equilibration of waters and the acid extraction of CO_2 from carbonates are indirect analytical procedures, involving temperature-dependent fractionation factors (whose values are not well defined) with respect to the original samples and which might be reevaluated on the normalized scale.

Table 5. Worldwide standards in use for the isotopic composition of hydrogen, carbon, oxygen, sulfur, and nitrogen

Element	Standard	Standard abbreviated
H	Standard Mean Ocean Water	SMOW
C	Belemnitella americana from the Cretaceous Peedde formation, South Carolina	PDB
O	Standard Mean Ocean Water	SMOW
S	Troilite (FeS) from the Canyon Diablo iron meteorite	CD
N	Air	N_2 (atm.)

1.6 General Remarks on Sample Handling

Isotopic differences between samples to be measured are often extremely small. Therefore, great care has to be taken to avoid any isotope fractionation during chemical or physical treatment of the sample.

To convert geologic samples to a suitable form for analysis, many different preparation techniques must be used. They have, nevertheless, one general feature in common: any preparation procedure providing a yield of less than 100% may produce a reaction product that is isotopically different from the original specimen, because the different isotopic species have different reaction rates.

A quantitative yield of a pure gas is usually necessary for the mass spectrometric measurement in order to prevent not only isotope fractionation during sample preparation, but also interference in the mass spectrometer. Contamination with gases having the same molecular masses and having similar physical properties may be a serious problem. This is especially critical with CO_2 and N_2O, on the one hand (Craig and Keeling 1963), and N_2 and CO, on the other. When CO_2 is used, interference by hydrocarbons and a CS^+ ion may also be a problem.

Contamination may result from incomplete evacuation of the vacuum system and/or from degassing of the sample. How gases are transferred, distilled, or otherwise processed in vacuum lines is briefly discussed under the different elements. All errors due to chemical preparation limit the overall precision of an isotope ratio measurement to 0.1‰ to 0.2‰, while modern mass spectrometer instrumentation enables a precision better than 0.02‰ for light elements other than hydrogen. Still larger errors must be expected when elements of very low concentration are extracted by chemical methods (e.g., carbon and sulfur from igneous rocks). Table 6 summarizes which gases are used for mass-spectrometric analysis of the various elements.

Table 6. Gases most commonly used in isotope ratio in mass spectrometry

Element	Gas
H	H_2
C	CO_2
N	N_2
O	$CO_2, (O_2)$
S	SO_2, SF_6
Si	SiF_4

Chapter 2   Isotopic Properties of Selected Elements

The foundations of stable isotope geochemistry were laid in 1947 by Urey's paper on the thermodynamic properties of isotopic substances and by Nier's development of the ratio mass spectrometer. Before going into details of the naturally occurring variations of stable isotope ratios, it is useful to discuss some general trends pertinent to the whole field of isotope geochemistry.

1. Detectable isotope fractionation occurs only when the relative mass differences between the isotopes of a specific element are large, i.e., measurable isotope fractionations should be detectable only for the light elements (in general up to a mass number of about 40).
2. All those elements that form solid, liquid, and gaseous compounds which are stable over a wide temperature range, are likely to have variations of isotopic composition. Generally, the heavy isotope is concentrated in the solid phase in which it is more tightly bound. Heavier isotopes tend to concentrate in molecules in which they are present in the highest oxidation state.
3. Isotopic variations in most biological systems can be best explained by assuming kinetic effects. During biological reactions (e.g., during photosynthesis, bacterial reactions, and other microbiological processes) the lighter isotope is very often enriched in the reaction product relative to the starting substances.

2.1 Hydrogen

Until 1931 it was assumed that hydrogen consisted of only one isotope. Urey et al. (1932a,b) detected the presence of a second, heavy, stable isotope, which was called deuterium. Way et al. (1950) gave the following average abundances of the stable hydrogen isotopes:

^{1}H: 99.9844%

^{2}D: 0.0156% .

Hagemann et al. (1970) reported the absolute abundance of deuterium in the SMOW standard to be 155.8 ppm. In addition to these two stable isotopes there is a third naturally occurring but radioactive isotope, ^3H, tritium, with a half-life of approximately 12.5 years.

The isotope geochemistry of the stable hydrogen isotopes is one of the most interesting, for several reasons:

1. Hydrogen has by far the largest relative mass difference between its two stable isotopes. This results in hydrogen showing the largest variations in stable isotope ratios of all elements.
2. Hydrogen is nearly omnipresent in the forms of H_2O, OH^-, H_2, and CH_4, even at great depths in the earth's mantle. Therefore, it is conceivable that hydrogen plays a major role, directly or indirectly, in many naturally occurring geologic processes.

In Fig. 7 the ranges of hydrogen isotope composition of some geologically important reservoirs are given. The isotope geochemistry of hydrogen has been reviewed by Taylor (1974a) and Friedman and O'Neil (1978).

2.1.1 Preparation Techniques and Mass Spectrometric Measurement

The determination of the D/H ratios is usually performed on H_2 gas. Water is converted to hydrogen by passage over hot zinc or uranium at about 750 °C, e.g., as described by Bigeleisen et al. (1952), Friedman

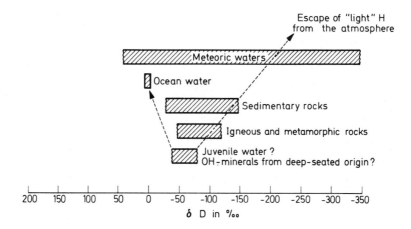

Fig. 7. D/H ratios of some geologically important materials (δD relative SMOW)

(1953), Craig (1961a), and Godfrey (1962). Most of the hydrogen generated from hydroxyl-bearing minerals is liberated in the form of water, but some is liberated as molecular hydrogen (Savin and Epstein 1970a). The resulting H_2 gas is converted in many laboratories to water by reaction with copper oxide. The water is then treated as described above.

A difficulty in measuring D/H isotope ratios is that along with the H_2^+ and HD^+ formation in the ion source, H_3^+ is produced as a by-product of ion-molecule collisions. Therefore, an H_3^+ correction has to be made. The relevant procedures have been evaluated by Schoeller et al. (1983), who also introduced a new alternative. The analytical error for hydrogen isotope data is usually given as $\pm 0.5\%o$ to $\pm 2\%o$ depending on different laboratories.

2.1.2 Standard

In the past there has been some confusion due to the fact that different laboratories have expressed their results in noncorresponding scales. To resolve this confusion, in 1976 a consultant meeting of the International Atomic Energy Agency in Vienna recommended that the zero point of the δD scale should be Vienna SMOW. The I.A.E.A. distributes a second water standard called SLAP (Standard Light Antarctic Precipitation) which has a δ-value of $-428\%o$.

2.1.3 Fractionation Mechanisms

2.1.3.1 *Vapor Pressure and Freezing-Point Differences*

The most effective processes that produce hydrogen isotope variations are those due to vapor pressure differences of water, and to a much smaller degree, those due to differences in freezing points. Because the vapor pressure of HDO is slightly lower than that of H_2O, the concentration of D is higher in the liquid than in the vapor phase.

The physical processes responsible for the fractionation of hydrogen isotopes in water and the distribution of the resulting fractionations in nature are the same as those applying to the fractionation of oxygen isotopes in water. Therefore, the fractionation of ^{18}O parallels that of D in most cases.

2.1.3.2 Equilibrium Exchange Reactions

Bottinga (1969a) and Richet et al. (1977) calculated the hydrogen isotope fractionations for gaseous hydrogen compounds. They demonstrated that very large fractionations on the order of several hundred per
mil occur in the systems water vapor-methane, H_2O-H_2, and H_2O-H_2S.
D/H fractionation factors between hydrous minerals and water have
been determined experimentally by Suzuoki and Epstein (1976), Sakai
and Tsutsumi (1978), Graham et al. (1980, 1984, 1986), Satake and
Matsuo (1984), and Liu and Epstein (1984). The resulting fractionation
curves (see Fig. 8) are much more complex than originally thought by
Suzuoki and Epstein (1976). As Fig. 8 demonstrates the forms of the
curves are extremely variable, even within one mineral group. Therefore, extrapolations to temperatures outside of the convenient experimental range are not justified. The linear relationship between $1/T^2$
and the mineral water fractionations proposed by Suzuoki and Epstein

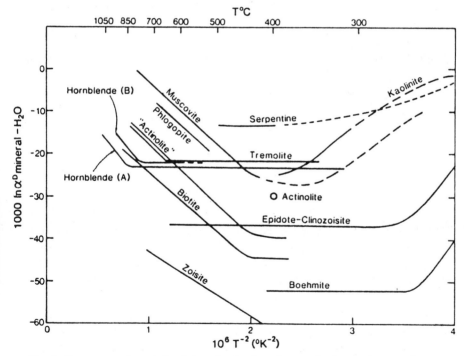

Fig. 8. Some experimentally determined mineral-water hydrogen isotope fractionation curves (Sheppard 1984)

(1976) for micas and amphiboles in the temperature range from 800°
to 450 °C is not followed by epidotes (Graham et al. 1980), by tremo-
lite (Graham et al. 1984), or by chlorite (Graham et al., in press). Furthermore, the near linear relationship between atomic mass/charge ratio
of the octahedrally coordinated cation and the mineral-H_2O fractionation noted by Suzuoki and Epstein (1976) has not been verified for
other minerals (Graham et al. 1980).

Hydrogen isotope fractionations in the temperature range below
400 °C are not well known, because the exchange rates between water
and mineral are slow. O'Neil and Kharaka (1976) reported that $\approx 30\%$
of hydrogen isotope exchange between kaolinite and water was reached
after 8 months of reaction at 200 °C. With a modified technique –
relative to conventional techniques – Liu and Epstein (1984) determined the fractionation for kaolinite-water in the temperature range from
200° to 350 °C. The relationship found by them is opposite to that
observed for temperatures above 400 °C. These data, combined with
data on natural samples (Lambert and Epstein 1980; Marumo et al.
1980), suggest that there is a reversal in isotope fractionation at temperatures around 200 °C.

One reason for this complicated behavior of hydrogen isotope fractionation might be that hydrogen occurs in some minerals in more than
one structural site (Graham et al., in press). In minerals where hydrogen
bonding occurs there is a qualitative relationship between the length
of the O–H–O bridge and the fractionation factor: the shorter the bond
length in the mineral, the more strongly the mineral concentrates pro-
tium. Satake and Matsuo (1984) demonstrated that besides the hydrogen bond effect, structural effects such as distortion of the Mg-octa-
hedron may also control the D/H fractionation.

When applying these experimental data to natural assemblages isotope equilibrium is a necessary prerequisite. However, in the case of
hydrogen it is especially difficult to establish whether isotope equilibrium is commonly preserved in high-temperature environments, because
the rate of hydrogen isotope exchange is – compared to oxygen – relatively rapid. Graham (1981) could demonstrate that for micas closure
temperatures for hydrogen are about 200 °C lower than those for oxygen. Rapid hydrogen transport may proceed by hydrolysis of Si–O
and Al–O bonds, thus supporting the idea that water appears to be essential for isotope exchange. Graham (1981) could further demonstrate
that hydrogen isotope exchange between coexisting hydrous minerals
proceeds by a quite different mechanism in the absence of hydrous

fluid than in its presence. The presence of water greatly facilitates diffusion rates by at least two orders of magnitude. The kinetics of the hydrogen exchange reaction may be quantified by the determination of activation energies for hydrogen diffusion in hydrous minerals (Graham 1981).

2.1.3.3 Other Fractionation Effects

It is well known that clays and shales may act as semipermeable membranes. This effect is also known as "ultrafiltration". Coplen and Hanshaw (1973) demonstrated that both hydrogen and oxygen isotope fractionations may occur during ultrafiltration in such a way that the residual water is enriched in the heavier isotopes, which is due to the preferential adsorption of the heavier isotopes at the clay minerals. This phenomenon has important implications in explaining the isotopic compositions of formation waters.

In salt solutions, isotope fractionation can occur between the water in the "hydration sphere" and the free water (Truesdell 1974). The influence of hydration on the D/H activity ratio is discussed briefly in Sect. 2.3. Very interesting fractionation effects have been observed between crystal water and the mother fluid. Barrer and Denny (1964) and Matsuo et al. (1972) reported that D is depleted in hydrated salts and enriched in the aqueous solution with which salts were in equilibrium. In the gypsum-water system, Matsubaya and Sakai (1973) gave, for instance, a fractionation factor of 0.980 at 25 °C. There are a few examples, such as ice-water, where deuterium is enriched in the solid phase (O'Neil 1968).

Appreciable hydrogen isotope fractionations seem probable in biochemical processes, e.g., during bacterial production of molecular hydrogen and methane (Krichevsky et al. 1961). Cloud et al. (1958) observed that hydrogen gas given off by a bacterial culture was depleted in deuterium by a factor of 20, rather than 3.7 expected if the hydrogen gas had been in isotopic equilibrium with water.

2.2 Carbon

Carbon is one of the most abundant elements in the universe, but it occurs in the earth as a trace element. The average carbon content of the crust and the mantle probably lies in the range of several hundred parts per million. Besides playing a key role in the biosphere, inorganic carbon also exists in a diversity of compounds with different oxidation states like in diamond and in carbon dioxide. This distribution of more oxidized carbon compounds of inorganic origin and of more reduced carbon in the biosphere is an ideal situation for naturally occurring isotope fractionations.

Carbon has two stable isotopes:

^{12}C = 98.89% (reference mass for atomic weight scale),
^{13}C =   1.11% (Nier 1950).

The naturally occurring variations of the carbon isotope composition is greater than 100‰. Heavy carbonates with a δ-value of more than +20 and light methane of around −90‰ have been reported in the literature (see also Fig. 11).

2.2.1 Preparation Techniques

The gas used in all $^{13}C/^{12}C$ measurements is CO_2, for which the following preparation methods exist:

1. a) *Carbonates* are reacted with 100% phosphoric acid at temperatures between 25° and 75 °C (depending on the carbonate) to liberate CO_2 (see also Sect. 2.3).
   b) Thermal decomposition.
2. *Organic compounds* are generally oxidized at ≈ 1000 °C in a stream of oxygen or by an oxidizing agent like CuO. Systems in use have been described, for instance, by Wedeking et al. (1983).

The determination of isotope abundances at specific positions within organic molecular structures can be of great interest. Special degradative techniques have been devised to produce CO_2 quantitatively from the positions of interest (DeNiro and Epstein 1977; Monson and Hayes 1982).

2.2.2 Standards

As the commonly used international reference standard PDB has been exhausted for several years there was a need for introducing new stan-

Table 7. δ^{13}C-values of NBS-reference samples relative to PDB. (After Coplen et al. 1983)

NBS-16	CO_2	-41.48
NBS-17	CO_2	- 4.41
NBS-18	Carbonatite	- 5.00
NBS-19	Marble	+ 1.92
NBS-20	Limestone (Solenhofen)	- 1.06
NBS-21	Graphite	-28.10
NBS-22	Oil	-29.63

dards. Coplen and Kendall (1982) prepared two new standards in the form of gaseous CO_2, which is particularly useful, because it avoids any errors during chemical preparation procedures. Table 7 summarizes the δ^{13}C-values of the presently available NBS standards.

With respect to the oil standard, one may ask whether this material can be regarded as a suitable standard substance, because in a recent intercalibration study Schoell et al. (1983) concluded that - 29.81 is the correct δ^{13}C-value.

2.2.3 Fractionation Mechanisms of Carbon Isotopes

The two main carbon reservoirs, organic matter and sedimentary carbonates, are isotopically quite different from each other because of the operation of two different reaction mechanisms:

1. a kinetic effect during photosynthesis, leading to a depletion of ^{12}C in the remaining CO_2, and concentrating the light ^{12}C in the synthesized organic material;
2. a chemical exchange effect in the system: atmospheric CO_2-dissolved HCO_3^-, which leads to an enrichment of C^{13} in the bicarbonate.

1. Carbon isotope fractionations by organisms grown in the laboratory have been reported by Park and Epstein (1960), Abelson and Hoering (1961), Smith and Epstein (1971), Seckbach and Kaplan (1973), Pardue et al. (1976), Wong and Sackett (1978), Fuchs et al. (1979), and Wong et al. (1979). Recent reviews by Deines (1980b) and O'Leary (1981) have summarized the biochemical background of carbon isotope fractionation during CO_2 uptake.

The main isotope-discriminating steps during biological carbon fixation are (1) the uptake and intracellular diffusion of CO_2 and (2) the first CO_2-fixing carboxylation reaction. Such a two-step model was first proposed by Park and Epstein (1960):

$$CO_{2(external)} \underset{k_2}{\overset{k_1}{\rightleftharpoons}} CO_{2(internal)} \overset{k_3}{\rightarrow} R-COOH .$$

From this simplified scheme follows that the diffusional process is reversible, while the enzymatic carbon fixation is irreversible. However, in specific cases such as aquatic plants, natural conditions are probably further complicated by additional fractionations (i.e., hydration of CO_2). Furthermore, this model nicely explains why organic substances derived from atmospheric CO_2 or oceanic HCO_3^- differ in their ^{13}C-content, namely that the isotopic composition depends upon the carbon source available.

The total isotope fractionation depends on which of the two steps becomes dominant or rate-controlling. Following O'Leary (1981) fractionations associated with k_1 and k_2 are roughly $-4\%o$. The fractionation of the irreversible enzymatic carboxylation reaction (k_3) is considerably larger, but may vary from $-17\%o$ to $-40\%o$ or even lower. The initial chemical product formed during the carboxylation reaction in the majority of plant families is a three-carbon molecule, phosphoglyceric acid. These species are therefore called C_3 plants and this type of photosynthesis is called the "Calvin cycle".

In contrast, carboxylation by phosphoenolpyruvate carboxylase yields a C_4 dicarboxylic acid as the first product of carbon fixation. These C_4 plants discriminate by only -2 to $-3\%o$ relative to bicarbonate which is the active species in this reaction (Reibach and Benedict 1977). This is one of the reasons for the small overall fractionation observed in the C_4 (or "Hatch-Slack") pathway.

Since the work of Park and Epstein (1960) and Abelson and Hoering (1961) it is well known that ^{13}C is not uniformly distributed among the total organic matter, but varies between lipids, carbohydrates, and proteins. Although the causes of these $\delta^{13}C$-differences are not entirely clear, kinetic isotope effects seem to be more plausible (DeNiro and Epstein 1977, Monson and Hayes 1982) than thermodynamic equilibrium effects (Galimov 1973). The latter author postulated that ^{13}C-concentrations at individual carbon positions within organic molecules are principally controlled by structural factors. Approximate calculations suggested that reduced $C-H$ bonded positions are systematically depleted in ^{13}C, while oxidized $C-O$ bonded positions are enriched in ^{13}C. Many of the observed relationships are qualitatively consistent with that concept, however, it is difficult to identify any general mechanism by which thermodynamic factors should be able to control chemical equilibrium within a complex organic structure. Experimental evidence presented by Monson and Hayes (1982) suggests that kinetic effects will be dominant in most biological systems.

2. Isotopic equilibria in the system $CaCO_3 - CO_2 - H_2O$ have been summarized by Usdowski (1982). The overall reaction of carbonate precipitation, either abiotically or by biological consumption, is given by:

$$Ca^{2+} + 2\,HCO_3^- \rightarrow CaCO_3 + H_2O + CO_2$$

with the largest ^{13}C fractionation between $CaCO_3$ and CO_2.

The temperature dependence for carbon solute species with respect to $CaCO_3$ is shown in Fig. 9. At low temperatures ($<50\,^\circ C$) the largest fractionation step occurs between $CO_{2(gas)}$ and HCO_3^-. Several experimental studies have attempted to determine this equilibrium fractionation factor and of these it is generally regarded that the values determines by Mook et al. (1974) are the most reliable. The isotope exchange takes place through the hydration-dehydration of CO_2 gas, which is a relatively slow process (Mills and Urey (1940). Besides the overwhelming importance of equilibrium effects, kinetic effects are also observed in the carbonate-CO_2-H_2O system (Usdowski et al. 1979; Turner 1982; Michaelis et al. 1985). In supersaturated solutions precipitation of calcite occurs without fractionation between solid and solution due to the fact that precipitation is faster than isotopic equilibration.

Other equilibrium fractionation reactions occur in the systems involving calcite, CO_2, graphite, and CH_4, which have been calculated

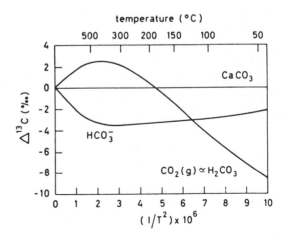

Fig. 9. Temperature dependence of carbon isotope fractionation for carbon solute species with respect to $CaCO_3$. The $\Delta^{13}C$-value is equal to the difference in $\delta^{13}C$-values between the solute species and $CaCO_3$ (note the crossover point on the H_2CO_3 curve at about $190\,^\circ C$ (Robinson 1975)

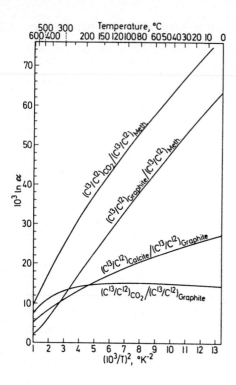

by Bottinga (1969a) (see Fig. 10). Of these, the calcite-graphite fractionation has become a useful geothermometer, especially at high temperatures (i.e., Valley and O'Neil 1981), discussed in more detail on p. 190).

2.2.4 Interactions Between the Carbonate-Carbon Reservoir and Organic Carbon Reservoir

In Fig. 11, δ^{13}C-variations of some important carbon compounds are schematically demonstrated. As has already been mentioned, the two most important carbon reservoirs on earth, the carbonate and the reduced carbon of biological origin are characterized by very different isotopic compositions: the carbonates being isotopically heavy with a mean δ^{13}C-value around 0‰ and the biogenically reduced carbon compounds being isotopically light with a mean δ^{13}C-value around $-25‰$. Furthermore, large ^{13}C/^{12}C-fractionation can be found between the possible decay products of the carbonates — CO_2 formed during decar-

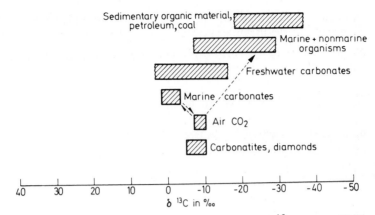

Fig. 11. $^{13}C/^{12}C$ ratios of some important carbon compounds ($\delta^{13}C$ relative PDB)

bonatization — and the decay products of the organic material, CH_4 and CO_2. In a closed system containing these different carbon species, variations in $\delta^{13}C$ can occur as a result of oxidation-reduction reactions. Carbonates with strongly negative $\delta^{13}C$-values are, therefore, usually interpreted as resulting from participation of organic matter carbon (see, for instance, Presley and Kaplan 1968; for interstitial waters, Hoefs 1970 and Sass and Kolodny 1972; for carbonate concretions, Kolodny and Gross 1974; for carbonates from the Mottled Zone, Israel, i.e., thermal metamorphism by combustion of organic matter). In all these cases the relative contribution of organic carbon has been estimated.

2.3 Oxygen

Oxygen is the most abundant element on earth. It occurs in gaseous, liquid, and solid compounds, most of which are thermally stable over large temperature ranges. These facts make oxygen one of the most interesting elements in isotope geochemistry. Oxygen has three stable isotopes with the following abundances (Garlick 1969):

$^{16}O = 99.763\%$
$^{17}O = 0.0375\%$
$^{18}O = 0.1995\%$.

Because of the higher abundance and the greater mass difference, the $^{18}O/^{16}O$ ratio is normally determined. Baertschi (1976) determined the absolute value for the $^{18}O/^{16}O$ ratio in Standard Mean Ocean Water (SMOW) as $(2005.20 \pm 0.45) \times 10^{-6}$. The $^{18}O/^{16}O$ ratio may vary by about 10% or in absolute numbers from about 1:475 to 1:525.

2.3.1 Preparation Techniques

In almost all cases CO_2 is the gas used in mass-spectrometric measurement. Different methods are used to liberate the oxygen from the various oxygen-containing compounds.

The oxygen in *silicates* and *oxides* is usually converted to CO_2 through fluorination with F_2, BrF_5, or ClF_3 in nickel tubes at 500° to 600 °C. Decomposition by carbon reduction at 1000 °C to 2000 °C is suitable for quartz and iron oxides, but not for all silicates (Clayton and Epstein 1958). The liberation of oxygen by F_2, BrF_5, or ClF_3 has been described by Taylor and Epstein (1962), Clayton and Mayeda (1963), and Borthwick and Harmon (1982). Unwanted product gases are removed from the oxygen by cold traps, and excess F_2 is removed by reaction with KBr to form KF. The oxygen is converted to CO_2 over a heated graphite rod.

Care must be taken to ensure quantitative oxygen yields. Low yields caused by inadequate reaction temperatures or reaction times result in anomalous $^{18}O/^{16}O$ ratios, high yields are often due to excess moisture.

Stepwise fluorination techniques are useful in studying isotopic gradients in mineral grains, such as quartz and feldspar which have been hydrothermally altered. Haimson and Knauth (1983) applied a partial fluorination technique to several hydrous silica samples by reducing the amount of fluorine. With this technique they tried to react away various amounts of water, organic matter, and other impurities leaving the stronger silicon-oxygen bonds unreacted. Hamza and Epstein (1980) described a stepwise fluorination procedure to isotopically analyze the oxygen of hydroxyls in several silicate minerals. They suggested that the difference in $\delta^{18}O$-values of the total mineral and the OH group can be used as a geothermometer.

Phosphates must be treated in a similar way first described by Tudge (1960), subsequently modified by Kolodny et al. (1983).

Sulfates are precipitated as $BaSO_4$, and then reduced at 1000 °C with carbon to CO_2 and CO. The CO is converted to CO_2 by sparking between platinum electrodes (Longinelli and Craig 1967). Care has to be taken that no memory effects occur (Sakai and Krouse 1971).

Table 8. Isotope fractionations for various carbonates occurring during CO_2 liberation with phosphoric acid at 25 °C (Rosenbaum and Sheppard 1986)

Carbonate	α	$10^3 \ln \alpha$
Calcite	1.01025	10.20
Aragonite	1.01034	10.29
Dolomite	1.01178	11.71
Siderite	1.01163	11.56

Carbonates are reacted with 100% phosphoric acid at various temperatures between 25° and 150 °C (McCrea 1950; Rosenbaum and Sheppard 1986).

The following reaction scheme:

$$3 \, CaCO_3 + 2 \, H_3 \, (PO_4) \rightleftharpoons 3 \, CO_2 + 3 \, H_2O + Ca_3 \, (PO_4)_2$$

shows that only two-thirds of the oxygen originally present in the carbonates is liberated. Since there are characteristic differences in the isotopic fractionation factors associated with the phosphoric acid liberation of CO_2 from various carbonates, this has to be considered when the isotopic composition of different carbonates is compared (see Table 8).

Wachter and Hayes (1985) demonstrated that careful attention must be given to the concentration of phosphoric acid. In their experiments best results were obtained by using 105% phosphoric acid and a reaction temperature of 75 °C. This high reaction temperature has the advantage of reducing the reaction time, but should not be applied when attempting to discriminate between mineralogically distinct carbonates by means of differential rates of phosphorolysis.

The $^{18}O/^{16}O$ ratio in *water* is usually determined by equilibration of a small amount of CO_2 with a surplus of water and analyzing the resulting CO_2. For this technique the exact value of the fractionation for the $CO_2 \rightleftharpoons H_2O$ equilibrium at a given temperature is of crucial importance. In addition, this fractionation factor enters also all $\delta^{18}O$-values of waters analyzed with respect to carbonate standards and all $\delta^{18}O$-values of silicates. Therefore, it is logical that a number of authors have experimentally determined this fractionation factor ranging from 40.7‰ to 42.4‰ at 25 °C. The more recent determinations by O'Neil et al. (1975) and Brenninkmeijer et al. (1983) suggest that the best value is

$$\alpha^{18}_{H_2O} (CO_2) = 1.04115 \pm 0.00005 .$$

It is also possible to quantitatively convert all water oxygen directly to CO_2 by reaction with guanidine hydrochloride. This technique was described by Dugan et al. (1985) and has the advantage that it is not necessary to assume a value for the H_2O-CO_2 isotope fractionation to arrive at a $\delta^{18}O$-value.

2.3.2 Standards

Two different δ-scales are in use: $\delta^{18}O_{(SMOW)}$ and $\delta^{18}O_{(PDB)}$, because of two different categories of users. The SMOW standard was originally a hypothetical water sample defined by Craig (1961b). Today, the International Atomic Energy Agency (IAEA) distributes two different water standards: Vienna SMOW and SLAP (Standard Light Antarctic Precipitation), with a normalized $\delta^{18}O$-value of $-55.5‰$ (Gonfiantini 1978, 1984). The original standard introduced for the paleotempera-ture determinations was the PDB standard, a Cretaceous belemnite from the Pee Dee formation, which has been long exhausted. The need of introducing additional stable isotope standards was recognized at an IAEA consultants' meeting in 1976. Table 9 gives the $\delta^{18}O$-values of these new standards (Coplen et al. 1983).

Table 9. $\delta^{18}O$-values of isotope standards relative to PDB now being distributed by the NBS. (After Coplen et al. 1983)

NBS 16	CO_2	-36.09
NBS 17	CO_2	-18.71
NBS 18	Carbonatite	-23.00
NBS 19	Marble	-2.19
NBS 20	Limestone (Solenhofen)	-4.14

It is relatively difficult to intercompare results between different kinds of samples, e.g., water and calcium carbonate. The conversion equation of $\delta^{18}O_{PDB}$ versus $\delta^{18}O_{SMOW}$ is

$$\delta_{SMOW} = 1.03086 \, \delta_{PDB} + 30.86‰$$

and for CO_2 samples

$$\delta_{SMOW} = 1.04143_{(\delta PDB-CO_2)} + 41.43\%o$$

and

$$\delta_{SMOW} = 1.4115_{(\delta SMOW-CO_2)} + 41.15\%o \ .$$

2.3.3 Fractionation Mechanisms

The $^{18}O/^{16}O$-ratio varies in nature by about 100‰. These variations result from both equilibrium and kinetic fractionation effects. Figure 12 presents a schematic diagram of the naturally occurring variations of oxygen isotopes.

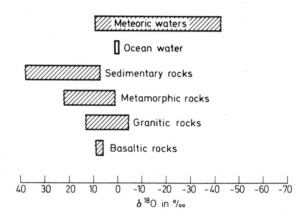

Fig. 12. $^{18}O/^{16}O$ ratios of important oxygen-containing compounds ($\delta^{18}O$ relative SMOW)

2.3.3.1 Equilibrium Exchange Reactions

An excellent consistency in the relative ^{18}O-contents of different minerals can be found in nature. Taylor (1967) made an attempt to arrange coexisting minerals according to their relative tendencies to concentrate ^{18}O (Table 10). This order of decreasing ^{18}O-content is due to a crystal-chemical relationship associated with the relative affinity for ^{18}O. The more highly polymerized the silicate, the greater is the tendency to concentrate ^{18}O, except for the OH group in silicates, which do not follow this relationship. Qualitatively the equilibrium isotope effect might be separated into a bond strength factor and a mass factor. The ^{18}O-rich minerals have the most strongly bonded oxygen and/or the

Table 10. Sequence of minerals in the order (bottom to top) of their increasing tendency to concentrate O^{18} during equilibrium oxygen isotopic exchange

Minerals	δ Value[a]
1. Quartz (tridymite)	15.0
2. Dolomite	14.2
3. K-feldspar, albite	13.0
4. Calcite	12.8
5. Na-rich plagioclase	12.5
6. Ca-rich plagioclase	11.5
7. Muscovite, paragonite	11.3
8. Augite, orthopyroxene, diopside (kyanite, glaucophane)	10.5
9. Hornblende (sphene, lawsonite)	10.0
10. Olivine, garnet (zircon, apatite)	9.5
11. Biotite	8.5
12. Chlorite	8.0
13. Ilmenite	5.5
14. Magnetite, hematite	4.5

[a] The δ-values given above are completely hypothetical, but they are reasonably typical of low- to middle-grade metamorphism of pelitic schists. The minerals in parentheses are less well placed in the sequence than are the major minerals.

oxygen is bonded to cations of the lowest atomic weight or highest ionic potential (charge to size ratio).

Each oxygen atom in quartz is very strongly bonded between 2 Si-atoms and Si-bonds are the strongest in the silicate structures. The Al–O bond is longer and, therefore, weaker than the Si–O bond and the atomic weight of Al is almost the same as that of Si. Therefore, all else being equal, feldspars have lower $^{18}O/^{16}O$ ratios than quartz and calcic plagioclases lower $^{18}O/^{16}O$ ratios than alkali feldspars.

The mass effect for carbonates with divalent cations is demonstrated in Table 11 (O'Neil et al. 1969). As seen in Table 11 the masses of the

Table 11. Oxygen isotope fractionation between various carbonates and water at $250\,^{\circ}C$

Carbonate	1000 ln α	Atomic weight of cation
$CaCO_3$	7.9	40.07
$SrCO_3$	7.4	87.63
$CdCO_3$	6.8	112.41
$BaCO_3$	6.2	137.37
$PbCO_3$	5.6	207.20

It should be mentioned, that the effects are much larger at lower temperatures than $250\,^{\circ}C$.

divalent cations have a relatively small effect on the ^{18}O-content of carbonates. However, carbonates, all being relatively ^{18}O-rich, seem to be primarily influenced by the strongest bond in their structure, namely the bond between the oxygen and the small, highly charged C^{4+} ion. For silicates this seems to imply that the mass effect, i.e., the substitutions of octahedral cations in silicate structures, will generally influence the fractionation behavior slightly compared to the $Al-Si-Fe^{3+}-O$ substitutions.

Kieffer (1982) presented a theoretical model with which oxygen isotope fractionations can be predicted. Her model confirms, in general, the relative order of ^{18}O-enrichment and defines the region in which the fractionation factors do not follow a $1/T^2$ trend. Furthermore, Kieffer (1982) was able to show that crossovers in fractionation factors generally do not occur in silicate minerals of comparable composition. Thus, she found no indication for a crossover between olivine and pyroxene at $\sim 1200\,^{\circ}C$, as has been proposed by Kyser et al. (1981).

On the basis of these systematic trends in the $^{18}O/^{16}O$-ratios of many minerals, it has become apparent that significant temperature information could be obtained up to temperatures of $1000\,^{\circ}C$ and even higher in natural assemblages, if calibration curves could be worked out for the various mineral pairs.

Table 12 summarizes the principal experimental methods that have been applied to determine oxygen fractionation factors. Various criteria have been used to establish the attainment of isotopic equilibrium or the percentage of exchange. For direct mineral-fluid exchange experiments, isotopic equilibrium can be approached from opposite directions. For those experiments where equilibrium is only approached, not attained, an extrapolation technique is used (Northrop and Clayton 1966). This technique assumes that the measured amount of exchange is directly proportional to the distance the system is off from isotopic equilibrium. An important modification of the Northrop and Clayton (1966) technique has been introduced by Matsuhisa et al. (1979) and Matthews et al. (1983a,b) by using a three-isotope exchange method as illustrated in Fig. 13. The initial $^{18}O/^{16}O$-fractionation for the mineral-water system is selected to be close to the assumed equilibrium. In contrast the initial $^{17}O/^{16}O$-fractionation is chosen to be very different from the equilibrium value. In this way the change in the $^{17}O/^{16}O$-fractionations monitor the extent of isotopic exchange, while $^{18}O/^{16}O$-fractionations closely bracket the equilibrium value enabling accurate determination of its value. Using the three-isotope

Table 12. Experimental calibration methods for oxygen-containing mineral systems. (After Sheppard 1984)

Method	System	Remarks	Reference
1. Direct mineral-fluid exchange experiments	Calcite-H_2O Quartz-H_2O	Recrystallization evident	O'Neil et al. (1969) Clayton et al. (1972) Matsuhisa et al. (1979)
2. Direct mineral-mineral exchange	Quartz-K-feldspar Enstatite-basalt magma	Both minerals in common solution Exchange via O_2 or CO_2 $1250°-1500°C$	Blattner and Bird (1974) Muehlenbachs and Kushiro (1974)
3. Crystallization a) Gel b) Glass	Quartz-H_2O Muscovite-H_2O Jadeite-H_2O	Synthesis from silica gel restricted to very limited T-range	Clayton et al. (1972) O'Neil and Taylor (1969) Matthews et al. (1983a)
4. Polymorphic transformation	Calcite-H_2O Quartz-H_2O	Aragonite-calcite Cristobalite-quartz	Clayton (1959) Matsuhisa et al. (1979)
5. Synthesis a) Cation exchange b) Complete	Albite-H_2O Muscovite-H_2O Muscovite-H_2O Magnetite-H_2O Magnetite-H_2O Quartz-H_2O Quartz-magnetite Dolomite-H_2O	K-feldspar + NaCl → Na Feldspar + KCl Paragonite + KCl → Muscovite + NaCl Kaolinite + KCl → Muscovite + HCl + H_2O $FeCO_3 + H_2O → Fe_3O_4 + CO_2 + H_2$ $Fe_2O_3 + H_2 + (H_2O) → Fe_3O_4 + H_2O$ Hydrothermal synthesis of silicic acid $(265-465°C)$ Fayalite + H_2O → Magnetite + quartz + H_2 Calcite → high Mg-calcite → dolomite solution and reprecipitation	O'Neil and Taylor (1967) O'Neil and Taylor (1969) O'Neil and Taylor (1969) O'Neil and Clayton (1964) Bertenrath et al. (1973) Matthews and Beckinsale (1979) Downs et al. (1981) Matthews and Katz (1977)

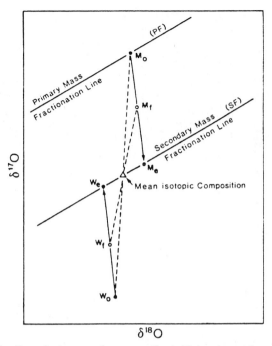

Fig. 13. Schematic diagram of the three-isotope exchange method. Natural samples plot on the primary mass fractionation line (*PF*). Initial isotopic compositions are mineral (M_0) and water (W_0) which is well removed from equilibrium with M_0 in $\delta^{17}O$, but very close to equilibrium with M_0 in $\delta^{18}O$. Complete isotopic equilibrium is defined by a secondary mass fractionation line (*SF*) parallel to *PF* and passing through the bulk isotopic composition of the mineral plus water system. Isotopic compositions of partially equilibrated samples are M_f and W_f and completely equilibrated samples are M_e and W_e. Values for M_e and W_e can be determined by extrapolation from the measured values of M_0, M_f, W_0, and W_f. (After Matthews et al. 1983a; Sheppard 1984)

technique Matthews et al. (1983a,b) provided a consistent set of temperature coefficients for silicate pairs (see Table 13).

For most minerals, the upper temperature limit of these experimental calibrations is around 700 °C. Mayeda et al. (1986) extended the experimental calibrations up to 1200 °C in dry systems, using calcite as the common exchange phase. Most of the experimentally determined equilibrium constants are in good agreement with the calculations of Kieffer (1982).

Isotope exchange reactions between minerals and fluids have been attributed to two major processes, diffusion and dissolution-precipitation. Available oxygen isotope data suggest that the exchange reactions

Table 13. Coefficient A for silicate-pair fractionations. (After Matthews et al. 1983a,b)

	Ab	Cc	Jd	Zo	An	Di	Wo	Mt
Qz	0.5	0.5	1.09	1.56	1.59	2.08	2.20	6.11
Ab	–	0.0	0.59	1.06	1.09	1.58	1.70	5.61
Cc	–	–	0.59	1.06	1.09	1.58	1.70	5.61
Jd	–	–	–	0.47	0.50	0.99	1.11	5.02
Zo	–	–	–	–	0.03	0.52	0.64	4.55
An	–	–	–	–	–	0.49	0.61	4.52
Di	–	–	–	–	–	–	0.12	4.03
Wo	–	–	–	–	–	–	–	3.91

$$(1000 \ln\alpha_{A-B} = \frac{A}{T^2} 10^6)$$

Abbreviations: Qz = quartz; Ab = albite; Cc = calcite; Jd = jadeite; Zo = zoisite; An = anorthite; Di = diopside; Wo = wollastonite; Mt = magnetite.

proceed in two steps, the first through surface controlled dissolution-precipitation reactions when the fluids and minerals are out of chemical equilibrium and when chemical equilibrium is attained through a diffusional mechanism (Matthews et al. 1983c; Cole et al. 1983). The most obvious general trend with respect to diffusion rates is the difference of several orders of magnitude between wet and dry conditions. Relatively high diffusion rates are observed in hydrothermal experiments probably due to lowering of the activation energies, whereas under dry conditions exchange rates are characterized by high activation energies and slow diffusion rates (Graham 1981; Freer and Dennis 1982, Dennis 1984; Giletti and Yund 1984).

2.3.3.2 Fractionations Due to Kinetic Processes

Oxygen isotope fractionations occurring during photosynthesis (Dole and Jenks 1944) and respiration (Lane and Dole 1956) are kinetic processes. This is discussed in more detail on p. 145.

2.3.3.3 Fractionations Due to Other Processes

Vapor Pressure Difference. Variations found in the isotopic composition of natural waters are due to vapor pressure differences. The light isotopic component has a higher vapor pressure than the heavy one, as has already been discussed.

Hydration. Oxygen isotope fractionation occurs during the hydration of most ions. The distinction between the activity isotope ratio and the concentration isotope ratio of water is rarely made and is only of academic interest in dilute solutions, since the difference between these ratios is within the precision of the method. However, Feder and Taube (1952), Taube (1954), Sofer and Gat (1972), and Truesdell (1974) have reported significant deviations of the $^{18}O/^{16}O$-activity ratio from the atom ratio of water in highly saline solutions. These effects have to be considered in the study of naturally occurring brines, such as oil field brines, hydrothermal solutions, and strongly evaporated water bodies, e.g., the Dead Sea.

2.3.4 Water-Rock Interaction

Oxygen isotope geochemistry is especially useful when applied to the study of water/rock interactions. The geochemical effect of such an interaction between water and rock or mineral is a shift of the oxygen isotope ratios of the rock and/or the water away from their initial values. Two end-member models can be considered:

1. Rock $\gg$ water. In this case, the $\delta^{18}O$-value of the rock remains unchanged and that of the fluid is modified. Examples are the positive $\delta^{18}O$-shifts up to 15‰ observed in geothermal waters (for instance, Craig 1963).
2. Water $\gg$ rock. The $\delta^{18}O$-value of the rock is modified and that of the fluid remains constant. Examples are the effects observed in submarine weathered basalts which have interacted with an infinitely large reservoir of seawater.

Besides the water/rock ratio, the other factor which affects the oxygen isotope systematics is the temperature of alteration. These parameters can vary independently, and, therefore, it is difficult to assess the relative importance of these two variables. Water rock ratios have often been calculated from oxygen isotope analyses of whole rocks and fluid, using a simplified, closed-system, material balance calculation (Taylor 1974, 1977):

$$W/R = \frac{\delta_{rock}f - \delta_{rock}i}{\delta_{H_2O}i - (\delta_{rock}f - \Delta)} ,$$

where $\Delta = \delta_{rock}f - \delta_{H_2O}f$. This model requires adequate knowledge of both the initial (i) and final (f) isotopic states of the system.

The above model bases on closed-system conditions, which assume continuous recirculation and cyclic reequilibration of the water with the rock. However, some of the heated water will be lost from the system by escape to the surface. In the open system, in which each increment of water makes only a single pass through the rock, the water/rock is lower than under closed-system conditions (Taylor 1977, 1978).

The intergrated W/R ratio is given by the equation:

$$W/R = \log_e [W/R_{closed\,system} + 1] .$$

In Fig. 14 comparisons are made between the "closed" and "open" system for the two specific parameters 500° and 300 °C and a $\delta^{18}O^i_{H_2O}$ of $-14‰$.

In actual hydrothermal systems, the true 500 °C curve would lie between the two 500 °C curves plotted in Fig. 14, probably close to the right-hand curve, as the single-pass system is an unrealistic one. It should be noted that such models only give minimum values of W/R because appreciable water may move through fractures without exchanging (i.e., after the wall rocks next to the fractures have become already markedly depleted in ^{18}O).

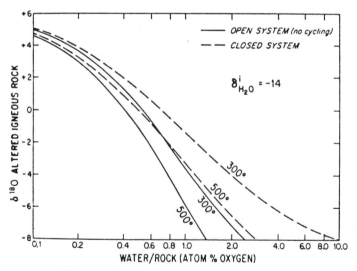

Fig. 14. $\delta^{18}O$-values of altered igneous rocks the calculated water/rock ratios for open and closed system conditions, assuming initial values of $\delta^i_{rock} = +6.5$ and $\delta^i_{H_2O} = -14‰$ (Taylor 1978)

2.4 Sulfur

Sulfur has four stable isotopes with the following abundances (Mac-Namara and Thode 1950):

^{32}S: 95.02%

^{33}S: 0.75%

^{34}S: 4.21%

^{36}S: 0.02%

Sulfur is present in nearly all natural environments: as a minor component in igneous and metamorphic rocks, mostly as sulfides; in the biosphere and related organic substances, like crude oil and coal; in ocean water as sulfate and in marine sediments as both sulfide and sulfate. It may be a major component: in ore deposits, where it is the dominant nonmetal, and as sulfates in evaporites. These occurrences cover the whole temperature range of geologic interest. Sulfur is bound in various oxidation states, from sulfides to elemental sulfur, to sulfates. From these facts it is quite clear that sulfur is of special interest in stable isotope geochemistry.

Thode et al. (1949) and Trofimov (1949) were the first to observe wide variations in the abundances of sulfur isotopes. Today, variations on the order of 150‰ have been found, e.g., the "heaviest" sulfates have δ^{34}S-values greater than +90‰ and the "lightest" sulfides have δ-values of around −65‰. For a schematic diagram, see Fig. 15.

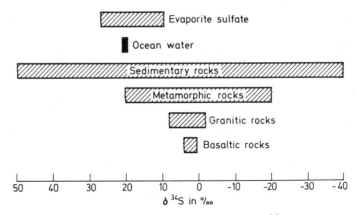

Fig. 15. ^{34}S/^{32}S ratios in some geologically important materials (δ^{34}S relative CD troilite)

The following papers have summarized the whole field or broader aspects of the naturally occurring variations: Thode (1970), Rye and Ohmoto (1974), Nielsen (1978, 1979), Ohmoto and Rye (1979).

2.4.1 Preparation Techniques

Some aspects concerning the chemical preparation of the various sulfur compounds have been discussed by Rafter (1957), Ricke (1964), and Robinson and Kusakabe (1975). The gas used in mass-spectrometric measurement is usually SO_2. Puchelt et al. (1971) and Rees (1978) described a method using SF_6 which has some distinct advantages, because (1) it is without any memory effect and (2) fluorine is monoisotopic avoiding any correction of the raw data.

Pure sulfides are concerted to SO_2 by reaction with an oxidizing agent, like CuO, Cu_2O, V_2O_5, and O_2. For any method used, it is particularly important to minimize the production of sulfur trioxide and sulfates. Combustions in vacuum with a solid oxidant minimize the presence of contaminant gases, particularly CO_2, and purification of the SO_2 is often unnecessary.

For the extraction of sulfates and total sulfur a suitable solvent and reducing agent are needed. Thode et al. (1961) used a reducing agent which was a mixture of HCl, H_3PO_2, and HJ. Tin(II)-phosphoric acid (Kiba solution) has been used by Sasaki et al. (1979). Sakai et al. (1978) and Ueda and Sakai (1983) described a method in which sulfate and sulfide disseminated in rocks are converted to SO_2 and H_2S, respectively. These authors showed that when sulfate and sulfide were attacked at 280 °C under vacuum by dehydrated phosphoric acid containing stannous ions, both SO_2 and H_2S can be obtained separately, but simultaneously, from the same specimen.

2.4.2 Standard

The reference standard commonly used is sulfur from troilite of the Canon Diablo iron meteorite, while Russian investigators refer to troilite from the Sikhote Alin meteorite.

2.4.3 Fractionation Mechanisms

There are two types of reactions producing mainly the naturally occurring sulfur isotope variations:

- a kinetic effect, during the bacterial reduction of sulfate to "light"
  H_2S, which gives by far the largest fractionations in the sulfur cycle;
- various chemical exchange reactions, e.g., between sulfate and sulfides, on the one hand, and between the sulfides themselves, on the
  other, where there is a definite order of concentrating ^{34}S.

1. The principal organisms which transform sulfate to H_2S are the
sulfate-reducing bacteria belonging to the genera Desulfovibrio and
Desulfatomaculum. These bacteria gain their energy for growth by
coupling anaerobic oxidation of hydrogen and organic matter to the
reduction of sulfate. Many of the environmental limitations on the sulfate-reducing bacteria were reviewed by ZoBell (1958) and are summarized in Table 14. A more recent review was given by Chambers and
Trudinger (1979).

Table 14. Some environmental limits of sulfate-
reducing bacteria (ZoBell 1958)

Factor	Limits
E_H	+350 to −500 mV
pH	4.2 to 10.4
Pressure	1 to 1000 atm
Temperature	0 to 100 °C
Salinity	< 1 to 30% NaCl

Harrison and Thode (1957a,b), Kaplan et al. (1960), Nakai and
Jensen (1964), Kemp and Thode (1968), McCready et al. (1974),
and McCready (1975) showed that fractionations of up to nearly 5%
could be achieved in the laboratory under a variety of conditions. In
the experiments cited above, such parameters as temperature, electron
source, sulfate concentration, and bacterial population density are
varied. The reaction chain during anaerobic sulfate reduction can be
described schematically as follows:

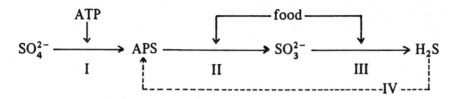

ATP = Adenosine triphosphate, APS = Adenosine-5'-phosphosulfate. For more details, see Goldhaber and Kaplan (1974).

Under normal conditions the rate-controlling step is reaction II, that is the breaking of the first S—O bond. However, extremely low sulfate concentration or extraordinarily high food supply makes reaction I rate-controlling and brings the net fractionation to zero (Kemp and Thode 1968, McCready 1975).

In contrast to the conditions in laboratory cultures, bacterial activity in natural environments is characterized by much slower growth rates and hence a much slower turnover of sulfur through the cells. Then reaction IV, the back reaction between the enzymatically bound sulfur species, may become important (Trudinger and Chambers 1973).

Two different model environments are representative for natural reduction of sulfate to sulfide:

a) The simplest model case is achieved in a body of stagnant water, which has become anoxic due to insufficient vertical mixing. Sulfate-reducing bacteria may grow rapidly under these conditions until the environment becomes poisoned due to H_2S production. Typical examples are the Black Sea and local oceanic deeps. In all these cases, the H_2S is extremely depleted in ^{34}S, while the sulfate consumption and change in $\delta^{34}S$ remain negligible (open-system environment, see also p. 182).

b) When H_2S is continually extracted from the system, for instance, by degassing or by precipitation of iron sulfide, the bacteria continue to reduce sulfate until no more food or sulfate is available. The extraction of the light sulfur isotope from the system changes the observed δ-patterns drastically following the Rayleigh equation, given on p. 10–11: the $\delta^{34}S$ of the residual sulfate increases with decreasing sulfate concentration and, consequently, the H_2S produced at a later stage also exhibits "heavier" $\delta^{34}S$-values. The change in $\delta^{34}S$ of residual sulfate and of produced H_2S with decreasing sulfate concentration follows the relation shown in Fig. 16. The curves indicate that extremely heavy sulfate may be found in the final stage of bacterial activity in a "closed-system" environment.

Fig. 16. Variations of δ^{34}S-values of sulfide-produced and residual sulfate in a closed system (Rayleigh-type fractionation). Assumed fractionation factor: 1.025, assumed starting composition of sulfate +10‰

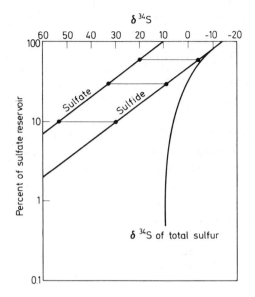

Besides the respiratory sulfate reduction other biological sulfur isotope fractionations occur in nature. Comparatively little is known on isotope effects associated with sulfur-oxidizing organisms. Fry et al. (1986) summarized the recent literature and demonstrated that isotope effects associated with bacterial oxidation of inorganic sulfur compounds are small.

During assimilation of sulfate by microorganisms, plant and animals sulfur isotope fractionation is in the range of −2 to −3‰ (Kaplan 1983). This is similar to the small effect measured during nitrogen fixation (Delwiche and Steyn 1970), but is in marked contrast to assimilated carbon, where a considerable ^{12}C-enrichment occurs. In the case of sulfur and nitrogen, fractionation only occurs in the initial activation step. Once sulfate or nitrogen has been transferred into the cell, it is completely reduced to the protein.

2. Under low temperature conditions (T < 50 °C) bacterial activity seems to be the only mechanism for the reduction of sulfate. However, at appreciably higher temperatures (> 250 °C) sulfate will be reduced to H_2S by reactions with Fe^{2+} components in rocks (Ohmoto and Rye 1979). Such a process is an important mechanism in environments where circulation of ocean water through hot volcanic rocks is established.

At these relatively high temperatures (> 250 °C) probably isotopic equilibrium seems to be established between sulfate and hydrogen sulfide. The isotope exchange between sulfate and sulfide may be written:

$$^{32}SO_4^{2-} + H_2^{34}S \rightleftharpoons {}^{34}SO_4^{2-} + H_2^{32}S \, .$$

The theoretical value for the exchange reaction is $\alpha \approx 1.075$ at 25 °C (Tudge and Thode 1950). Therefore, if this exchange takes place, although no mechanism is yet known, it should lead to sulfides being depleted in ^{34}S by up to 75‰ relative to sulfates. Sakai (1957) has extended this calculation up to temperatures of 1000 °C. Thode et al. (1971) have measured the equilibrium constant K between SO_2 and H_2S in the temperature range 500° to 1000 °C and compared the results with theoretical calculations. Robinson (1973) measured the fractionation between H_2S and HSO_4^- and found fractionation factors slightly lower than Sakai's theoretical values. Sakai (1957) first suggested that isotope fractionation between different metallic sulfides would bring about a slight variation in the isotope ratios during their deposition. Theoretical studies of fractionations between sulfides have been done by Sakai (1968) and Bachinski (1969), who reported the reduced partition function ratios and the bond strength of sulfide minerals and their relationship to isotope fractionation. Similar to oxygen in silicates, a relative order of ^{34}S-enrichment in coexisting sulfide minerals can be arranged (Bachinski 1969; Table 15).

Instead of directly comparing the equations for the fractionation factors suggested by various investigators, Ohmoto and Rye (1979) critically examined all the available experimental raw data in terms of (1) attainment of equilibrium, (2) uncertainties in the measurements, (3) minimum or maximum fractionation factors when equilibrium was not attained, and (4) compatibility with the fractionation factors

Table 15. Equilibrium isotope fractionation factors of important sulfides with respect to H_2S. The temperature dependence is given by the equation $10^3 \ln \alpha = A/T^2 + B$. (After Ohmoto and Rye 1979)

Mineral	Chemical composition	A
Molebdynite	MoS_2	0.45
Pyrite	FeS_2	0.40
Sphalerite	ZnS	0.10
Pyrrhotite	FeS	0.10
Chalcopyrite	$CuFeS_2$	−0.05
Covelline	CuS	−0.40
Galena	PbS	−0.63
Chalcosite	Cu_2S	−0.75
Argentite	Ag_2S	−0.80

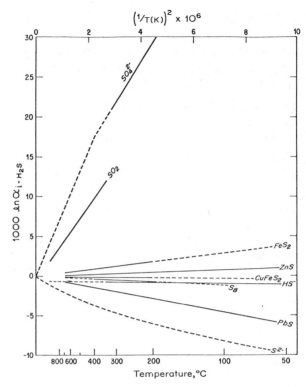

Fig. 17. Equilibrium fractionation factors among sulfur compounds relative H_2S (*solid lines* experimentally determined, *dashed lines* extrapolated or theoretically calculated) (Ohmoto and Rye 1979)

estimated from other sets of experiments. Figure 17 and Table 15 give a summary of what Ohmoto and Rye (1979) believe to be the best fractionation factors relative to H_2S.

3. As already mentioned in Sect. 2.1 both hydrogen and oxygen isotopic compositions of residual water may become heavier during the passage through clay-rich sediments (ultrafiltration). A similar phenomenon was observed by Nriagu (1974) on $\delta^{34}S$-values of sulfate in solution. He demonstrated that ^{32}S was preferentially absorbed into sediments and the remaining sulfate was enriched in ^{34}S up to 6‰. The fractionation factor was dependent on the sulfate concentration in the solution and on the amount of sulfate absorbed in the sediment. This process could play an important role for the sulfur isotopic composition in formation and connate waters.

2.4.4 Experimental Determination of Sulfide Systems

In the last years more and more experimental determinations of the distribution of sulfur isotopes in sulfides under "hydrothermal" conditions have become available. Some of the more recent investigations are those by Kiyosu (1973), Czamanske and Rye (1974), and Hubberten (1980).

Two essentially similar approaches have been used. One approach is to have both sulfides present in the equilibrium vessel but to keep them physically separated and to effect isotope exchange between them via transport of sulfur vapor. The second approach uses hydrothermal solutions instead of a gas phase. The problem involved in this technique comes from the difficulties involved in the exchange reactions between a mineral pair.

The experimental determinations of equilibrium constants performed so far do not agree very well with each other. For example, sulfur fractionations in sphalerite-galena pairs can vary considerably depending on which experimental curve is used. For a fractionation of 2‰ the calculated temperatures from the three experimental curves are 319° ± 15 °C (Czamanske and Rye 1974), 286° ± 19 °C (Grootenboer and Schwarcz 1969), and 360° ± 30 °C (Kajiwara and Krouse 1971). Rye (1974) has argued that the Czamanske and Rye (1974) curve gives the best agreement with filling temperatures of fluid inclusions over the temperature range from 370° to 125 °C. Mineral pairs may give geologically reasonable temperatures as long as the two minerals were formed in equilibrium with the ore-bearing solution which in turn was uniform in temperature and in physicochemical conditions such as f_{O_2} and p_H . Sphalerite-galena pairs in many deposits give reasonable temperatures even when these minerals are not contemporaneous, suggesting that during the deposition of sphalerite and galena the conditions have not changed very much. Examples for which galena-sphalerite pairs agree well with the filling temperatures of fluid inclusions have been reported from Providencia (Mexico), Casapalca (Peru), and Creede (Colorado, USA) by Rye (1974). With the sphalerite-galena pair temperatures can be determined to within ±40 °C, even taking into account the uncertainty in the calibration curves. This sulfur isotope geothermometer is especially useful on massive or metamorphosed ore deposits where other methods such as filling temperatures in fluid inclusions fail.

Pyrite-galena pairs, on the other hand, are not very suitable for a temperature determination, because pyrite often seems to precipitate over larger portions of ore deposition than galena and the chemistry of the solutions and the temperature might have changed considerably over this period.

2.5 Selenium

Because selenium is, to some extent, chemically similar to sulfur, one might except to find some analogous fractionations of the selenium isotopes in nature.

Six stable selenium isotopes are known with the following abundances:

^{74}Se: 0.87% ^{78}Se: 23.52%

^{76}Se: 9.02% ^{80}Se: 49.82%

^{77}Se: 7.58% ^{82}Se: 9.19%

^{82}Se/^{76}Se ratios have been determined by Krouse and Thode (1962), Rees and Thode (1966), and Rashid et al. (1978). Selenium is extracted in its elemental form from natural samples and then fluorinated to SeF_6 which is introduced into the mass spectrometer.

From theoretical calculations Krouse and Thode (1962) deduced that ^{76}Se and ^{82}Se differ in their chemical properties to the extent that isotope fractionations up to 60‰ are predicted for ^{76}Se/^{82}Se exchange processes, provided that mechanisms are available for such exchanges.

In this connection it is interesting to note that in addition to sulfate-reducing bacteria, other anaerobic bacteria are known to reduce selenates and selenites. Rashid et al. (1978) showed that numerous organisms fractionate selenium isotopes during SeO_3^{2-} reduction. For instance, six different Salmonella species were found to reduce $^{76}SeO_3^{2-}$ faster than $^{82}SeO_3^{2-}$ with δ^{82}Se values ranging from -5‰ to -40‰.

2.6 Nitrogen

More than 99% of the known nitrogen on or near the earth's surface is present as atmospheric N_2 or as dissolved N_2 in the ocean. Only a minor amount is combined with other elements, mainly C, O, and H. Nevertheless, this small part plays a decisive role in the biological world. Since nitrogen occurs in various oxidation states and in gaseous, liquid, and solid forms, it is a highly suitable element for the search of natural variations in its isotopic composition.

Nitrogen consists of two stable isotopes, ^{14}N and ^{15}N. Atmospheric nitrogen, determined by Nier (1950), has the following composition:

^{14}N: 99.64%

^{15}N: 0.36%

Much progress has been achieved during the last years concerning nitrogen isotope geochemistry (Sweeney et al. 1978; Letolle 1980; Kaplan 1983). One important application is the problem whether it is possible to distinguish fertilizer nitrogen from compounds of natural origin. Artificial fertilizers are systematically depleted in ^{15}N compared with soil organic matter due to their synthesis from air. Therefore, one might expect that $\delta^{15}N$-values of cultivated soil organic matter may show a shift towards lighter values. So far, none of the detailed studies in progress have confirmed this shift (Letolle 1980).

Preparation procedures have been described by Bremner and Keeney (1966) and Ross and Martin (1970). The conversion of organic nitrogen is done by the Kjeldahl procedure. Nitrate is reduced to ammonium and then oxidized to N_2 by the use of lithium hypobromide. Overall analytical precision can be better than ±0.2‰.

As has been mentioned earlier, isotope ratios are measured in the form of N_2. The standard gas used is atmospheric nitrogen (Mariotti 1983). It is imperative that N_2 be free of carbon monoxide, which gives an interfering peak in the mass spectrum.

Basic biochemical reactions of nitrogen are:

1. Nitrogen Fixation

$$N_2 + 3 H_2O \xrightarrow{\text{nitrogenase}} 2 NH_3 + 3/2 O_2$$

In the nodules of the roots of plants many bacteria can convert molecular nitrogen into nitrogen compounds. However, the high energy needed to break molecular nitrogen makes natural nitrogen fixation a very inefficient process.

2. Nitrification. When organic matter decays in the soil, bacteria utilize the complex nitrogen-containing molecules to form ammonia, which is oxidized by nitrifying bacteria first to NO_2^- and subsequently to NO_3^-. Conversion of ammonia into nitrate can thus be written as:

$$NH_3 \quad + 3/2\ O_2 \ \rightarrow \ HNO_2 + H_2O$$
$$KNO_2 + 1/2\ O_2 \ \rightarrow \ KNO_3\ .$$

Bacteria such as Nitrosomonas can accomplish the first oxidation while Nitrobacter the second. Under normal aerobic conditions, nitrate is the most common form of combined nitrogen in the ocean. Nitrate can be utilized by vascular plants.

3. Denitrification. Denitrification by organisms such as Pseudomonas denitrificans or Thiobacillus denitrificans results in the conversion of nitrate to nitrogen gas and seems to be the only significant mechanism to convert combined nitrogen into N_2. Denitrification takes plase in poorly aerated soil and in stratified anaerobic water bodies. Denitrification supposedly balances the natural fixation of nitrogen. If it did not occur atmospheric nitrogen would be exhausted in less than 100 million years.

A model for denitrification may involve two consecutive steps:
a) Uptake of substrate into the cell with little or no isotope fractionation.
b) Reduction of the substrate with breaking of an $N-O$ bond which produces a large isotope effect. It is generally agreed that biological denitrification occurs as the following reaction sequence $NO_3^- \rightarrow NO_2^-$ $\rightarrow NO \rightarrow N_2O \rightarrow N_2$. The denitrification step $NO_2^- \rightarrow N_2O$ was investigated by Mariotti et al. (1982). The fractionation factor changed from 10 to 30‰, the largest values were obtained under lowest reduction rates. Generally, the same factors that influence isotope fractionations during bacterial sulfate reduction are also operative during bacterial denitrification.

The fractionation factors for the three main processes involved in biogenic utilization are summarized in Table 16. As Table 16 demonstrates, isotope fractionations during nitrogen fixation are small compared to bacterial nitrification and denitrification. Equilibrium processes are ammonia volatilization from ammonia ion and the solution of nitrogen gas. The first process only seems to have a significant isotope effect.

Since in igneous rocks the main source of nitrogen seems to be present as ammonium, temperature dependence of the $NH_4^+ - NH_3$ isotope exchange reaction is of great importance. Experimental data by

Table 16. Nitrogen isotope fractionation factors for the major biological processes occurring in nature [calculated fractionation factors among gaseous nitrogen compounds by Richet et al. (1977) are excluded]

Kinetic Isotope Effects

Nitrogen fixation	Atm. $N_2 \to$ fixed nitrogen	$\alpha = 1.000$	Hoering and Ford (1960)
		$\alpha = 1.004$	Delwiche and Steyn (1970)
Nitrification	$NH_4^+ \to NO_2^-$	$\alpha = 1.02$	Miyake and Wada (1971)
		$\alpha = 1.035$	Mariotti et al. (1981)
Denitrification	$NO_3^- \to N_2$	$\alpha = 1.02$	Miyake and Wada (1971)
			Wellmann et al. (1968)
			Delwiche and Steyn (1970)
		$\alpha = 1.01-$ 1.03	Mariotti et al. (1982)

Equilibrium Isotope Effects

Ammonia volatilization

$NH_{4(aq)}^+ \rightleftharpoons NH_{3(g)}$ 25 $^\circ$C	$\alpha = 1.034$	Kirschenbaum et al. (1947)

Solution of nitrogen gas

$N_{2(soln.)} \rightleftharpoons N_{2(g)}$ 0 $^\circ$C	$\alpha = 1.00085$	Klots and Benson (1963)

Nitzsche and Stiehl (1984) indicate fractionation factors of 1.0143 at 250° and of 1.0126 at 350 $^\circ$C.

Other fractionation effects affecting the ammonium ion may be ion exchange in soils. Delwiche and Steyn (1970) studied the fractionation between aqueous solution and kaolinite clays, however, a general conclusion on the efficiency of this process in natural systems cannot yet be made.

Still another process causing isotope fractionations might be diffusional migration of natural gas. Stahl et al. (1977) concluded that migration of nitrogen within porous rocks results in decreasing ^{15}N-contents with increasing migration distance. Into the same direction point the large differences found in mantle-derived rocks, which may be explained by differential degassing of mantle nitrogen. This point will be discussed later on p. 85.

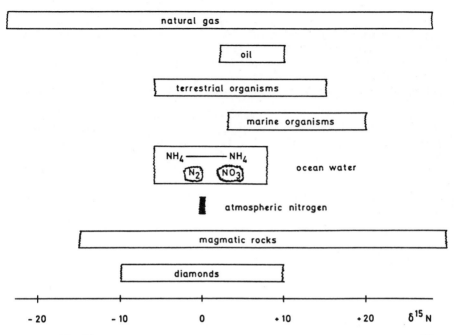

Fig. 18. $^{15}N/^{14}N$ ratios of some geologically important nitrogen compounds ($\delta^{15}N$ relative atmospheric nitrogen)

These various fractionation mechanisms lead to considerable differences in $^{15}N/^{14}N$ ratios of natural substances (see Fig. 18).

2.7 Silicon

Silicon has three stable isotopes with the following abundances (Bainbridge and Nier 1950):

^{28}Si: 92.27%
^{29}Si: 4.68%
^{30}Si: 3.05%

We might expect small fractionations, small, because there are no redox reactions: Si is always bound to oxygen. Furthermore, no geologically important gaseous silicon compound is known and the liquid components are of minor importance.

Douthitt (1982) has summarized the literature data and added 132 measurements on terrestrial materials. The total variation range is

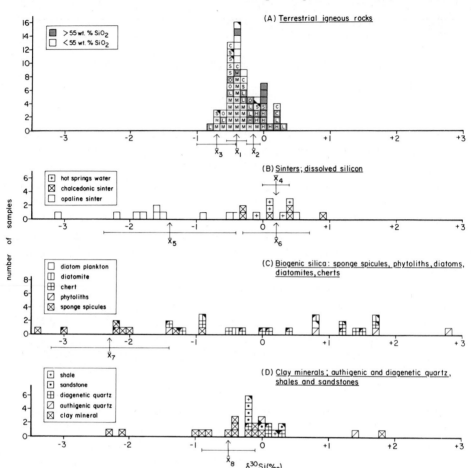

Fig. 19. Histogram of δ^{30}Si-values for various terrestrial samples (δ^{30}Si-values relative CalTech Rose Quartz Standard) (Douthitt 1982)

6.2‰ (see Fig. 19). However, δ^{30}Si-values show little variation (1.1‰) in igneous rocks and minerals. Coexisting minerals exhibit small, systematic, silicon isotope fractionations. In a general way δ^{30}Si increases with the silicon contents of igneous rocks and minerals. Relatively large fractionations occur in opaline sinters, biogenic opal, clay minerals, and authigenic quartz. A kinetic isotope fractionation of ~3.5‰ has been postulated by Douthitt (1982) to occur during the low precipitation of opal and possibly, poorly ordered phyllosilicates. This fractionation coupled with a Rayleigh precipitation model is capable of explaining most nonmagmatic δ^{30}Si-variations.

2.8 Boron

Boron has two stable isotopes (Bainbridge and Nier 1950):

^{10}B: 18.98%

^{11}B: 81.02%

Boron is well known to be a highly mobile element geochemically and, therefore, variations in its isotopic composition might be expected. The large relative mass difference between ^{10}B and ^{11}B (only hydrogen and oxygen have larger relative mass differences) and large chemical isotope effects (Bigeleisen 1965) make boron a very promising element to look for isotope variations. Because boron is difficult to handle in mass-spectrometric measurements, data on boron isotope variations, reported in the past, partly contradict each other. However, recent studies by Spivack and Edmond (1986) and Swihart et al. (1986) obviously have overcome these experimental difficulties, thus a renewed interest in boron isotope variations is expected for the very near future.

Since earlier studies by Schwarcz et al. (1969) it is well known that there is a tremendous kinetic isotope effect during the absorption of dissolved ^{10}B onto clay minerals. The magnitude of this isotope effect depends on the specific clay minerals, the solution temperature, and the boron concentration. This isotope effect is thought to be responsible for the observed ^{11}B-enrichment of ocean water by about 4% relative to crustal material. It is also reflected in marine and nonmarine evaporite borates (Swihart et al. 1986). Thus, it should be possible to use boron isotopic compositions as a tracer for boron of marine origin.

Spivack (1985) demonstrated that the mean $\delta^{11}B$ of unaltered mid-ocean ridge basalts is $-3.6 \pm 0.5‰$ relative to NBS 951 Boric Acid Standard. Altered basalts have $\delta^{11}B$-values between 0 and 9‰. Hydro-thermal solutions from midoceanic ridges fall on a mixing line between ocean water and fresh basalt (Spivack 1985).

As shown by Kanzaki et al. (1979) the boron isotopic composition of high-temperature fumarolic products should be closely related to that of the parent magma. Based on such fumarolic products, Nomura et al. (1982) proposed an average andesite value of about 6‰ for Japanese island arc rocks.

2.9 Chlorine

Chlorine has two stable isotopes with the following abundances (Boyd et al. 1955):

^{35}Cl: 75.53%
^{37}Cl: 24.47%

Older measurements by Hoering and Parker (1961) reported no detectable differences in the isotopic composition lying outside the analytical precision. By using methyl chloride as the gas being introduced into the mass spectrometer, Kaufmann et al. (1984) reported isotopic variations up to 1.3‰ in natural chlorides. Dissolved chloride in groundwaters shows both enrichment and depletion with respect to seawater chloride. Kaufmann et al. (1984) suggested that variations in the chlorine isotope ratio may be significant in diffusion-controlled systems. And, indeed, Desaulniers et al. (1986) observed a preferential upward diffusion of ^{35}Cl relative to ^{37}Cl in slow flowing groundwater systems.

In brines from deep aquifers Kaufmann et al. (1986) observed a positive correlation between δ^{37}Cl-values and chloride concentrations. They suggested that diffusion processes may be responsible for this apparent relationship.

2.10 Calcium

Isotopic variations of the alkali and alkaline earth elements have been reported in the literature, however, the results are often contradictory and will not be repeated here. We are still unable to give a well-defined picture of the possible naturally occurring differences. Calcium is the only element which has been recently analyzed with a sophisticated technique (Russel et al. 1978). Calcium has six stable isotopes in the mass range of 40 to 48 with the following abundances (Bainbridge and Nier 1950).

^{40}Ca: 96.97%
^{42}Ca: 0.64%
^{43}Ca: 0.145%
^{44}Ca: 2.06%
^{46}Ca: 0.0033%
^{48}Ca: 0.185%

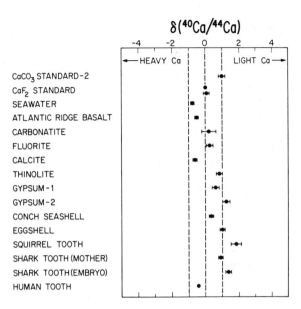

Fig. 20. $^{40}Ca/^{44}Ca$ ratios of some terrestrial samples [δ-values relative to a CaF_2 standard (Russell et al. 1978)]

Due to the large mass difference between ^{40}Ca and ^{48}Ca (the percent mass difference is the next highest after ^2H and ^1H) and due to the low abundance of all Ca isotopes other than ^{40}Ca, Ca isotope abundance measurements are especially subjected to instrumental fractionation effects. By using a double-spike technique and by using a mass-dependent law for correction of instrumental mass fractionation Russell et al. (1978) were able to demonstrate that differences in the $^{40}Ca/^{44}Ca$ ratio are clearly resolvable to a level of 0.5‰. Russel et al. (1978) found (1) no samples which have Ca fractionated by more than 2.5‰, (2) meteorites, lunar, and terrestrial samples show the same small range of $^{40}Ca/^{44}Ca$, and (3) small Ca isotope fractionations (∼2.5‰) are definitely present in nature. Ca of biological origin does not show fractionation effects larger than observed for nonbiogenic samples. In Fig. 20 the variation range for the terrestrial samples analyzed by Russel et al. (1978) are given.

Chapter 3   Variations of Stable Isotope Ratios in Nature

3.1 Extraterrestrial Materials

3.1.1 Meteorites

Formerly, it was generally agreed that the initial material of the solar system was isotopically homogeneous, but in recent years it has become quite clear that this is not the case. Traces of presolar matter, which have survived in some primitive meteorites, can be recognized by their anomalous isotopic composition, which lies outside the range normally found on the earth and in the solar system. Shima (1986) has summarized the extremes of isotopic variations in extraterrestrial materials.

During the past few years tremendous progress has been achieved in this field. A recent summary of the vast amount of new data has been presented by Pillinger (1984). For carbon and nitrogen isotope studies this success is partly due to the introduction of a new mass-spectrometric measurement technique similar to that used for noble gas techniques, namely the static mass spectrometer. Details for the experimental procedure were published by Gardiner and Pillinger (1979). With this technique sensitivities at the nanogram and even the picogram level are possible with precisions of the order $\pm 1\%_o$ or even better (Wright et al. 1983). A limitation of the static method is that it is not immediately applicable to hydrogen, oxygen, and sulfur because of instrument background, sample degradation, or adsorption problems.

In meteorites and other extraterrestrial materials a totally new class of isotope effects has to be considered, namely isotope fractionation due to spallogenic effects. This includes all effects caused by the interaction of primary and secondary particles, for example, protons, neutrons, α-particles, and mesons from both galactic radiation and solar cosmic radiation. The reaction products of such elemental interactions with cosmic particles cover the total range in mass number up to, and even slightly above, those of the target elements themselves and they are characterized by an approximately equal production rate for the different isotopes of a given element. Consequently, a spallogenic admixture manifests itself most pronouncedly in an enhanced abundance of the rare isotopes.

3.1.1.1 Oxygen

It was the work of Clayton et al. (1973a) on oxygen isotope variations in carbonaceous chondrites which caused a revision of theories regarding solar system formation. Once oxygen isotope anomalies had been discovered, attempts to characterize them and to establish the extent of their occurrence followed by the Chicago group. Excess ^{16}O occurs in all samples of carbonaceous chondrites (C2, C3, and C4) (Clayton et al. 1976, 1977, Clayton and Mayeda 1984; see Fig. 21). Almost 5% excess ^{16}O was found in spinel, pyroxene, and sometimes olivine, smaller amounts of about 1% in melilite, feldspars, and grossular. Clayton et al. (1977) proposed that the first material condensing from the solar nebula had an oxygen isotope composition of $\delta^{18}O - 40$ and $\delta^{17}O - 43‰$. The initial condensates later underwent a partial exchange close to the terrestrial fractionation.

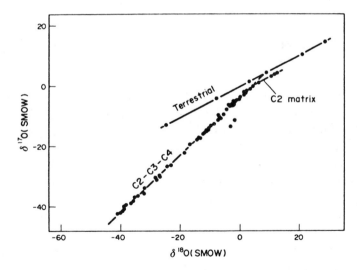

Fig. 21. Three isotope plots for terrestrial samples and some anhydrous phases in C2, C3, and C4 carbonaceous chondrites. The maximum observed ^{16}O excess is 5% (Clayton et al. 1977)

The supposition that measurement of three isotopes can distinguish between nuclear and chemical/physical fractionation effects was questioned by Thiemens and Heidenreich (1983), who found a mass-independent fractionation process during the photochemical dissociation of ozone. The mechanism postulated for the dissociation process also works with other gases such as CO_2, however, it may, of course, be irrelevant with respect to understanding the internal isotopic systematics in meteorites.

Whatever the cause of ^{16}O-isotope variations in meteorites might be, their presence has permitted a classification into seven categories (Clayton and Mayeda 1978). In the order of increasing ^{16}O-contents, they are: (1) L and LL chondrites; (2) H chondrites, (3) E chondrites and aubrites; (4) eucrites, howardites, and diogenites; (5) ureilites; (6) C2 carbonaceous chondrites hydrous matrix; (7) C2, C3, and C4 carbonaceous chondrites anhydrous minerals.

These characteristic ^{18}O-contents are obviously due to the different oxygen isotope compositions of the parent bodies from which the different types of meteorites were derived. There is a peculiar group of achondrites, the so-called SNC meteorites, which are considered to have originated from a large planet, perhaps Mars. Clayton and Mayeda (1983) demonstrated that the SNC achondrites and eucrites each have characteristic oxygen isotope compositions being near the terrestrial isotope composition. The two groups could not have been derived from the same oxygen reservoir and thus probably had different parent bodies.

Clayton and Mayeda (1978) have tried further to relate the above groups to the various iron meteorite classes through the oxygen isotope analysis of minor oxygen-bearing phases in iron meteorites. The aim is to elucidate the genetic links between stones and irons which could have been originally part of the same planetary body.

3.1.1.2 The Volatiles Hydrogen, Carbon, Nitrogen

Before considering the wealth of information currently being obtained from the analysis of individual components, the bulk isotopic composition of carbonaceous chondrites is briefly discussed. Kerridge (1985) analyzed δD, $\delta^{13}C$, and $\delta^{15}N$-values for 26 carbonaceous chondrites and mainly confirmed the ranges observed by earlier workers. In addition, Kerridge (1985) could demonstrate that several chondrites reveal unusual compositions which are not only useful for classification of carbonaceous chondrites, but also strengthen the conclusion that some specimens such as Al Rais and Renazzo are clearly different from other carbonaceous chondrites.

One of the basic questions in the interpretation of meteoritic δD, $\delta^{13}C$, and $\delta^{15}N$-values is the distinction of nuclear from mass-dependent processes. By using stepwise combustion techniques Kerridge (1983) observed systematic patterns of isotopic heterogeneity for H, C, and N in the insoluble organic fraction from the Orgueil and Murray carbonaceous chondrites. The insoluble organic matter contains a number of components with distinct origins, either nucleogenetic or chemical, or both, which were mixed together with little or no equilibration. Kerridge's data cannot be reconciled with a single mass fractionation process acting upon a single precursor composition.

Hydrogen. Considerable efforts have been undertaken in recent years to analyze D/H ratios in meteoritic materials (Robert et al. 1978, 1979a,b; Kolodny et al. 1980; McNaughton et al. 1981, 1982; Becker and Epstein 1982; Robert and Epstein 1982; Fallick et al. 1983; Kerridge 1983; Yang and Epstein 1983).

The D/H ratio of our solar system cannot be obtained directly from a study of the sun, because deuterium is most easily destroyed in thermonuclear reactions. Therefore, the analysis of meteorites is the next best choice, especially of primitive chondrites. In these meteorites hydrogen is bound in hydrated minerals and in organic matter. The most exciting result from these studies is that exceptional deuterium enrichments occur – up to a δD-value of 10000‰ (Yang and Epstein 1983) – in special deuterium-carrier phases associated with the organic matter. Pillinger (1984) suggested that ion-molecule reactions which seem to operate independently of temperature appear to be the only feasible fractionation mechanisms to be responsible for the extreme deuterium enrichments in the organic matter. The δD-values in the phyllosilicates of carbonaceous chondrites are relatively constant at about 110‰ (Yang and Epstein 1982).

Carbon. The $^{12}C/^{13}C$-ratio varies greatly in stars from 4 to > 100 in late-type stars and even the mean interstellar ratio of 60 ± 8 or 67 ± 19 (Penzias 1980) differs appreciably from the terrestrial ratio of 89. Thus, exotic carbon in meteorites should be easily recognizable by its anomalous isotopic ratio. In a direct search for exotic carbon Swart et al.

(1983) measured several carbonaceous chondrites enriched in anomalous noble gases. The Murchison and Allende chondrites contain up to 5 ppm C that is enriched in ^{13}C by up to 1100‰. The $^{12}C/^{13}C$-ratio is approximately 42 compared to 89 for terrestrial carbon.

Part of the carbon isotope variation of the total carbon in carbonaceous chondrites ($\delta^{13}C$-values from – 30 to +65‰) is related to the presence of ^{13}C-enriched carbonate. The ^{13}C-enrichment in the carbonate has been attributed to kinetic isotope effects in Fischer-Tropsch-type reactions in the formation of organic compounds in these meteorites (Lancet and Anders 1970). Yuen et al. (1984) have questioned the Fischer-Tropsch model on the ground that individual hydrocarbons (C_1 to C_5) of increasing molecular weight show decreasing ^{13}C-abundances as might be expected for a kinetically controlled process rather than thermochemical equilibrium. Higher molecular weight alkanes (greater than C_{17}) follow the opposite trend which implies that they were produced by cracking (Gilmour and Pillinger 1983).

Using stepped heating techniques to resolve contamination and to distinguish between different carbon phases, Carr et al. (1983), Grady et al. (1983), and Grady et al. (1985) demonstrated consistent carbon isotope patterns for ordinary chondrites, enstatite chondrites, and ureilite achondrites which have aided in establishing intergroup/class relationships.

Nitrogen. ^{14}N and ^{15}N are synthesized in two different astrophysical processes: ^{14}N during hydrostatic hydrogen burning and ^{15}N during explosive hydrogen and helium burning (Prombo and Clayton 1985). Thus, it can be expected that nitrogen should be isotopically heterogeneous in interstellar matter.

What was considered by Kaplan (1975) to be a wide range of ^{15}N-values in meteorites has continuously expanded (Kung and Clayton 1978; Robert and Epstein 1980, 1982; Grady et al. 1983; Lewis et al. 1983; Prombo and Clayton 1985). The lowest $\delta^{15}N$-value directly measured so far, is – 326‰ in a fraction of an Allende acid residue (Lewis et al. 1983), the highest $\delta^{15}N$-value is 973‰ in a whole-rock sample of the stony-iron meteorite of Bencubbin (Prombo and Clayton 1985). Two possibilities exist to explain the ^{15}N-enrichments: either by extreme chemical or photochemical fractionation processes, which would require very low temperatures below 40 K in a presolar molecular cloud or by nucleosynthetic reactions, which would require failure of homogenization in nova explosions (Prombo and Clayton 1985).

There are three groups of rare achondrites, the so-called SNC meteorites, which differ from other achondrites. It has been suggested that these meteorites may have originated on Mars. This idea has been supported by Becker and Pepin (1984), who found a nitrogen component trapped in a glass inclusion, but not present in the surrounding basaltic matrix, with a $\delta^{15}N$-value of at least as high as 190‰. Such a δ-value can be explained with dilution of a Martian atmospheric component ($\delta^{15}N$: 620±160‰) by either terrestrial atmosphere absorbed on the samples or by indigenous nitrogen from the rock.

3.1.1.3 Sulfur

The most intense research on sulfur isotopes was in the 1960's (Thode et al. 1961; Jensen and Nakai 1962; Hulston and Thode 1965; Monster et al. 1965; Kaplan and Hulston 1966).

Troilite is the most abundant sulfur compound of iron meteorites and has $\delta^{34}S$-values from 0‰ to 0.6‰ relative to Cañon Diablo troilite (Kaplan and Hulston 1966). Stony meteorites also contain a wide variety of sulfur compounds. Monster et al. (1965) and Kaplan and Hulston (1966) separated the various sulfur constituents. The results are summarized in Table 17.

In contrast to most terrestrial environments, sulfate in the Orgueil meteorite is isotopically the lightest sulfur compound. Monster et al. (1965) suggested that kinetic isotope effects in a sulfur-water reaction may be responsible for the genesis of sulfur compounds in meteorites. There is no evidence for biological activity having occurred in meteorites, because biogenic sulfate would be characterized by an enrichment in the heavy isotope. Measurements of more than two isotopes of sulfur might help in identifying genetic relationships between mete-

Table 17. Distribution and isotopic composition of sulfur in the Orgueil carbonaceous chondrite. (After Monster et al. 1965)

Form of sulfur	Sulfur (wt.%)	$\delta^{34}S$
$MgSO_4 \times 7\ H_2O$	2.1	– 1.3
Elemental sulfur	1.8	1.5
Troilite sulfur	0.8	2.6
Other forms of sulfur	0.3	–
Total sample	5.0	0.4

orites in a similar way to that already known from oxygen isotope studies. Hulston and Thode (1965) and Kaplan and Hulston (1966) have measured the $^{33}S/^{32}S$ and $^{36}S/^{32}S$ ratios in meteoritic material. Their data give strong evidence against variations due to inhomogeneities in the processes of nucleosynthesis and due to cosmic ray effects. The occurrence of an excess of ^{33}S and ^{36}S from spallation processes has been demonstrated by Hulston and Thode (1965) in the iron (non-troilite) phase. McEwing et al. (1983) found a slight ^{34}S excess in metal spheroids from the rim of the Canyon Diablo crater which they interpreted as an impact-induced fractionation effect.

3.1.2 Tektites

Tektites are derived from terrestrial rocks, which were hit by asteroids or comets (Bentor 1986). They consist of a silica-rich glass (average 75% SiO_2) resembling obsidian, yet distinct from obsidians in composition and texture.

   Taylor and Epstein (1964, 1966a, 1969) have shown that the oxygen isotope composition of tektites usually ranges from 8.9‰ to 11.8‰. Various tektite groupings based on chemical composition and geographic occurrence all show a systematic increase in ^{18}O with decreasing SiO_2 content. These systematic correlations, as noted by Taylor and Epstein (1969), are caused probably by vapor fractionation of the tektite material during impact.

3.1.3 The Moon

Three different kinds of material are recognized on the lunar surface: (1) crystalline rocks of different composition, (2) brecciated rocks (rock fragment and fine-grained debris) from meteorite impact, and (3) fines or soils with grain sizes down to $< 1\ \mu m$. The crystalline rocks represent deep-seated lunar materials which are generally very poor in volatiles, such as carbon and nitrogen, but rich in sulfur. The brecciated rocks represent an intermediate group, while the dust and fines are heavily influenced through the bombardment of the solar wind.

   Starting with the oxygen isotope composition of the common rock-forming minerals in lunar igneous rocks, one can say that they vary little from one sampled locality to another (Epstein and Taylor 1970,

1971, 1972; Onuma et al. 1970a; Clayton et al. 1973b). The $\delta^{18}O$ of pyroxene is always between 5.3‰ and 5.8‰, olivines are between 4.9‰ and 5.1‰, and plagioclase between 5.6‰ and 6.4‰. The small range of variation implies that the source regions of the rocks are very similar in oxygen isotope composition. Assuming that the source regions are composed largely of a mixture of olivine, pyroxene, and plagioclase, the lunar interior should have a $\delta^{18}O$-value of 5.5‰ ±0.2‰. Lunar igneous rocks are, therefore, essentially identical to many mafic and ultramafic rocks on earth. The fractionations observed among coexisting minerals indicate temperatures of crystallization of ≈ 1000 °C or higher, similar to values observed in terrestrial basalts (Onuma et al. 1970a,b). Furthermore, it seems to imply that the water content of lunar basalts is negligible.

In comparison with other terrestrial rocks, the range of observed ^{18}O-values is narrow. For instance, the group of terrestrial plagioclase exhibits a variation which is at least ten times greater than that for all lunar rocks (Taylor 1968). This difference may be attributed to the much greater role of low-temperature processes in the evolution of the earth's crust and to the presence of water on the earth.

The question of the presence of water on the moon is very important in understanding the origin and conditions of its formation. Small amounts of water have been found in lunar soils (Epstein and Taylor 1970, 1971, 1972; Friedman et al. 1970, 1974). This water is typically present in concentrations between 120 and 360 ppm. Epstein and Taylor (1970, 1971) argued that the absolute concentration and the isotope data strongly suggest that terrestrial atmospheric water vapor is the dominant source of this "lunar water", meaning that the water is largely a result of terrestrial contamination.

3.1.3.1 Sulfur

The most notable feature of the sulfur isotope geochemistry of lunar rocks is their uniformity and proximity to the Canon Diablo meteorite standard. The range of published values is between – 2‰ and +2.5‰, however, as noted by Des Marais (1983), the actual range is likely to be considerably narrower than 4.5‰ due to systematic discrepancies either between laboratories or between analytical procedures. The average $\delta^{34}S$-value of all measurements is 0.55‰, but it is not yet clear whether this small ^{34}S enrichment relative to the Canon Diablo standard reflects the real sulfur isotope composition of lunar rocks or whether it reflects an artifact. The very small range in $\delta^{34}S$-values supports the idea that the very low oxygen fugacities on the moon prevent the formation of SO_2 or sulfate and thus eliminating exchange reactions between oxidized and reduced sulfur species.

3.1.3.2 Nitrogen and Carbon

As shown by Des Marais (1983) nitrogen and carbon abundances are extremely low in lunar rocks. Des Marais (1983) presented compelling evidence that all lunar rocks are contaminated by complex carbon compounds during sample handling. This carbon which is released at relatively low combustion temperatures, exhibits low $\delta^{13}C$-values between – 23‰ and – 31‰, whereas the carbon liberated at higher temperatures has higher $\delta^{13}C$-values. However, we must ask if the extremely low C- and N-contents of lunar materials indicate that the moon is strongly depleted in these elements. Perhaps the losses of carbon and nitrogen in rocks which crystallized at depth were less extensive. Another complication for the determination of the indigenous isotope ratios of lunar carbon and nitrogen arises from spallation effects (which results from the interaction of primary and secondary particles of cosmic ray origin with the lunar surface). These spallation effects lead to an increase in the $\delta^{13}C$- and $\delta^{15}N$-values depending upon cosmic ray exposure ages of the rocks.

   Summarizing, enrichments of the heavy isotopes on the surfaces of the lunar fines are found in ^{13}C, ^{15}N, ^{34}S, and ^{41}K (Barnes et al. 1973). This suggests that all enrichments are due to the same processes, most probably the influence of the solar wind. Detailed interpretation of their isotopic variations is difficult due to the lack of knowledge of the isotopic composition of the solar wind and due to uncertainties of the mechanisms for trapping. However, it seems obvious that the lighter isotopes have been preferentially lost from the moon because of kinetic effects involved in the vaporization-condensation processes.

3.1.4 Mars

Measurements from the Viking mission have shown that the Martian atmosphere consists mainly of CO_2 with traces of N_2, Ar, O_2, CO,

Table 18. Comparison of the isotope ratios in the Martian and terrestrial atmosphere. (After Owen et al. 1977)

Isotope ratio	Mars	Earth
$^{12}C/^{13}C$	90	89
$^{16}O/^{18}O$	500	499
$^{14}N/^{15}N$	165	277
$^{40}Ar/^{36}Ar$	3000	292

and O. The relative abundance of oxygen and carbon isotopes seems to be similar to values measured for the terrestrial atmosphere, however, a strong enrichment in ^{15}N by about 75% relative to the earth has been found (Biemann et al. 1976; Nier et al. 1976; Owen et al. 1977). This ^{15}N-enrichment is attributed to selective escape, implying a higher initial nitrogen abundance during the early stages of Martian history. In Table 18 a comparison of the isotope ratios in the Martian and terrestrial atmosphere is made (Owen et al. 1977).

3.1.5 Venus

The mass spectrometer on the Pioneer mission, which entered the Venus atmosphere on December 9, 1978 measured the atmospheric composition relative to CO_2, the dominant constituent. The $^{13}C/^{12}C$ and $^{18}O/^{16}O$ ratios have been found to be close to the earth value, while the $^{15}N/^{14}N$ ratio is within 20% of that of the Earth (Hoffman et al. 1979).

One of the major problems related to the origin and evolution of Venus is that of its "missing water". There is no liquid water on the surface of Venus today and the water vapor content in the atmosphere is probably not more than 220 ppm (Hoffman et al. 1979). This means that either Venus was formed of material very poor in water or whatever water was originally present has disappeared, possibly by the escape of hydrogen into space. By measuring the D/H ratio Donahue et al. (1982) tested this idea. The D/H ratio they measured was $1.6\pm0.2 \times 10^{-2}$. The 100-fold enrichment of deuterium relative to the Earth is consistent with an outgassing process, however, the magnitude of this process is difficult to understand.

3.1.6 Interplanetary Dust

In the last few years a new type of extraterrestrial material has become available for isotope investigations: micrometeorites collected as stratospheric dust particles by aircraft at 20-km altitude. Ion microprobe measurements of D/H ratios by McKeegan et al. (1985) gave δD-values between – 386 to +2534‰. The hydrogen isotope composition is heterogeneous on a scale of a few microns demonstrating that the dust is unequilibrated. Elemental and molecular ion signals in different dust fragments show that a carbonaceous phase, not water, is the carrier of the hydrogen. Carbon isotope ratios are similar to terrestrial values within the limits of analytical uncertainties.

3.1.7 The Galaxy

The galactic distributions of some stable isotope ratios have been derived from the study of interstellar molecules by means of their radio spectra (see Table 19, after Penzias 1980). Two observations are especially important: (1) the heavy isotopes are enriched substantially in the galaxy relative to the Earth and (2) there is a distinct difference in isotopic composition between the galactic center and the galactic plane.

Table 19. Estimated isotope abundances and corresponding $\approx 90\%$ confidence uncertainties in our interstellar/solar system. (After Penzias 1980)

	$^{13}C/^{12}C$	$^{18}O/^{16}O$	$^{17}O/^{16}O$	$^{14}N/^{15}N$
Galactic plane	1.3 ± 0.2	0.7 ± 0.3	1.2 ± 0.4	1.4 ± 0.2
Galactic center	3.9 ± 1.0	2.0 ± 0.6	3.2 ± 1.0	2.0 ± 0.6

3.2 The Isotopic Composition of Mantle Material

Considerable geochemical and isotopic evidence has accumulated supporting the concept that many parts of the mantle have experienced a complex history of partial melting, intrusion, crystallization, recrystallization, deformation, and alkali metasomatism. The result of this complex history is that the mantle is chemically and isotopically heterogeneous.

Heterogeneities in radiogenic isotopes are relatively easy to detect because the processes which produce basaltic melts and a refractory residue do not fractionate radiogenic isotopes and if they do, these effects are corrected for by measurement of the nonradiogenic isotopes. Heterogeneities in stable isotopes are more difficult to detect: stable isotope ratios are affected by the various partial melting-crystal fractionation processes because they are governed by the temperature-dependent fractionation factors between residual crystals and partial melt and between cumulate crystals and residual liquid.

Sources for information about the composition of the mantle are from the direct analysis of unaltered ultramafic xenoliths brought rapidly to the surface in explosive volcanic vents. Due to the rapid transport these peridotite nodules are in many cases chemically fresh and by most workers considered as the best sample available from the mantle. The other source of information about the mantle is from basalts, which represent partial melts from the mantle. The problem with basalts is that they do not necessarily represent the mantle composition if partial melting causes isotope fractionations relative to the precursor material. Partial melting of Ca-Al-containing peridotites would result in extraction of various basaltic magmas as the Ca-Al-rich minerals were dissolved, leaving behind refractory residues dominated by olivine and orthopyroxene which may differ in the isotopic composition from the original materials. With respect to the volatiles it has to be kept in mind that during partial melting the volatiles will be enriched in the melt and depleted in the parent material. At present, knowledge of isotope fractionations related to this process is more or less nonexistent, but probably this effect should be small at the very high temperatures present in the mantle.

Let us first consider which isotope fractionation processes might have modified the original mantle isotope composition in the samples analyzed. Both, extrapolation of experimental data and theory demand that isotope fractionation factors in the magmatic temperature range should be small ($\alpha \approx 1.000$), but relatively little is known about the exact isotope fractionation factors at very high temperatures ($> 1000\ ^{\circ}C$). For gases, Stern et al. (1968) predicted isotopic reversals to occur, even more complex behavior is expected for liquids and solids.

With respect to volatiles a further complication has to be kept in mind. Assuming that the volatiles are stored in various accessory minerals present in the mantle, the volatiles will be preferentially depleted in the parent material during partial melting and enriched in the melt.

This process alone may cause isotope fractionation. During further ascent of the melts the volatiles are degassed preferentially, which in turn might be accompanied by isotope fractionation. Two different examples might exemplify the possible phenomena. Javoy et al. (1978) observed experimentally a difference of 4‰ between CO_2 and carbon dissolved in a basaltic melt at 1200 °C. The effect of degassing upon the sulfur isotope composition in basalts has been discussed by Sakai et al. (1982). Normally ^{32}S will be enriched in the melt due to the preferential degassing of SO_2 enriched in ^{34}S. At especially high oxygen fugacities loss of SO_2 will cause an enrichment of ^{34}S in the melt, because the sulfate/sulfide ratio is significantly increased.

In summary, the assumption that partial melting of peridotitic mantle material will not produce any measurable fractionations between the magma and the residual phases may be true only to a first approximation. In detail, the effects might be very complex and might cause some small but measurable shifts in the isotopic composition relative to the precursor material. With this in mind, we will now discuss the various evidences of oxygen, hydrogen, carbon, sulfur, and nitrogen isotope compositions of the mantle.

3.2.1 Oxygen

Basalts. There is general agreement that unaltered midocean ridge tholeiitic basalts and ocean island tholeiites have relatively constant $\delta^{18}O$-values with an average value of 5.7 ± 0.5‰ (Muehlenbachs and Clayton 1972a; Pineau et al. 1976a; Kyser et al. 1982). Ocean island alkalic basalts are enriched in ^{18}O by approximately 0.5‰ (Kyser et al. 1982; Harmon and Hoefs 1984; see Fig. 22). These differences in ^{18}O-contents, although not well understood, may be interpreted as indicating a heterogeneous mantle source.

Many island-arc and continental basalts have $^{18}O/^{16}O$ ratios similar to those of their oceanic counterparts, but $\delta^{18}O$-values in excess of 7‰ have been reported from a variety of localities, which will be discussed in more detail under "magmatic rocks".

Spinel and Garnet Lherzolites. Figure 22 illustrates the range in $\delta^{18}O$-values for spinel and garnet lherzolites which is between 5.0 and 7.0, the spinel lherzolites show a pronounced maximum around 5.7 (Javoy 1980; Kyser et al. 1981, 1982).

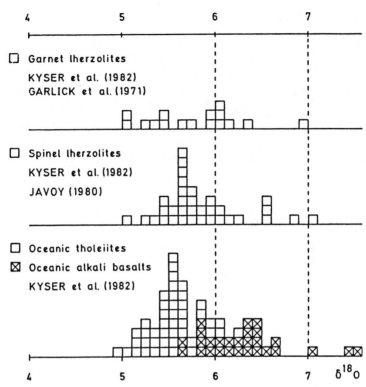

Fig. 22. Oxygen isotope variations in spinel and garnet lherzolites and in oceanic tholeiites and oceanic alkali basalts

When we look at separated minerals the ^{18}O-spread becomes even larger (Javoy 1980; Kyser et al. 1981, 1982). To explain this large spread Kyser et al. (1981) described a temperature-dependent fractionation between coexisting pyroxene and olivine. This interpretation was, however, questioned by Gregory and Taylor (1986), who noted that olivines had a much larger spread (4.4 to 7.5‰) than pyroxenes (5.6 to 6.5‰). They attributed this phenomenon to secondary mantle heterogeneities produced by nonequilibrium processes. This intermineral disequilibrium in the xenoliths does not necessarily demonstrate that the mantle in the basalt source regions will show similar disequilibrium, but is suggestive that this may indeed be the case.

Of special significance with respect to the oxygen isotope composition of mantle-derived rocks is the process of mantle metasomatism in which metasomatic fluids rich in Fe^{3+}, Ti, K, LREE, P, and other LIL elements tend to react with peridotite mantle-forming micas, amphiboles, and other accessory minerals. The origin of metasomatic fluids is likely to be either (1) exsolved fluids from an ascending magma or (2) fluids or melts derived from subducted, hydrothermally altered oceanic crust.

Hydrothermally altered oceanic crust can show large $\delta^{18}O$-variations from 2.0 to ~20‰ (Spooner et al. 1974; Magaritz and Taylor 1976; Gregory and Taylor 1981). During subduction, homogenization processes operate within the ocean crust, which reduce the overall oxygen isotope variation. If the typical scale of homogenization were a few tens of meters within the basalt-sediment section, $\delta^{18}O$-values in melts produced from this volume should be around +10‰. Because such high $\delta^{18}O$-values have not been observed in mantle-derived rocks, the homogenization process must be more effective and operate on a larger scale.

3.2.2 Hydrogen

The concept of juvenile water has a long tradition and has influenced thinking in various fields in igneous petrology and ore genesis. Juvenile water is defined as water that originates by degassing from the mantle and that has never been in contact with the Earth's surface. To analyze the isotopic composition of juvenile water two different approaches have been made: firstly, the direct analysis of water in and from mantle-derived materials, secondly the analysis of OH-bearing minerals such as micas and amphiboles of deep-seated origin (see Fig. 23).

Water. The difficulty in analyzing the water content of basaltic and other mantle-derived materials lies in the fact, that degassing, mixing, and contamination with other water sources may have modified the primary composition. In addition, the fine-grained and glassy matrix material of volcanic rocks is susceptible to low temperature hydrous alteration processes. Altogether this complicates the interpretation of δD-values in basaltic rocks (Kyser and O'Neil 1984). Therefore, it is desirable to analyze separated hydrous minerals rather than whole-rock samples.

OH-Bearing Minerals. Micas and amphiboles in rapidly quenched volcanic rocks will probably preserve their primary isotopic composition,

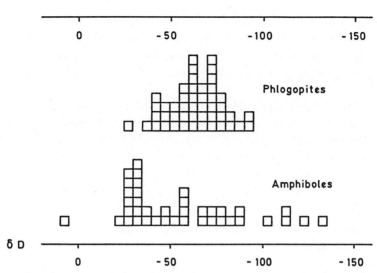

Fig. 23. Deuterium isotope variations in mantle-derived minerals. (Data source: Sheppard and Epstein 1970; Boettcher and O'Neil 1980; Kuroda et al. 1975; Javoy 1980; Graham et al. 1982)

however, because of the limited knowledge on fractionation factors and unknown temperatures of final isotope equilibration between the minerals and water, the calculation of the isotopic composition of water is rather crude and uncertain. Assuming that the temperature of isotopic equilibration lies in the range between 850° and 1000 °C, the fractionation factor between phlogopite and water is around 1.000 (Suzuoki and Epstein 1976), whereas in the case of amphiboles the fractionation factor is about 0.985 (Graham et al. 1984). From these estimates and the phlogopite data shown in Fig. 23, it is quite clear that the hydrogen isotope composition of mantle water should lie, in general, between – 50 and – 80, a range which was first proposed by Sheppard and Epstein (1970) and subsequently by several other authors. Furthermore, Fig. 23 demonstrates that the mantle is heterogeneous with respect to hydrogen isotopes.

The δD-range for amphiboles is much larger than that for phlogopites (Boettcher and O'Neil 1980); the reasons for this are not well understood. The lowest δD-values are from K-richterites, whereas pargasites are relatively D-rich by comparison. One interesting interpretation for the deuterium-enriched amphiboles has been offered by Graham et al. (1982), namely the derivation from seawater via subduction.

3.2.3 Carbon

The presence of CO_2 in the upper mantle has been well documented through the following observations:

1. CO_2 is a quantitatively significant constituent in volcanic gases associated with basaltic eruptions (see discussion in Sect. 3.4.1).
2. The eruption of carbonatite and kimberlite rocks testifies to the storage of CO_2 in the upper mantle. Experimental petrologists agree, for instance, that kimberlitic liquids can be produced by a small degree of partial melting of garnet lherzolite containing H_2O and CO_2 (Eggler and Wendtlandt 1979; Wyllie 1979).

Besides as CO_2, mantle carbon may exist as methane, graphite, and diamond.

Figure 24 summarizes data from fluid-inclusion CO_2, from ocean ridge basalts, from kimberlites and carbonatites, and from diamonds. Most $\delta^{13}C$-values are between -8 and $-3‰$. Although the mean isotopic composition of selected carbonatite complexes is indistinguishable from that of kimberlite pipes as well as that of most diamonds, at a single locality the $\delta^{13}C$-values of carbonatites, kimberlites, and diamonds may vary in the per mil range for different complexes or pipes. Deines et al. (1984) have convincingly documented this heterogeneity

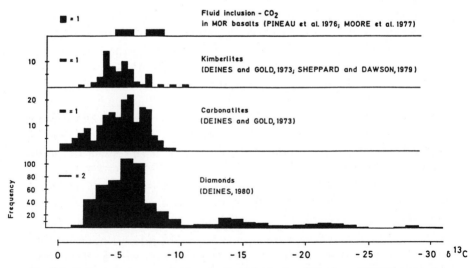

Fig. 24. Carbon isotope variations in fluid inclusion CO_2 from MORB, in kimberlites, carbonatites, and diamonds

by comparing diamonds from the Finsch and Premier kimberlite. Maybe the most puzzling fact is the large variation range observed in diamonds (see Fig. 24). While maybe more than 90% of all diamonds analyzed fall in the restricted δ^{13}C-range between -8 and $-3‰$, some of them are extremely depleted in ^{13}C with δ-values lower than $-30‰$ (Galimov 1985). By discussing different models of diamond genesis Deines (1980a) demonstrated that carbon isotope fractionations during the formation of diamonds should be on the order of a few per mil. Therefore, the observed large range of δ^{13}C-values should reflect mainly carbon isotope variations in the mantle source.

3.2.4 Sulfur

Basalts. As previously mentioned degassing of SO_2, which is the predominant sulfur species at basaltic temperatures, may cause some isotope fractionation in subaerial basalts. Therefore, we will concentrate on the sulfur isotope composition of submarine basalts; where fractionation effects due to degassing should be minimized. As shown in Fig. 25, deep-sea basalts from the Atlantic and Pacific Ocean (data from Kanehira et al. 1973; Grinenko et al. 1975; Hubberten and Puchelt 1980; Sakai et al. 1982; Hubberten 1984) reveal that

1. in general, the primary sulfide sulfur fraction has slightly negative δ^{34}S-values between -1.2 and $0‰$;
2. the sulfate-sulfur fraction, which may amount up to 36%, varies between 2.3 and 5.5‰ (Sakai et al. 1982) and
3. the total sulfur exhibits values in the range between 0.4 to 1.2‰, pointing to small, but nevertheless distinct, heterogeneities.

Sulfur from Peridotites and Mantle Xenoliths. Heilmann and Lensch (1977) analyzed sulfides and whole-rock sulfur from peridotites of the Ivrea Zone. The different bodies showed slightly different δ^{34}S-values with mean values of 0.5, 0.8, and 1.9‰, respectively (see Fig. 25).

In xenoliths of upper mantle origin sulfides are relatively rare, but they occur in some ultramafic and eclogitic xenoliths from kimberlites. Bulk rock lherzolites and garnet pyroxenite xenoliths from the Obnazhennaya kimberlite give δ^{34}S-values from 0.5 to 2.1‰ (Grinenko and Ukhanov 1977), while eclogites from the Premier and Roberts Victor kimberlite yields values between 0.2 to 2.1‰ (Tsai et al. 1979).

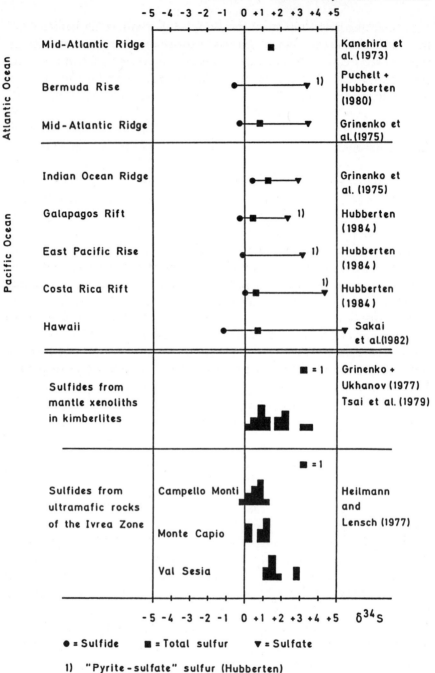

Fig. 25. Sulfur isotope variations in ridge basalts and in sulfides from kimberlites and peridotites

In conclusion, the observed spread of $\delta^{34}S$ mean values between 0.0 and +2‰ indicates mantle heterogeneities and favor a slight enrichment in ^{34}S relative to meteoritic sulfur. The reasons for this enrichment are not at all clear, but might be due to the loss of volatiles preferentially enriched in ^{32}S.

3.2.5 Nitrogen

The data base for evaluating the nitrogen isotope composition of the mantle, although increased in the past few years, is still rather limited. Becker and Clayton (1977) were the first to report a large range in $\delta^{15}N$-values between 6 and 20‰ in mantle-derived rocks. In the meantime more $\delta^{15}N$-values became available and the majority of them show distinct negative $\delta^{15}N$-values (Javoy and Pineau 1986). In diamonds, for instance, which may contain 1.7 to 3.7 vol% N_2 (Melton and Giardini 1974) an inverse correlation between N concentrations and ^{15}N contents exists (Javoy et al. 1984): low ^{15}N contents are correlated with high N concentrations. In basaltic glasses Exley et al. (1986b) observed low N-contents between 0.2 and 2 ppm and $\delta^{15}N$-values between – 4 and 14‰ (average 7.5‰) with no correlation between both parameters. These findings suggest that the primary mantle is depleted in ^{15}N relative to the atmosphere and that the residual nitrogen in the mantle may be enriched in ^{15}N due to degassing processes.

3.3 Igneous Rocks

As was stated in the introductory section, igneous rocks show relatively small differences in isotopic composition because of their high temperature of formation. However, especially due to secondary alteration processes, the variation in isotopic composition is sometimes larger than expected from their temperature of formation. The importance of $^{18}O/^{16}O$ measurements to some of the classical problems of igneous petrology is well established (see, e.g. Taylor 1968).

Provided an igneous rock has not been affected by subsolidus isotope exchange or hydrothermal alteration, its oxygen isotope composition will be determined by:

1. the $^{18}O/^{16}O$ ratio of the source region in which the magma was generated;

2. the temperature of magma generation and crystallization;
3. the mineralogical composition;
4. the evolutionary history of a magma including such additional processes as isotope exchange, assimilation of country rocks, magma mixing, etc.

In the following sections some of these points are discussed in more detail.

3.3.1 Differences Between Volcanic and Plutonic Igneous Rocks

Relatively large differences in O-isotope composition are observed between fine-grained, rapidly quenched volcanic rocks and their coarse-grained plutonic equivalents (Taylor 1968; Anderson et al. 1971). Fractionations in mafic plutonic rocks are on the average about twice as great as for the corresponding fractionations in mafic extrusive rocks. This difference may result from the retrograde exchange or postcrystallization exchange reactions of the plutonic rocks with a fluid phase. This interpretation is supported by the fact that basaltic and gabbroic rocks from the lunar surface have the same "isotopic temperatures" corresponding to their initial crystallization. Due to the absence of water on the moon, no retrograde exchange took place.

3.3.2 Fractional Crystallization

Because fractionation factors between liquid and solid are small at the relatively high temperatures of magmatic melts, fractional crystallization seems to play only a minor role in influencing the oxygen isotopic composition of magmatic rocks. Matsuhisa (1979) reported that within a lava sequence from Japan $\delta^{18}O$ increased by approximately 1‰ from basalt to dacite. Muehlenbachs and Byerly (1982) analyzed an extremely differentiated suite of volcanic rocks at the Galapagos spreading center and showed that 90% fractionation only enriches the residual melt by about 1.2‰. On Ascension Island Sheppard and Harris (1985) observed a difference of nearly 1‰ in a volcanic suite ranging from alkali basalt to obsidian.

In summary, modelling for closed-system, crystal fractionation predicts an ^{18}O-enrichment of about 0.4‰ per 10 wt% increase in SiO_2 content (Sheppard and Harris 1985; Harmon and Hoefs, in press).

3.3.3 Assimilation of Crustal Rocks

Because the various surface and crustal environments are characterized by different and distinctive isotope compositions, stable isotopes provide a powerful tool of discriminating between the relative role of mantle and crust in magma genesis. This is especially true when stable isotopes are considered together with radiogenic isotopes, because variations within these independent isotopic systems arise from unrelated geologic processes. A mantle melt that has been affected by upper crustal contamination will exhibit increases in $^{18}O/^{16}O$ and $^{87}Sr/^{86}Sr$ ratios that correlate with increase in SiO_2 and decrease in Sr content. In contrast, a mantle melt which evolves only through differentiation unaccompanied by interaction with crustal material, will mainly reflect the isotopic compositions of the source region, independent of variations in chemical composition. In this latter case, correlated stable and radiogenic isotope variations would be an indication of variable crustal contamination of the source region, i.e. crustal material which has been recycled into the mantle via subduction. In many igneous provinces a positive correlation between O- and Sr-isotope compositions has been observed. One of the most striking examples is shown in Fig. 26.

The feature presented in Fig. 26 is generally interpreted to reflect a mixing between a mantle-derived melt with low $^{18}O/^{16}O$ and low $^{87}Sr/^{86}Sr$ ratios and a crustal component enriched in both ^{18}O and ^{87}Sr. This mixing is viewed as a combined fractional crystallization-assimilation process (AFC process, Taylor 1980) occurring at relatively shallow depths during magma residence in a magma chamber. Thus, a

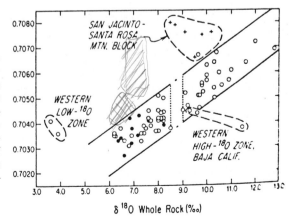

Fig. 26. Whole-rock $\delta^{18}O$-values versus initial $^{87}Sr/^{86}Sr$ ratios for plutonic igneous rocks, Peninsular Ranges batholith. *Solid circles* indicate gabbros. The $\delta^{18}O$ gap between 8.5 and 9.0‰ divides the western group of granodiorites and tonalites from the eastern group (Taylor and Silver 1978)

two-component mixing process is an oversimplification for crustal contamination and at least three end members have to be considered: the magma, the country rocks, and the cumulates.

Modelling by Taylor (1980) and James (1981) has demonstrated that it is possible to distinguish between the effects of source contamination as well as crustal contamination. Magma mixing and source contamination are two-component mixing processes which obey two-component hyperbolic mixing relations, while crustal contamination is a three-component mixing process, involving the magma, the crustal contaminant, and the cumulates.

3.3.4 Secondary Alteration Processes

It is now generally recognized that the emplacement of a hot magma into relatively permeable, water-saturated country rocks commonly initiates the development of a hydrothermal circulation system. As the magma cools and crystallizes, these hydrothermal systems alter the recently solidified igneous rocks. The amount of interaction between igneous rocks and the hydrothermal fluid depends upon the permeability of the affected rocks, the water/rock ratio of the system, and the chemical and isotopic differences between the rock and the fluid. In subaerial regions the circulating fluid is dominantly meteoric water, while in submarine environments the fluid is ocean water. Since both types of water are very different in isotopic composition the observed alteration effects are also different. Therefore, these two different environments are discussed separately.

3.3.4.1 Ocean Water Interaction

Evidence of altered oceanic crust comes from samples dredged from the seafloor, from DSPD drilling sites, and from the study of ophiolite complexes, which presumably represents pieces of old oceanic crust. Two types of alteration can be distinguished: low-temperature weathering may markedly enrich the groundmass of basalts, but does not affect the isotopic composition of phenocrysts. It is well known that volcanic glasses are readily altered to more stable phases, generally smectites or other clay minerals. This low-temperature alteration leads to an ^{18}O-enrichment by 1 to 3‰ of the whole rock, but may increase to about 10‰ (Muehlenbachs and Clayton 1972a,b, 1976). The amount of ^{18}O-enrichment correlates with the H_2O content: the higher the water content, the higher the $\delta^{18}O$-values.

At temperatures in excess of about 300 °C hydrothermal systems beneath the midocean ridges have depleted deeper parts of the oceanic crust by 1 to 2‰. Similar findings have been reported from ophiolite complexes, see, for instance, the oxygen isotope study of the Oman ophiolite by Gregory and Taylor (1981). At Oman maximum $\delta^{18}O$-values occur in the uppermost part of the pillow lavas and decrease through the sheeted dike complex. Normal mantle values are observed at the base of the dike complex and below that depth down to the MOHO $\delta^{18}O$-values are lower than mantle values by about 1 to 2‰. The oxygen isotopic composition of oceanic crust subducted into the mantle in the geologic past will depend upon the variation in the oxygen isotope composition of seawater during earth history. Muehlenbachs and Clayton (1976) have concluded that the ^{18}O-enrichment is balanced by the ^{18}O-depletion in the oceanic crust. Gregory and Taylor (1981) reached the same conclusion which implies that the $\delta^{18}O$ of the oceans has not changed much with time. This point will be also discussed again on p. 142.

3.3.4.2 Interactions Between Meteoric Groundwaters and Igneous Intrusions

Mainly through the work of H.P. Taylor and co-workers it has become well established that many epizonal igneous intrusions have interacted with meteoric groundwaters on a very large scale. The interaction and transport of large amounts of meteoric water through hot igneous rocks produces a depletion of ^{18}O in the igneous rocks by as much as 10 to 15‰ and a corresponding shift in the $\delta^{18}O$-values of the water. Igneous complexes that are abnormally low in ^{18}O characteristically display the following geologic, petrologic, and isotopic features (after Taylor 1974b).

1. The intrusions are emplaced into highly jointed volcanic rocks that are permeable to groundwater movement.
2. The feldspars are commonly depleted in ^{18}O to a greater degree than the other coexisting minerals and the feldspars, particularly the alkali feldspars, commonly show a "clouding" or a turbidity.

3. The primary igneous pyroxenes and olivines are usually altered to amphibole, chlorite, Fe-Ti oxides, and/or epidote.
4. Intergrowths of turbid alkali feldspars and quartz are quite common.
5. Miarolitic cavities are locally present in the intrusives.
6. The OH-bearing minerals have abnormally low δD-values relative to "normal" igneous rocks.

Low ^{18}O-igneous rocks have now been observed in many localities, the best documented are in the Skaergaard intrusion in Greenland, in the Scottish Hebrides, and in portions of the Idaho batholith, the southern California batholith, and the Coast Range batholith, British Columbia (Taylor and Forester 1971, 1979; Magaritz and Taylor 1976, 1986; Criss and Taylor 1983, and others).

These types of hydrothermal systems seem to represent the "fossil" equivalents of the deep portions of modern geothermal systems such as occur at Wairakei, New Zealand, Steamboat Springs, Nevada, and Yellowstone Park, Wyoming. The enormous scale of interaction between meteoric groundwaters and epizonal igneous intrusions is best documented in the Challis Volcanic Field, Idaho, where thousands of square kilometers are involved (Criss et al. 1984). The amounts of meteoric water needed to produce these effects are large but not unreasonable, however, considering the tens of thousands of years during which such hydrothermal systems probably persist. Assuming normal amounts of rainfall, only about 5% of the annual precipitation has to be added to the geothermal system, to produce the observed O-isotope shifts.

Recently Ferry (1985a,b) has integrated the isotope data of Taylor and Forester (1971) and Forester and Taylor (1977) on Skye with chemical and mineralogical changes observed in the gabbros and granites. Both isotopic and petrologic data point to pervasive flow of fluid through gabbros at 500° to 1000 °C and through granites at 450° to 550 °C. The difference between the behavior of the granites and gabbro can be attributed to the difference between the solidus temperatures of the two rock types.

For the granitic rocks Ferry (1985b) postulated that during cooling an interconnected set of channelways along grain boundaries between quartz and feldspar developed due to the volume decrease of quartz during β- to α-transformation. These channelways would allow access to hydrothermal fluids that could then chemically and isotopically react with minerals preferentially at grain boundaries.

3.3.5 Basaltic Rocks

Before discussing the primary isotopic composition of volcanic rocks, secondary factors that can modify the primary composition are discussed briefly. Because of their glass content and very fine grain size volcanic rocks are very susceptible to low-temperature ^{18}O-enrichment processes such as hydrothermal alteration, hydration, and weathering. The whole-rock $\delta^{18}O$-values of all whole-rock samples must be regarded with suspicion, unless the lavas were very recently erupted or unless the results have been checked by $\delta^{18}O$-analysis of the phenocrysts, which, however, is often not possible. Although there is no way to exactly correct for this O-enrichment effect, a crude estimate can be made by determining the water (and carbon dioxide) content of the rocks to be analyzed (Taylor et al. 1984; Ferrara et al. 1985). Although it is difficult to estimate the primary water content of basaltic magmas, there is some general consensus that primary magmas should not have more than 0.5%, certainly not more than 0.9% as determined for some Hawaiian alkali basalts (Moore 1970). Thus, any water content > 0.5% or > 0.9% should be of secondary origin, and has to be corrected for when the primary $\delta^{18}O$-value of the rock is evaluated. Such corrections are schematically demonstrated in Fig. 27.

Basaltic rocks from different tectonic settings exhibit characteristic differences in oxygen isotope composition. This was first suggested by Kyser et al. (1982) and subsequently verified in an extensive compilation by Harmon and Hoefs (1984) (see Table 20). MORB basalts have

Table 20. Comparison of O-isotope variations of basalts erupted in different tectonic settings. (Data from Harmon and Hoefs 1984)

Tectonic setting	Number of analyses	$\delta^{18}O$ range	$\bar{x}\,\delta^{18}O \pm 1\,\sigma$
Midocean ridges	(67)		
Tholeiitic basalts		+5.4 to +6.6	+5.78 ± 0.23
Oceanic islands	(110)		
Tholeiitic basalts		+5.0 to +5.8	+5.38 ± 0.23
Alkali basalts		+5.5 to +8.2	+6.34 ± 0.62
Continental intraplate settings	(92)		
Tholeiitic basalts		+6.4 to +9.0	+6.91 ± 0.59
Alkali basalts		+5.9 to +8.2	+6.88 ± 0.89
Oceanic island arcs	(165)	+5.0 to +9.7	+6.31 ± 0.79
Continental margin arcs	(323)	+5.2 to +8.8	+6.71 ± 1.08

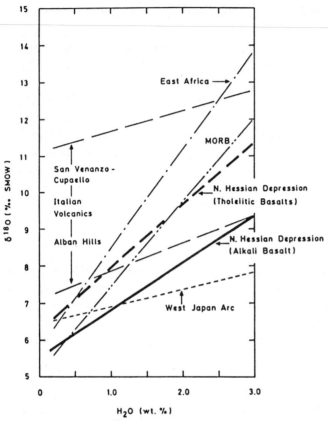

Fig. 27. Plot of $\delta^{18}O$ versus H_2O which compares the effect of secondary water uptake on the oxygen isotope composition of volcanic rocks and glasses of different chemical character (Harmon et al. 1987)

the lowest average $\delta^{18}O$-values (5.82±0.3) with the smallest overall O-isotope variation. The ^{18}O-enrichment observed for continental volcanic rocks compared to their oceanic counterparts of equivalent bulk composition either implies that the subcontinental upper mantle is enriched in ^{18}O due to mantle metasomatism (Holm and Munksgaard 1982) or that contamination by ^{18}O-rich continental crust is an important process in the magma genesis of continental magmas (e.g., Taylor et al. 1979). Although there is strong evidence that the subcontinental mantle shows some enrichment in radiogenic isotopes, evidence for an ^{18}O-enrichment due to metasomatic processes is much weaker. In many cases where ^{18}O-enriched volcanics have been observed these can be best explained by assuming different mechanisms of crustal interaction in different tectonic settings. Examples of such crustal interaction − besides others − have been described from the Andes (Harmon et al. 1984), from the Lesser Antilles (Davidson 1985), and from Antarctica (Hoefs et al. 1980).

The highest ^{18}O-enrichments are found in K-rich leucite-bearing rocks. Taylor et al. (1984) argued that these extreme ^{18}O-enrichments are due to secondary subsolidus processes, because these K-rich volcanic rocks containing leucite are especially susceptible to hydration and other low-temperature alteration processes.

3.3.6 Granitic Rocks

On the basis of their $^{18}O/^{16}O$ ratios, Taylor (1977, 1978) subdivided granitic rocks into three groups: (1) "normal ^{18}O-granitic rocks" with $\delta^{18}O$-values between 6 and 10‰, (2) high ^{18}O-granitic rocks with $\delta^{18}O$-values $> 10‰$, and (3) low ^{18}O-granitic rocks with $\delta^{18}O$-values $< 6‰$.

Although this is a somewhat arbitrary grouping it allows some generalizations.

Many plutonic granites throughout the world have relatively uniform $\delta^{18}O$-values between 6 and 10‰. Granitoids at the low ^{18}O-end of the normal group have been described in island-arc areas where continental crust is absent, such as the Koloulo Igneous Complex of Guadalcanal (Chivas et al. 1982), which are considered to be entirely mantle-derived. Similar $\delta^{18}O$-values have been measured in granites which could be derived by partial melting of lower crustal granulitic rocks. Granites at the high end of the normal granitic rock $\delta^{18}O$-range may have formed by partial melting of crust that contained both a sedimentary and a volcanic fraction. Furthermore, it is interesting to note that many of the "normal" granites are of Precambrian age in which metasediments quite often have $\delta^{18}O$-values below 10‰ (Longstaffe and Schwarcz 1977).

Granitic rocks with $\delta^{18}O$-values higher than 10‰ require derivation from some type of ^{18}O-enriched sedimentary or metamorphic rock. For instance, such high ^{18}O-contents are observed in some Caledonian granitoids in Scotland (Halliday et al. 1980), in Hercynian granites of western Europe (Michard-Vitrac et al. 1980; Hoefs and Emmermann 1983), in Damaran granites, SW Africa (Haack et al. 1982), and in

granites from the Himalaya (Blattner et al. 1983). All these granites
are easily attributed to anatexis within a heterogeneous crustal source,
containing a large metasedimentary component.

Granitic rocks with $\delta^{18}O$-values lower than 6‰ cannot be derived
by any known differentiation process from "normal" basaltic magmas.
Excluding those low ^{18}O-granites which have exchanged with ^{18}O-
depleted meteoric-hydrothermal fluids under subsolidus conditions, a
few primary low ^{18}O-granitoids have been observed. These granites obviously inherited ^{18}O depletions when still predominantly liquid, prior
to their primary crystallization as low ^{18}O-magmas. Examples are described from Iceland (Hattori and Muehlenbachs 1982), from southern
Nevada (Lipman and Friedman 1975), and from Yellowstone (Hildreth
et al. 1984). In all cases, the ^{18}O-depletions can be attributed to the
interactions of magmas with ^{18}O-depleted country rocks in high-level
magma chambers.

3.3.7 Granitic Pegmatites

Stable isotope data on granitic pegmatites are mostly available from
oxygen isotope studies (Taylor et al. 1979; Taylor and Friedrichsen
1983; Longstaffe 1982). The interpretation of these data is complex
because of the small size of these bodies and their volatile, rich nature.
Rather surprisingly, only a relatively small spread in $\delta^{18}O$-values is observed, the $^{18}O/^{16}O$ of quartz, for example, varies from 8.6 to 11.8‰
(Taylor and Friedrichsen 1983). The whole-rock isotopic composition
appears to reflect the composition of the melt. Isotope exchange with
the wall rocks may occur, but the extent is difficult to assess, because
the country rocks intruded by granitic pegmatites often have similar
$\delta^{18}O$-values (Longstaffe 1982). Oxygen isotope fractionations among
pegmatitic minerals yield temperatures for crystallization between 540°
and 750 °C. Hydrogen isotope data have shown that interaction with
meteoric waters is atypical of pegmatite crystallization and emplacement (Taylor et al. 1979; Taylor and Friedrichsen 1983).

3.3.8 Hydrogen in Igneous Rocks

H-isotopes provide the primary guide to the source of the water present
in a magma or in volcanic rocks. δD-values also indicate to what extent

vapor-phase exsolution or assimilation may have affected a magma prior to eruption and crystallization (Nabelek et al. 1983; Taylor et al. 1983).

It is well known that volcanic rocks are quite susceptible to post-eruptive, subsolidus alteration. Kyser and O'Neil (1984) demonstrated that D/H ratios and water contents in fresh submarine basalt glasses from midoceanic ridges can be altered by (1) outgassing, (2) addition of seawater at magmatic temperature, and (3) low-temperature hydration. Extrapolations to possible unaltered δD-values indicate the primary δD-value of most tholeiite and alkali basalts near – 80 ± 5‰ .

The H-isotope composition of subaerial volcanic rocks can be modified by slow inward diffusion of meteoric water into interstitial glass and fine-grained groundmass at low temperature. Such an uptake should produce a decrease in δD-values concomitant with increasing H_2O content. Kyser and O'Neil (1984) observed that the outer rims of volcanic glasses can undergo low-temperature hydration by hydroxyl groups having δD-values of – 100‰ .

In addition to low temperature hydration, exsolution of water from a magma can deplete the magma in deuterium (Nabelek et al. 1983; Taylor et al. 1983). In contrast, a D-enrichment would occur if H_2 or CH_4 is lost from the melt because these gases are strongly depleted in D relative to the melt at magmatic temperatures (Richet et al. 1977).

The hydrogen isotopic composition of biotite and hornblende from granitic rocks in most cases is within the range between – 90‰ and – 50‰ (Godfrey 1962; Sheppard and Epstein 1970; Kuroda et al. 1974; Taylor 1974b). However, quite a significant number of hornblendes and biotites, mostly from shallow intrusions, have lower and more variable δD-values (down to – 180‰). This variation seems to be a consequence of the interaction of meteoric groundwaters of different isotopic composition with magmas or hot igneous rocks. The D/H ratios of igneous rocks and minerals are much more sensitive to such processes than are the $^{18}O/^{16}O$ ratios because igneous rocks contain about 60 atom% oxygen and it thus requires exchange with a very large amount of H_2O to change their $^{18}O/^{16}O$ ratios appreciably. The water/rock ratios in these hydrothermal convective systems are typically so low that over wide areas the only evidence for meteoric water exchange is given by δD-values. Figure 28 illustrates the changes in δD$_{(biotite)}$ and $\delta^{18}O_{(feldspar)}$ due to isotope exchange between a granodiorite and high-latitude meteoric water as a function of the water/rock ratio. Such L-shaped patterns are characteristic of hydrothermal alteration (Taylor 1978).

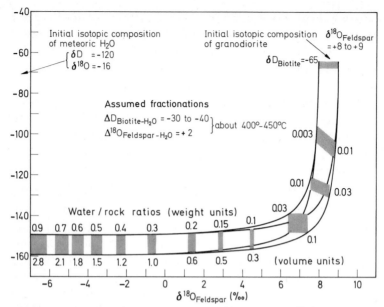

Fig. 28. Plot of calculated values of $\delta D_{(biotite)}$ and $\delta^{18}O_{(feldspar)}$ that would be obtained during hypothetical meteoric-hydrothermal alteration of a typical granodiorite at varying water/rock ratios, assuming the initial conditions shown in the diagram (Taylor 1978)

3.3.9 Carbon and Sulfur in Igneous Rocks

Earlier measurements by Hoefs (1973) and Fuex and Baker (1973) demonstrated that reduced carbon in igneous rocks was isotopically depleted relative to "mantle" $\delta^{13}C$-values. Two models have been developed to explain this duality: Pineau and Javoy (1983) suggested that the isotopically light carbon is the residue after partial outgassing from the magma. By employing a stepwise combustion technique to extract carbon Des Marais and Moore (1984) and Mattey et al. (1984) considered the isotopically light carbon, which was released at low combustion temperatures, to be surficial organic contamination. At very high temperatures isotopically heavy carbon with typical mantle values was released from midocean ridge basalts and glasses. However, island-arc glasses yielded lower $\delta^{13}C$-values which might be explained by mixing of two different carbon compounds in the source regions: a MORB-like carbon and an organic carbon component from subducted pelagic sediments. More measurements from other tectonic settings are needed to confirm this relationship.

Sulfur. Midocean ridge basalts and submarine Hawaiian basalts have a very narrow range in sulfur isotope composition around zero (Sakai et al. 1982, 1984). In subaerial volcanic rocks the variation in δ^{34}S-values is larger and generally shifted towards positive values. One of the reasons for this larger variation is certainly degassing of sulfur. The effects of this process on the sulfur isotope fractionation depends on the ratio of sulfate to sulfide in the magma which is directly proportional to the fugacity of oxygen (Sakai et al. 1982). Especially enriched in ^{34}S (up to δ^{34}S-values of +20‰) are arc-volcanic rocks (Ueda and Sakai 1984; Harmon and Hoefs 1986). This overall enrichment is considered to be mainly a product of mantle heterogeneities resulting from the recycling of marine sulfate during subduction.

3.4 Volcanic Gases and Hot Springs

In the fluid phase of magmatic melts, H_2O is the main constituent, followed by CO_2 and SO_2. Most workers agree that typical concentrations in basaltic and andesitic magmas are in the ranges: H_2O from 35 to 90 mol%, CO_2 from 5 to 50 mol%, and SO_2 from 2 to 30 mol%. The ultimate origin of these volatiles — whether juvenile in the sense that they originate from primary mantle degassing or recycled during subduction processes — is an important question which might be solved by detailed isotope studies in suitable cases.

There are three sources of information on the isotopic composition of the fluid and gaseous phase related to magmatic processes: (1) volcanic gases, (2) fluid inclusions, and (3) hot springs.

3.4.1 Volcanic Gases

The chemical compositions of volcanic gases vary significantly with distance from volcanic vent, eruption history, temperature of collection, and sampling techniques. Volcanic gases are frequently subject to several sources of contamination. It is relatively simple to recognize and correct modifications from atmospheric contamination. More problematic are those which involve usual magmatic constituents such as meteoric water. The complexities involved have been discussed by Anderson (1975) among others.

Water. A long-standing problem is the source of the water vapor observed in many volcanic eruptions. How much is derived from the magma itself and how much is recycled meteoric water? Deuterium measurements can help to solve this problem. As will be discussed below, in most cases meteoric surface waters are the dominant source. However, in a few cases a magmatic water component is not excluded. Fumarolic water from Surtsey, Iceland gave a δD-value of $-53‰$ (Arnason and Sigurgeirsson 1968). Viglino et al. (1985) determined the isotopic composition of fumarole condensates from Augustine Volcano, an active stratovolcano in the Aleutian Arc. The isotopic data for the condensates form a linear $\delta D - \delta^{18}O$ array from low-temperature fluids ($< 100\ °C$), which are essentially local meteoric water ($\delta D \sim -150$, $\delta^{18}O - 19$), to high temperature ($> 450\ °C$) fluids collected at the volcano summit, which are enriched in both D and ^{18}O ($\delta D \sim -35$, $\delta^{18}O \sim -13.5$). These authors suggest that the isotopically enriched condensates are "magmatic" fluids released into the hydrothermal system during the 1976 eruption.

CO₂. CO_2 is the second most abundant magmatic gas species. Its importance for determining the composition of mafic magmas in the source region and for governing the vapor phase at elevated pressure became recognized recently. The volcano where gases have been collected and analyzed for the longest time (since 1912) is Kilauea in Hawaii. Gerlach and Thomas (1986) observed a constant $\delta^{13}C$-value of $-3.2‰$, which is heavier than common estimates for mantle carbon. These authors argued that the analyzed $\delta^{13}C$-value of the CO_2 reflects the $\delta^{13}C$-value of the parental magma. They attributed the observed ^{13}C-enrichment to be a typical feature of Hawaiian hot-spot volcanism. However, glasses from Loihi seamount — some 30 km south of Kilauea — contain indigenous carbon with typical mantle values between $-5.8‰$ and $-7.1‰$ (Exley et al. 1986a). $\delta^{13}C$-values of CO_2 from many other volcanoes around the world have been also interpreted to represent "mantle carbon". Sometimes, a ^{13}C-enrichment may be due to local contamination from carbonate-rich sediments in the volcano's basement (Allard 1979).

Sulfur. Sulfur is a major component of magmatic volatiles. The equilibrium proportions of H_2S, S_2, SO_2, and SO_3 depend upon temperature, pressure, oxygen fugacity, and bulk composition (Anderson 1975). In high-temperature gases SO_2 is the dominant gas phase. $\delta^{34}S$-

values of SO_2 collected from fumaroles between 1971 and 1979 at Kilauea have remained relatively constant at 0.9‰ (Sakai et al. 1982). However, elemental sulfur may show quite different and much more variable $\delta^{34}S$-values. The origin of sulfur and the fractionation behavior of different sulfur components is discussed below.

3.4.2 Fluid Inclusions

There are two different methods whereby gases may be extracted from rocks: (1) decrepitation by heating in vacuum and (2) crushing and grinding in vacuum. Both techniques may cause serious analytical difficulties. The major disadvantage of the thermal decrepitation technique is that although the amount of gas liberated is higher than by crushing at elevated temperatures, the different compounds present may begin to exchange isotopically or even to react with each other. This is especially crucial for hydrogen and carbon isotope measurements, when water and methane or carbon dioxide and methane coexist with each other. Crushing in vacuum avoids isotope exchange processes, however, during crushing large new surfaces are created which easily absorb some of the liberated gases. These absorption processes may also cause some fractionation effects. Both techniques preclude separating the different generations of inclusions in a sample. Therefore, in general, the results obtained represent the average isotopic composition of all generations of inclusions.

H-isotope data are particularly important in defining the origin of inclusion fluids and in detecting the mixing of different fluids if two or more sources are involved. The $\delta^{18}O$-value of inclusion waters in silicate is likely to have been modified through isotope exchange with the host mineral. Rye and O'Neil (1968) analyzed the ^{18}O-content of fluid inclusions in minerals from the Providencia ore deposit and found that the water present in oxygen-bearing minerals such as quartz and calcite indicated that the inclusions had exchanged ^{18}O with their host minerals. Only the $\delta^{18}O$-values of water in sphalerites with a very narrow $\delta^{18}O$-range between 5.8 and 6.2‰ were found to represent the primary isotopic composition.

Since the possibilities of isotope exchange are much more restricted for hydrogen that for oxygen, δD-values of magmatic waters should represent their primary isotopic composition.

3.4.3 Hot Springs

One of the principal conclusions drawn from stable isotope studies of
fluids in geothermal systems is that most hot spring waters (perhaps
95% or more) are meteoric waters derived from local precipitation
(Craig et al. 1956; Craig 1966; Clayton et al. 1968; Clayton and Steiner
1975; Truesdell and Hulston 1980). Most hot spring waters have deuterium contents similar to those of local precipitation, but are usually enriched in ^{18}O by isotope exchange with the country rock at elevated temperatures. This so-called oxygen-isotope shift is demonstrated
in Fig. 29. However, in areas where the D/H ratios of the local meteoric
waters are similar to "magmatic" water values, stable isotope techniques cannot distinguish conclusively between the meteoric-hydrothermal and the magmatic-hydrothermal solutions.

The magnitude of the oxygen isotope shift depends on the original
$\delta^{18}O$-value of both water and rock, the mineralogy of the rock, the
temperature, the water/rock ratio, and the time of contact. Waters
from systems containing carbonate rocks – having originally $\delta^{18}O$ be-

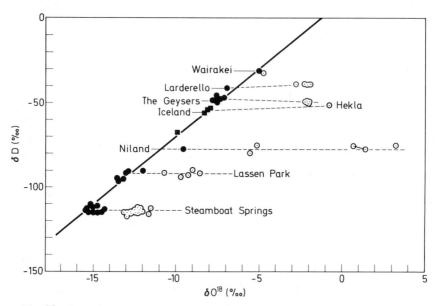

Fig. 29. Observed isotopic variations in near-neutral, chloride-type geothermal
waters and in geothermal steam. ■ Local meteoric waters or slightly heated, near-
surface groundwaters; ● hot spring geothermal water; ⊙ high temperature, high
pressure, geothermal steam. Niland = Salton Sea Geothermal Area. (After Craig
1963)

tween 20 and 30‰ — high temperatures and low water/rock ratios exhibit large oxygen shifts. One such example is the Salton Sea, where large $\delta^{18}O$-shifts in the fluid ($\approx$ +13‰) have indicated approximately equal amounts of water and rock (by mass) (Clayton et al. 1968). In the Broadlands-Wairakei area, New Zealand, much smaller $\delta^{18}O$-shifts accompany the hydrothermal alteration (Clayton and Steiner 1975), implying that the ratio of the mass of the water to the mass of the hydro-thermally altered rock is large. In the Valles Caldera, New Mexico, Lambert and Epstein (1980) observed an even larger fluid/rock ratio than in the Wairakei geothermal system.

A second class of geothermal waters consists of seawater. Sakai and Matsubaya (1974) showed that such thermal water systems at ocean coasts are mixtures of heated oceanic and local meteoric waters. They are characterized by $\delta^{18}O$- and δD-values intermediate between ocean and local meteoric waters. Similar conclusions may hold true for hot spring waters in Iceland.

What is sometimes neglected are the effects of boiling: loss of steam from a geothermal fluid can cause isotope fractionations. ^{16}O is con-centrated in the vapor at all temperatures below the critical tempera-ture, while hydrogen is increasingly concentrated in the vapor below 221 °C (Truesdell et al. 1977).

Quantitative estimates of the effects of boiling on the isotopic com-position of water can be made using known temperature-dependent fractionation coefficients and assumptions as to the extent to which the generated steam remains in contact with liquid water during the boiling process (Truesdell and Hulston 1980). If steam stays with as-cending water and separates when it reaches the surface, maximum isotope effects are found. If steam separation is continuously minimum isotope effects are observed.

3.4.4 Origin of Carbon and Sulfur in Geothermal Fluids

Most $\delta^{13}C$-values of CO_2 from geothermal waters fall in the range be-tween – 5‰ and –1‰ (Craig 1953; Hulston and McCabe 1962; Lyon 1974a,b; Panichi et al. 1977). Because those $\delta^{13}C$-values are close to the ratio for deep-seated carbon (see p. 82), these values might pos-sibly indicate a magmatic origin. However, high temperature leaching of carbonate rocks, or mixing between decarbonation CO_2 and CO_2 derived from carbonaceous matter could also produce such $\delta^{13}C$-values.

The CH_4 content in these fluids rarely exceeds 1% and thus is generally less than the CO_2 content. Therefore, the isotopic composition of
methane is probably the result of high-temperature equilibration with
CO_2 and thus not indicative of the source of the methane.

Elucidation of the origin of geothermal sulfur is complicated by the
fact that besides H_2S, SO_2, and sulfate, elemental sulfur can form in
appreciable amounts. A further complication arises from the fact that
contamination may be very important. Sakai et al. (1980) demonstrated
that seawater sulfate plays a decisive role in the Reykjanes geothermal
systems of Iceland. The same is true for the solfataras of the Santorini
volcano in Greece (Hubberten et al. 1975). For the active volcano White
Island, New Zealand Giggenbach and Robinson (1976) found $\delta^{34}S$-
values between +1‰ and +9‰, the low values being observed during
high-temperature periods. The high values measured for the low-temperature periods are considered to be due to the kinetically controlled
precipitation of isotopically light elemental sulfur in the fumaroles.
Native sulfur, which usually has higher $\delta^{34}S$-values than 0‰, could
fractionate into heavy sulfate and light H_2S through reactions such as

$$4 S + 4 H_2O \rightarrow 3 H_2S + H_2SO_4 .$$

3.4.5 Thermometers in Geothermal Systems

Although there are many isotope exchange processes occurring within
a geothermal fluid which have the potential to provide thermometric
information, only a few have been generally applied, because of suitable
exchange rates for achieving isotopic equilibrium (Hulston 1977;
McKenzie and Truesdell 1977; Panichi et al. 1977; Sakai 1977; Panichi
and Gonfiantini 1978, Truesdell and Hulston 1980; Giggenbach 1982).

In a single geothermal system several different reservoirs may exist
which generally increase in temperature with depth. The presence of
such reservoirs at different temperatures may be indicated by different
isotope thermometers equilibrating at different rates. The reaction rates
of these exchange reactions determine whether they will equilibrate in
deep geothermal reservoirs or if they will reequilibrate in shallower
reservoirs. Some isotopic reactions appear not to equilibrate at temperatures below 350 °C, whereas others equilibrate so rapidly that only
the temperature of collection is indicated (see Table 21).

Figure 30 summarizes the calculated fractionation factors as a function of temperature for some gases of geologic interest (after Richet

Table 21. Isotope temperature and rates of exchange to establish equilibrium for the hydrothermal fluid at Wairakei, New Zealand (Hulston 1977)

Element	Species	Isotope temperature	Rates of exchange
C	$^{13}CH_4 - ^{12}CO_2$	350 °C	$10^2 - 10^5$ y
S	$H^{34}SO_4 - H_2{}^{32}S$	350 °C	10^3 y
O	$S^{16}O_4 - H_2{}^{18}O$	280 °C	1 y
H	$H_2 - HDO$	260 °C	1–2 weeks
	Drill hole temperature	260 °C	

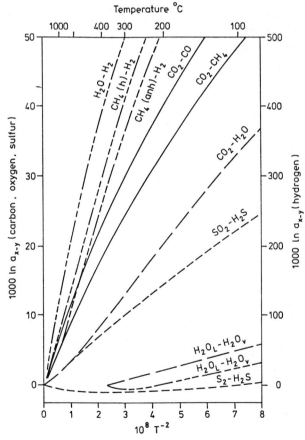

Fig. 30. Calculated equilibrium hydrogen, carbon, oxygen, and sulfur isotope fractionation factors as a function of temperature for some gases of geological interest. (After Richet et al. 1977, in Sheppard 1984)

et al. 1977, from Sheppard 1984) and the following paragraphs discuss the details of the more common geothermometers.

CO₂–CH₄. This thermometer was the first to be applied (Craig 1953; Hulston and McCabe 1962) on the basis of the exchange reaction:

$$CO_2 + 4 H_2 \rightleftharpoons CH_4 + 2 H_2O .$$

These early workers postulated close agreement between actually measured underground temperatures and isotope temperatures. However, more recently it became apparent that calculated temperatures are frequently much higher than those measured at the base of the geothermal wells (Ellis 1979; see also Table 21).

Giggenbach (1982) demonstrated that the observed $\Delta_{CO_2-CH_4}$ values are likely to represent frozen-in compositions attained after minimum residence times of 20 Ka at 400 °C or 10 Ma at 300 °C.

Hydrogen Isotope Geothermometers. At least five hydrogen-containing gases (H_2O, H_2S, H_2, CH_4, and NH_3) occur in major amounts in geothermal fluids. Of the various possible geothermometers only two have been widely used: (1) hydrogen-water and (2) hydrogen-methane. Arnason (1977) measured δD-values of hydrogen gas and water from drill holes and hot springs of Iceland. He found excellent agreement with measured downhole temperatures in most areas. In other regions variable degrees of reequilibration between the reservoir and the sampling point seem to occur (Lyon 1974a,b; Kiyosu 1983). Kiyosu (1983) observed that the hydrogen from low-temperature fumarolic gases (~100 °C) in northeastern Japan is lighter than from hot-water dominated geothermal gases from Iceland, which suggests that the hydrogen in low-temperature volcanic gases have equilibrated with water vapor at temperatures between 200°–400 °C.

The exchange for H_2-CH_4 is probably slower than that for H_2-H_2O. This has been found by a comparison of temperatures in Yellowstone Park and Broadlands, New Zealand samples (Lyon 1974b).

Oxygen Isotope Geothermometers. These geothermometers have the basic disadvantage that by far the largest oxygen reservoir is contained in the water itself and boiling during ascent produces isotope fractionation. Nevertheless, the most "well-behaved" geothermometer appears to be the sulfate-water system (McKenzie and Truesdell 1977; Sakai 1977; Truesdell and Hulston 1980). The exchange reaction is both sufficiently rapid to attain equilibrium and sufficiently slow to retain the indication of temperatures above 300 °C. Another oxygen geothermometer is the $CO_2 - H_2O$ pair, which has been applied with apparent success at Lardarello, Italy (Panichi et al. 1977). In those uncommon systems where both steam and water can be sampled separately the water-steam geothermometer can be useful (Truesdell and Hulston 1980).

Sulfur Isotope Geothermometers. Large isotope fractionations occur between oxidized and reduced sulfur species, namely between H_2S and HS^- and between SO_2, SO_4^{2-}, and HSO_4^-. In laboratory studies the exchange between H_2S or HS^- and HSO_4^- and SO_4^{2-} has been found to be very fast under acid conditions and very slow under alkaline conditions (Thode et al. 1971; Robinson 1973, 1978). Thus, the application of sulfur isotope geothermometers requires an estimate of the pH of the fluid. Using the $SO_4^{2-} - H_2S$ fractionation factors of Robinson (1973), the temperature in the Wairakei geothermal system is about 350 °C (see Table 21).

3.5 Ore Deposits

Stable isotope analyses provide the best geochemical data available for discerning the origin of the ore fluid, the sulfur, and the carbon species involved in ore deposition.

Of particular importance has been the ability to identify sources of water in hydrothermal ore deposits by their hydrogen and oxygen isotope ratios. The major isotopic characteristic of hydrothermal fluids is that, at a given locality, there is commonly a wide range in $\delta^{18}O$, but a narrow range in δD-values. We, thus, find the relationship for hot springs, shown in Fig. 30, where waters have undergone an ^{18}O-enrichment at more or less constant H-isotope compositions.

3.5.1 Origin of Ore Fluids

Ore fluids may be generated in a variety of ways. With respect to their isotopic composition five end-member components may be defined (see Fig. 31):

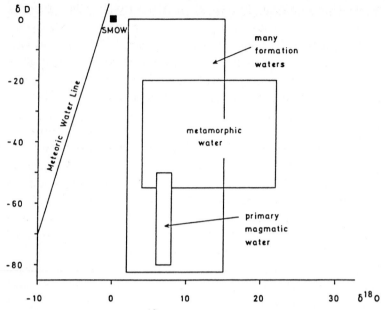

Fig. 31. Plot of δD versus $\delta^{18}O$ of waters of different origin

1. meteoric waters;
2. ocean water;
3. formation waters;
4. metamorphic waters;
5. magmatic waters.

Waters (1), (2), and (3) can be obtained as actual samples and measured isotopically. Waters (4) and (5) normally cannot be measured directly, except for fluid inclusions present in certain minerals such as quartz, but can be calculated from the analysis of hydroxyl minerals if the fractionation factors between the mineral and water and the temperature of formations are known.

1. Heated meteoric waters are a major constituent of the ore-forming fluids in many ore deposits and may become dominant in the latest stages of ore deposition. O'Neil and Silberman (1974), for instance, demonstrated that epidermal Au—Ag vein deposits of the Great Basin, Nevada are probably exclusively deposited from meteoric waters. The majority of vein minerals from these deposits have very low $\delta^{18}O$-values, whereas the D-values of fluid inclusions are *uniformly* low with

a range from -90 to $-140\%_0$ identical to that of recent spring waters in the area.

2. Many volcanogenic massive sulfide deposits are formed in submarine environments from heated oceanic waters. This concept gains support from the recently observed hydrothermal systems at ocean ridges, where the actually measured fluids show a slightly modified isotopic composition relative to $0\%_0$. Bowers and Taylor (1985) have modelled the δD- and $\delta^{18}O$-values of the evolving hydrothermal solution. At low temperatures the $\delta^{18}O$ of the fluid decreases relative to ocean water because the alteration products in the oceanic crust are ^{18}O-rich. At around 250 °C the solution is back to its initial value of seawater. Further reaction with basalt at 350° increases the modified seawater to $2\%_0$. The δD of the solution increases at all temperatures because mineral-water fractionations are nearly all less than zero. At 350 °C the δD-value of the solution is $2.5\%_0$. The best documented example for the role of ocean water during ore deposition is for the Kuroko-type deposits (see the extensive monograph by Ohmoto and Skinner 1983: Econ. Geol. Monogr. 5).

3. Connate and formation waters tend to increase in salinity as well as in temperature as they descend. During descent their ability to extract and transport metal increases as their chlorinity increases. One type of ore deposit which is associated with connate waters is the Mississippi-Valley-type ore deposit (Pinckney and Rye 1972; Heyl et al. 1974). More details about the isotopic composition of formation waters is discussed on p. 129.

A special type of water has been introduced by Charef and Sheppard (1986): "organic water" which is defined to be water whose deuterium content at least is derived from transformations of organic compounds during thermal maturation. Its D/H ratio is predicted to be very similar to that of the organic precursor material as fractionations during biodegration and oxidation reactions are generally quite small. Charef and Sheppard (1986) postulated that organic water should have δD-values below $-90\%_0$. Such organic waters may play an important role during the dewatering of sedimentary basins, transporting probably both metals and sulfur.

4. Metamorphic waters are generated by dehydration reactions. This process occurs not only in metamorphic rocks, but also in altered wall rocks accompanying many ore deposits. Typical ore deposits formed by the action of metamorphic fluids are skarn deposits. Combined petrologic and stable isotope studies by Taylor and O'Neil (1977),

Brown et al. (1985), and Bowman et al. (1985) have helped to resolve the complexities of formation. For instance, Bowman et al. (1985) concluded that meteoric water was unimportant during the main stage of skarn formation and that decarbonation of marble could supply the needed ^{18}O-rich fluid to mix with magmatic water. Into the category of metamorphic waters belong also the hydrothermal fluids that deposited gold-bearing quartz veins. These fluids are generally heavy, have $\delta^{18}O$-values between 8 and 14 and δD-values between -10 and $-50‰$ (Kerrick 1980; Marshall and Taylor 1981; Böhlke and Kistler 1986), and appear to owe their characteristics to deep crustal sources.

5. The isotopic composition of magmatic water has already been discussed on p. 98f. The classical theory of a hydrothermal solution arising from a cooling magma remains a reasonable explanation for many ore deposits. Fluids associated with porphyry copper and molybdenum deposits show the clearest evidence for a dominant magmatic water compound (Sheppard et al. 1971; Rice et al. 1985). Some involvement of nonmagmatic waters is indicated during ore-emplacement stages and became dominant during postore stages. Isotopic evidence supports a genetic model in which a magmatic-hydrothermal system is surrounded by a cooler meteoric-hydrothermal system that collapses inward and downward as the magmatic-hydrothermal system cools.

3.5.2 Wall Rock Alteration

Information on the origin and genesis of ore deposits can also be obtained by analyzing the alteration products in wall rocks. Delineation of hypogene and supergene ore deposits is possible by stable isotope analysis of the ores and associated alteration products. Because temperatures and isotopic compositions of the hydrothermal fluids are variable, hydrothermal clays exhibit a large scatter of points on a plot of δD versus $\delta^{18}O$ (see Fig. 32). All supergene clays plot either on the "Kaolinite Line" or slightly to the left of it. This is explained by slightly higher temperatures of formation than those occurring during surface weathering. Note that not only do supergene clays correlate with the Meteoric Water Line, but hypogene clays also correlate. This indicates that meteoric water was also a dominant constituent in the hypogene-hydrothermal system.

Whole rock $\delta^{18}O$-data in alteration halos can be used as an exploration guide for volcanogenic massive sulfide deposits (Green et al. 1983).

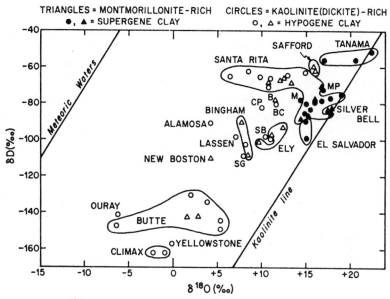

Fig. 32. δD versus $\delta^{18}O$ for hypogene and supergene clay minerals from various ore deposits (Taylor 1974a)

According to the generally accepted model for this type of deposit, the ore-forming fluid passes through the submarine sediment-volcanic section, emerges into the ocean, and precipitates the ore at or near the ocean floor. The conduit of altered rocks is thus subjected to large water fluxes, producing intense hydrothermal alteration and generally causing a depletion in $\delta^{18}O$-values. Thus, fossil hydrothermal conduits can be outlined by following the zones of ^{18}O-depletion. If a set of alteration zones has been mapped from field and petrographic data, $\delta^{18}O$-values might indicate zones of greatest water fluxes and of temperatures most appropriate for ore deposition. Oxygen isotope data are especially valuable in rock types which do not show diagnostic alteration mineral assemblages as well as those in which the assemblages have been obliterated by subsequent metamorphism (Beaty and Taylor 1982). In another example, Criss et al. (1985) found excellent spatial correlations between low $\delta^{18}O$-values and economic mineralizations. Such empirical spatial associations of mineralization with zones having anomalous low $\delta^{18}O$-values may be a useful guide for exploration of hydrothermal ore deposits.

3.5.3 Sulfur Isotope Composition of Hydrothermal Ore Deposits

A huge amount of literature exists on the sulfur isotope composition in hydrothermal ore deposits. Some of these data are discussed in the first and second editions of this book and, therefore, are not repeated here. Of the numerous papers on the subject the reader is referred to extensive reviews by Rye and Ohmoto (1974), Ohmoto and Rye (1979), and Nielsen (1978, 1979, 1985a). The basic principles to interpret the meaning of δ^{34}S-values in sulfidic ores were elucidated by Sakai (1968), which have been subsequently extended and clarified by Ohmoto (1972).

The major factors which control the sulfur isotope composition in hydrothermal ore deposits are:

1. Temperature which determines the fractionations between sulfur-bearing minerals. This point has been already discussed on p. 54 and is not further elucidated here.
2. The source of the sulfur, which can only be traced on the basis of the isotopic composition of the *total* sulfur in an ore deposit. $\delta^{34}S_{\Sigma S}$ can be grouped into three categories:
   a) deposits with δ^{34}S-values near zero should derive their sulfur from igneous sources, including sulfur released from magmas and sulfur leached from sulfides in igneous rocks;
   b) deposits with δ^{34}S-values near 20‰ should derive their sulfur from ocean water;
   c) deposits with δ^{34}S-values between 5 and 15‰ may receive their sulfur from local country rocks or from mixtures of (a) and (b).
3. Ohmoto (1972) has demonstrated that the p_H-value of the ore-forming fluid and the proportions of oxidized and reduced sulfur species may be of crucial importance when interpreting δ^{34}S-values in hydrothermal ore deposits.

Let us first consider the effect of p_H increase due to the reaction of an acidic fluid with carbonatic host rocks. At $p_H = 5$ practically all of the dissolved sulfur is undissociated H_2S, whereas at $p_H = 9$ the dissolved sulfide is almost entirely dissociated. Since H_2S concentrates ^{34}S relative to the dissolved sulfide ion, a p_H increase means as increase in the δ^{34}S of precipitated sulfides. Thus, when the p_H of an ore fluid increases from 5 to 9 during migration through the site of ore deposition, an increase in δ^{34}S should be observed between early precipitated and later precipitated sulfide minerals. At 250 °C the

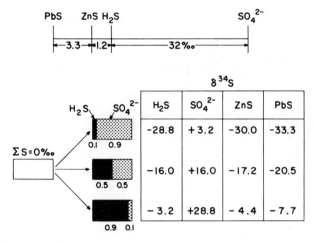

Fig. 33. Variation of $\delta^{34}S$ of H_2S, sulfate, and sulfide minerals with variation in the H_2S/SO_4^{2-} ratio of the hydrothermal solution (T = 200 °C, $\delta^{34}S_{\Sigma S}$ = 0‰) (Rye and Ohmoto 1974)

H_2S, HS^-/S^{2-} ratios at p_H values of 6, 7, and 8 are approximately 80:20:0, 35:35:30, and 5:25:70, respectively, and the corresponding $\delta^{34}S$-increase is ~1‰, ~3.5‰, and ~5‰.

An increase in oxygen fugacities has a much stronger effect on the $\delta^{34}S$-values than a p_H change, because of the large isotope fractionation between sulfate and sulfide. An illustration of how the sulfur isotopic composition of sulfur-containing phases are affected by the chemistry of fluids is given in Fig. 33. It is assumed that the isotopic composition of the total sulfur is 0‰, the temperature 200 °C, and that the ratio of H_2S to SO_4^{2-} changes from 1/9 to 9/1. $\delta^{34}S$-values of each aqueous sulfur species varies as a function of the H_2S/SO_4^{2-} ratio (e.g., $\delta^{34}S_{H_2S}$ from −28.8‰ to −3.2‰), but differences between $\delta^{34}S$-values of coexisting sulfur phases remain constant.

Figure 34a summarizes the stability fields of some important sulfur species at 250 °C in a p_H–f_{O_2} diagram. The transition field between the predominance of reduced and oxidized sulfur species is restricted to one decade of the f_{O_2} value. The $\delta^{34}S$-values at points A, B, C, and D are presented in Fig. 34b which indicates that changes of f_{O_2} have drastic effects on the sulfur isotopic composition. Magmatic sulfur with a total sulfur isotope composition of zero may yield very light sulfides, which are otherwise typical for biogenic sulfur (situation A), whereas a hydrothermal fluid with sulfur of seawater origin may yield

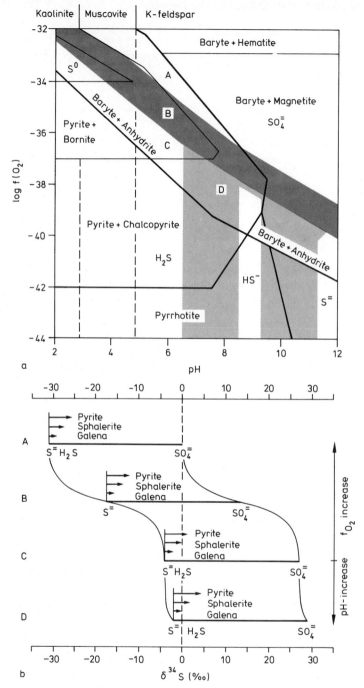

Fig. 34. a Stability fields of sulfur phases in a p_H-log f_{O_2} diagram at 250 °C for $\Sigma S = 0.1$ and $Ba^{2+} = 10^{-3}$ mol kg^{-1} (Nielsen 1985a). **b** δ^{34}S-values of sulfides and sulfate at 250 °C. Letters refer to those in **a** (Nielsen 1985a)

sulfides typical for magmatic sulfur (situation C and D). Figure 34 indicates clearly that the source of sulfur in hydrothermal fluids can only be traced when one has a knowledge of the $p_H - f_{O_2}$ relationships in an ore deposit.

Up to this point chemical and isotopic equilibrium among the ore-forming fluid and the precipitating sulfides and sulfate has been tacitly assumed. However, there is increasing evidence that isotopic disequilibrium is more common than originally assumed (Shelton and Rye 1982). Under such conditions of isotopic disequilibrium δ^{34}S-values do not respond to changes in the chemical environment of ore deposition as described above.

It is a well-known fact that at temperatures below $\sim$350 °C sulfide-sulfate mineral pairs do not attain isotopic equilibrium during the coprecipitation of sulfate and sulfide minerals (Ohmoto and Lasaga 1982; Shelton and Rye 1982). Obviously, sulfide minerals approach isotopic equilibrium with an H_2S-bearing fluid more quickly than aqueous sulfate and H_2S equilibrate. A similar relationship is commonly observed for aqueous sulfate-sulfide pairs in geothermal fluids (Robinson 1978). Ohmoto and Lasaga (1982) have computed the rate constants from experimental data and demonstrated that they are strongly dependent on temperature and p_H, but in a very complex manner. They suggested that mixing of sulfide-rich solutions with sulfate-rich solutions at or near the depositional sites may explain many of the observed fractionations.

In the following section a few examples of hydrothermal sulfide ore deposits of general interest are discussed.

Recent Sulfide Deposits at Midocean Ridges. Recent sulfide deposits have been identified on the East Pacific Rise off Mexico at 21° N and at the Galapagos Rift at 86° W. There are two possible primary sources for the sulfur in these vents: reduction of seawater sulfate and a mantle source, either as a direct magmatic contribution or indirectly during the leaching of the basaltic host rocks in the hydrothermal system. Studies by Styrt et al. (1981), Arnold and Sheppard (1981), Skirrow and Coleman (1982), Kerridge et al. (1983), and Zierenberg et al. (1984) show that the δ^{34}S-values are enriched in ^{34}S relative to a mantle source, implying the addition of sulfide derived from seawater. δ^{34}S-values from the East Pacific Rise deposits range from 1.3 to 4.1‰, which may be interpreted to indicate that the sulfides received a 10% seawater sulfate contribution to 90% derived from the leaching of basalt.

At the Galapagos Rift Skirrow and Coleman (1982) found somewhat higher δ^{34}S-values between 5.4 and 6.3‰ implying that the seawater contribution was more dominant than at the East Pacific Rise. Thus, there is agreement that both sources must be involved in the formation of these deposits, however, the relative importance of the two sources is still under debate.

Porphyry Copper Deposits. From δD- and δ^{18}O-measurements it has been concluded that porphyry copper deposits show the clearest affinity of a magmatic water imprint (Taylor 1974a). Many porphyry type copper deposits have been investigated for their sulfur isotope composition, but only Field and Gustafson (1976) and Shelton and Rye (1982) have taken a detailed look at the relationship between the sulfur isotopic composition of coexisting sulfates and sulfides. The majority of the δ^{34}S-values of the sulfides and sulfates in porphyry deposits fall between -3 and $+1$‰ and between $+8$ and $+15$‰, respectively. The calculated sulfate-sulfide equilibrium isotope temperatures lie typically between 450° and 650 °C and agree with temperatures estimated from other methods. Thus, the sulfur isotope data and temperatures seem to support the magmatic origin of the sulfur in porphyry deposits. This view was, however, questioned by Shelton and Rye (1982) who argued that sulfur isotope disequilibria between aqueous sulfate and H_2S may be an important feature of porphyry copper ore formation and may reveal remobilization of earlier precipitated ores as well as record changes in the chemistry of the ore-forming fluid.

Vein-Type Deposits. A wide spectrum of ore deposits of a different nature occurs in this category. Rye and Ohmoto (1974) have demonstrated the difficulty in interpreting the genesis of an ore body from the observed δ^{34}S-ranges. The only meaningful classification seems to be related to the temperature of the ore deposition. Vein deposits, due to their vertical extent, frequently exhibit a pronounced isotope zoning. Differences in $^{34}S/^{32}S$ ratios may result from temperature differences and from an increase in f_{O_2} or p_H toward the upper level. In many of these deposits the sulfur appears to have been derived from igneous sources and the sulfides were precipitated under $p_H - f_{O_2}$ conditions where H_2S was the dominant sulfur species.

Stratabound Deposits. Stratabound deposits are those confined to a single or a small number of stratigraphic units. They can be divided into two main categories, those enclosed in predominantly marine sedimentary rocks (carbonates and shales) and those enclosed in predominantly marine volcanic rocks or volcanoclastic sediments. Stratabound deposits in volcanic rocks are generally thought to be related to submarine volcanism and formed at or near the ocean water-rock interface. Reviews by Sangster (1968, 1976) and Nielsen (1985b) showed that the δ^{34}S-values of sulfides in these volcanic deposits are typically positive with a relatively narrow spread and that generally sulfide is depleted by about 15‰ in ^{34}S relative to the contemporaneous seawater sulfate.

The sulfur isotope data for most volcanic stratabound deposits are best explained by a mechanism in which ocean water became an ore-forming fluid by various degrees of reduction of its sulfate while in contact with hot volcanic rocks. The δ^{34}S-values of sulfate minerals, when present in the volcanic stratabound ores, are larger than those of contemporaneous seawater, as would be expected when seawater sulfate is partially reduced at temperatures above 250 °C in equilibrium with sulfides.

Biogenic Deposits. The discrimination between bacterial and inorganic sulfur in ore deposits on the basis of δ^{34}S-values is rather complex. The best criterion to distinguish between both types is the internal spread of δ-values. If individual sulfide grains at a distance of only a few millimeters or centimeters exhibit large and nonsystematic differences in δ-values, then it seems to be safe to assume an origin involving bacterial sulfate reduction. Irregular δ-variations are obviously due to bacteria growing in reducing microenvironments around individual particles of organic matter. In contrast, inorganic sulfate reduction would need a considerable supply of thermal energy by an ascending hot fluid. Such an environment is not consistent with bacterial reduction in a small-scale, closed system.

Unfortunately, this discrimination can only be successfully applied in deposits that have not been overprinted by later metamorphic events, which tend to homogenize the nonequilibrium δ-differences over very short distances. Two types of deposits, where the internal S-isotope variations fit the expected scheme of bacterial reduction, but where the biogenic nature was already known earlier from conventional geological analysis, are the "sandstone-type" uranium mineralization in the Colorado Plateau (Warren 1972) and the Kupferschiefer in Central Europe (Marowsky 1969).

Metamorphosed Deposits. It is generally assumed that metamorphism reduces the isotopic variations in a sulfide ore deposit. Recrystallization, liberation of sulfur in fluid and vapor phases, such as the breakdown of pyrite into pyrrhotite and sulfur, and diffusion at elevated temperatures should tend to reduce initial isotopic heterogeneities. However, Willan and Coleman (1983) were able to demonstrate that sulfide and sulfate minerals have not reequilibrated over distances of ~1 cm during greenschist or amphibolite facies metamorphism. They concluded that the variations found in sulfide deposits of the Dalradian metamorphic terrain, Scotland represent primary variations of the sulfur source.

3.5.4 Hydrothermal Carbonates

Similar to sulfur, the isotopic composition of carbon in hydrothermal carbonates depends not only on the δ^{13}C-value of the total carbon in the ore-forming fluid, but also on the oxygen fugacity, the pH, the temperature, the ionic strength of the fluid, and on the total concentration of carbon (Ohmoto 1972; Rye and Ohmoto 1974). The oxygen fugacity affects the oxidation states of the carbon species. If the oxygen fugacity is about 10^{-38} atm, most of the carbon is oxidized and the reduced carbon is negligible. However, at lower oxygen fugacities the reduced carbon increases in abundance and has a dramatic effect on the δ^{13}C-values of the coexisting carbonate species because the reduced carbon is strongly enriched in ^{12}C. Similarly, at a fixed f_{O_2} value, the δ^{13}C of the oxidized carbon varies with p_H because of the changing abundances of the HCO_3^- and CO_3^{2-} ions in aqueous solution.

Carbon in hydrothermal systems could originate from three different sources:

1. marine limestones which have average δ^{13}C-values near zero;
2. deep-seated carbon with a range in δ^{13}C-values between $-8‰$ and $-5‰$ (see p. 82);
3. organic carbon from sedimentary rocks with δ^{13}C-values typically lower than $-20‰$.

Considering the carbon isotope composition in more detail, Fig. 35 shows the variation of δ^{13}C-values of carbonate gangue minerals for a number of hydrothermal ore deposits. The different generations of carbonates are arranged from early to late as shown by arrows in Fig. 35. Most early carbonates have δ^{13}C-values between -7 and $-5‰$ which

Fig. 35. δ^{13}C-values of carbonates from various hydrothermal ore deposits. Different generations of carbonate are arranged from early to late (Rye and Ohmoto 1974)

Carbonates
• Fluid inclusion

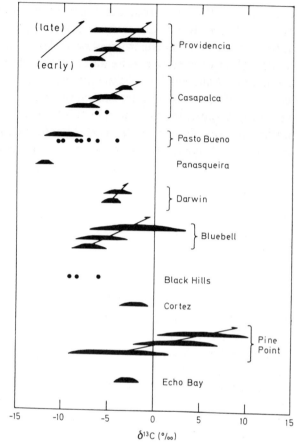

may indicate a deep-seated origin of the carbon. However, such δ-values can also be generated by simple mixing between carbonate-derived and organically-derived CO_2. Late-stage carbonates may show an enrichment in the heavy isotope relative to the main-stage carbonates as a result of (1) cooling of the ore fluid, (2) decreasing CO_2/CH_4 ratios in the fluid, and/or (3) increasing contribution of CO_2 from other sources.

3.6 Hydrosphere

First, some definitions concerning water of different origin are given.

The term *"meteoric"* applies to water that has gone through the meteorological cycle, i.e., evaporation, condensation, and finally precipitation. All continental surface waters, such as rivers, lakes, and glaciers, fall into this general category. Because meteoric water may seep into the underlying rock strata, it will also be found at various depths of the lithosphere. The *ocean*, although it continuously receives the continental run-off of meteoric waters as well as rain, is not regarded as being of meteoric nature. *Connate* water is "fossil" water, which has been trapped in the sediments at the time of burial. *Formation* water is present in rocks immediately before drilling and may be a useful nongenetic term for waters of unknown age and origin.

3.6.1 Meteoric Water

The natural water cycle has been compared to a multiple-stage distillation column with reflux of the condensate to the reservoir (Epstein and Mayeda 1953; Siegenthaler 1979). The oceans correspond to the reservoir, and the ice fields at the poles correspond to the highest stages of the column. In all processes concerning the evaporation and condensation of water the hydrogen isotopes are fractionated in proportion to the oxygen isotopes, because a corresponding difference in vapor pressures exists between H_2O and HDO in one case and $H_2{}^{16}O$ and $H_2{}^{18}O$ in the other. Therefore, the hydrogen and oxygen isotope distributions are correlated in meteoric waters. Craig (1961a) first defined the following relationship:

$$\delta D = 8 \, \delta^{18}O + 10 \, ,$$

which describes the interdependence of H- and O-isotope ratios in meteoric waters.

This relationship, shown in Fig. 36, is generally described as the "Meteoric Water Line". Neither the numerical coefficient 8 nor the constant 10, also called the deuterium excess "d", are really constant. Both vary with the conditions of evaporation, vapor transport, and precipitation and thus offer insight into climatic processes. Based on the "deuterium excess", Leguy et al. (1983) were able to separate rain waters from the Negev desert into three groups. These authors postulated that the two extreme groups, one has "deuterium excess" values above 22, the other has values less than 10‰, have quite different origins. The intermediate group, to which the majority of cases belong, seem to represent some combination of the two extreme groups.

Thus, the study of the deuterium excess "d", introduced originally by Dansgaard (1964), promises to be a source of additional climatic

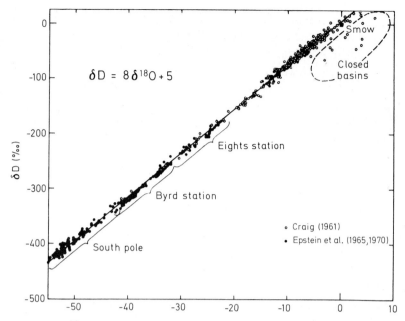

Fig. 36. δD versus $\delta^{18}O$ for various meteoric surface waters (Taylor 1974a)

information. Harmon and Schwarcz (1981) noted a shift of about 10‰ in the "d" value for meteoric waters in North-Central North America between interglacial and glacial times and Jouzel et al. (1982) interpreted the decrease of "d" in an Antarctic ice core from about 9‰ in the recent interglacial to 4.5‰ during full glacial conditions as indicating a 10% increase in relative humidity during glacial times.

Atmospheric precipitation generally follows a Rayleigh process at liquid-vapor equilibrium. The atmospheric Rayleigh process also explains why, at higher altitudes and latitudes, fresh waters become progressively lighter isotopically, whereas tropical waters show very small depletions relative to ocean water. Gat (1980) has pointed out some serious problems with this simple Rayleigh distillation model, namely that:

1. the meteorological pattern of air movement is not consistent with a gradual poleward movement of low-latitude air masses;
2. the latitudinal "distillation column" is certainly not closed, additional evaporation over middle- and high-latitude oceans, as well as over continents occurs;

3. the mathematical solution of a simple Rayleigh-type process does not predict a straight line, such as the meteoric water line, especially in its terminal stages (i.e., polar snow);
4. according to the Rayleigh model, the isotopic composition of precipitation depends only on the composition of the vapor and the temperature of condensation. However, evaporation and isotope exchange with water vapor *after* condensation also affects the isotopic composition.

Nevertheless, the distillation model describes the isotopic composition of rain water in qualitative terms, if not in quantitative terms.

As already mentioned several times, when water evaporates from the surface of the ocean, the water vapor is enriched in H and ^{16}O because $H_2{}^{16}O$ has a higher vapor pressure than HDO and $H_2{}^{18}O$. Under equilibrium conditions at 25 °C the fractionation factors for evaporating water are 1.0092 for ^{18}O and 1.074 for D (Craig and Gordon 1965). However, under natural conditions, the actual isotopic composition of water vapor is significantly more negative than the predicted equilibrium values, which is due to kinetic effects (Craig and Gordon 1965). Vapor leaving the surface of the ocean cools as it rises and rain forms when the dew point is reached. During the outraining of the moist air mass, the vapor is continuously depleted in the heavy isotopes, because the rain leaving the system is enriched in ^{18}O and D. If the air mass moves poleward and becomes cooler, additional rain will form having less ^{18}O than the first rain. This relationship is schematically shown in Fig. 37.

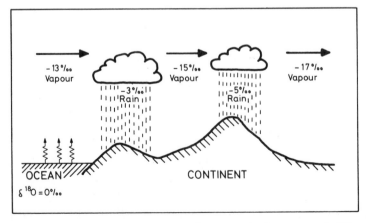

Fig. 37. Schematic fractionation in the atmospheric water cycle (Siegenthaler 1979)

The global distribution of D and ^{18}O in rain has been monitored since 1961 through a network of stations (Yurtsever 1975). From this extensive data base it can be deduced how geographic factors influence the isotopic composition of precipitation. Besides the dominant temperature effect, a latitude, an altitude, and an amount effect can be distinguished (Dansgaard 1964).

Marine precipitation at any given latitude is more ^{18}O-rich than continental precipitation. In the absence of any land masses, $\delta^{18}O$-values of precipitation can be expected to be near $-1\%o$ at the equator and to diminish to about $-5\%o$ at about $30°N$ or S latitude. At latitudes greater than $30°$ the rate of ^{18}O-exchange with latitudes increases by about a factor of 4. In continental areas, the latitude effect is roughly $0.5\%o$ per degree latitude (Dansgaard 1964). The altitude effect, which also favors the light isotopes, depends on local climate and topography, but gradients of $0.15-0.5\%o$ $\delta^{18}O/100$ m are typical (Gat 1980).

Besides a latitude and an altitude effect there is an "amount effect" in areas of high rainfall. Dansgaard (1964) observed that for each 100 mm increase in rainfall the value of $\delta^{18}O$ decreased by about $1.5\%o$. This "amount effect" is ascribed by Dansgaard (1964) to deep cooling of the air in heavy rainfall with only slight enrichments possible in later evaporation.

Our knowledge of the isotopic variations in precipitation will certainly be increased when short-term variations are analyzed from local stations. Especially under midlatitudinal weather conditions such short-term variations arise from varying contributions of tropical, polar, marine, and continental air masses and these in conjunction with other weather data should provide important climatic information.

Isotope variations in discrete meteorological events can be studied by analyzing hailstones, because they keep a record on the internal structure of a cloud. Jouzel et al. (1975) concluded that hailstones grow during a succession of upward and downward movements in a cloud.

Deviations from the Meteoric Water Line tend to lie on the right-hand side of the line, i.e., toward low δD and/or high $\delta^{18}O$ values (Craig 1961a; Clayton et al. 1966). These deviations are typically observed in lakes and other water bodies subjected to intense evaporation. In contrast to condensation, evaporation mostly takes place under kinetic conditions, especially when the relative humidity is significantly less than 100%. Thus, in dry climates such as in the Sahara Desert, $\delta^{18}O$-values as high as $+31\%o$ and δD-values as high as $+129\%o$ have been measured (Fontes and Gonfiantini 1967).

Snow and Ice Stratigraphy. The isotopic composition of snow and ice deposited in the polar regions and at high elevations in mountains depends primarily on the temperature. Snow deposited during the summer has less negative $\delta^{18}O$- and δD-values than snow deposited during the winter. A good example of the seasonal dependence has been given by Deutsch et al. (1966) on an Austrian glacier, where the mean δD-difference between winter and summer snow was observed to be $-14‰$. Systematic ^{18}O- and D-measurements have been used to study flow patterns of glaciers, snow accumulation rates, and climatic variations. Isotope profiles through a glacier should give lighter isotopic compositions at depth than near the surface, because deep ice may have originated from locations upstream of the ice-core site, where temperatures may be colder.

Annual accumulation rates of snow and firn have been also determined. For example, Epstein et al. (1965) found an average annual accumulation rate of 7 cm water at the South Pole during the time interval from 1958–1963. This estimate is in good agreement with results obtained by conventional stratigraphic methods and by radioactive dating methods. The isotopic composition of snow and firn at one site may record changing climatic conditions. Detailed studies over a time scale of 1000 years at different sites show considerable diversity in the records (Johnson et al. 1972; Dansgaard et al. 1975; Paterson et al. 1977). Because the isotopic record is based on distinct snowfall events, these isotopic differences between the different sites may result from storms of varying trajectories and may not be affected by changes in the mean temperature. Whether the different sites correlate or not, it can also be that each site records its own paleotemperature history.

The seasonal fluctuations of $\delta^{18}O$ and δD in snow and ice are gradually eliminated due to homogenization processes, such as melting and refreezing of water percolating downward through snow or firn. Judy et al. (1970) demonstrated, for example, that δD-values of individual snowfalls ranged from -230 to $-106‰$, whereas the δD-range in the snowpack some months later was reduced to only -182 to $-158‰$. They attributed this homogenization to recrystallization by vapor transport within the snowpack.

Although the seasonal variations of the annual layers are gradually obliterated, the absolute values record climatic conditions primarily in terms of the mean air temperatures. Therefore, continuous ice cores drilled from the continental ice sheets in Greenland and Antarctica contain a climatic record over more than the last 100000 years (Dansgaard et al. 1969, 1982; Johnson et al. 1972; Lorius et al. 1985). In Fig. 38 the δ^{18}O-record of two cores from Greenland and Antarctica are compared. Because of the difficulty in establishing absolute time scales for these cores, detailed comparisons between the δ^{18}O-records of the ice sheets in Greenland and Antarctica are unreliable. Nevertheless, there are remarkable similarities in both cores. The upper portions in both cores are rather constant and then at around 10000 years δ^{18}O-values decrease to about $-40‰$, suggesting significantly lower

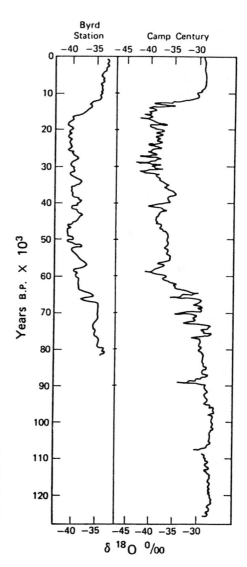

Fig. 38. Variations of δ^{18}O in ice cores from Byrd Station, Antarctica and Camp Century, Greenland. (The more negative δ^{18}O-values from about 70000 to 12000 years before present reflect colder climatic conditions during the last ice age) (Faure 1977)

average air temperatures. Towards the bottom of the core the $\delta^{18}O$-values increase again, implying a warmer climate.

3.6.2 Groundwater

In temperate and humid climates the isotopic composition of groundwater is similar to that of the precipitation in the area of recharge (Gat 1971). This is strong evidence of direct recharge to an aquifer. The variation of $\delta^{18}O$ and δD with altitude can be used for estimating the altitude of unknown recharge areas of artesian waters (Stahl et al. 1974).

According to Gat (1971) the main mechanism which can cause variations between precipitation and recharged groundwater are:

1. recharge from partially evaporated surface water bodies;
2. recharge that occurred in past periods of different climate when the isotopic composition of precipitation was different from that at present;
3. isotope fractionation processes resulting from differential water movement through the soil or the aquifer or due to exchange reactions with geologic formations.

In semiarid or arid regions, evaporation losses before and during recharge shift the isotopic composition of groundwater towards heavier values. Furthermore, transpiration of shallow groundwater through plant leaves may also be an important evaporation process. In deserts, evaporation becomes the dominant process in influencing isotopic composition. Gat and Dansgaard (1972) and Gat and Issar (1974) have demonstrated clearly that the isotopic composition of paleowaters (remnants of meteoric waters of past cooler climatic periods) can be distinguished from more recently recharged groundwaters which have been evaporated. Because rain is scarce and irregular in deserts, direct recharge to groundwaters appears to be negligible. However, in groundwaters of the Sinai desert, Gat and Issar (1974) demonstrated that direct rain recharge to aquifers is widespread.

3.6.3 Ocean Water

The isotopic composition of ocean water has been discussed in detail by Redfield and Friedman (1965), Craig and Gordon (1965), and

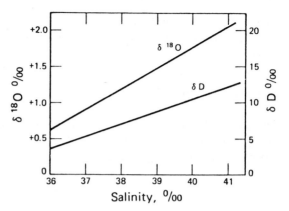

Fig. 39. Relationship between $\delta^{18}O$, δD, and salinity of water in the Red Sea due to preferential loss of $H_2{}^{16}O$ during evaporation (Faure 1977)

Broecker (1974). Ocean water with 35‰ salinity exhibits a very narrow range in isotopic composition, less than 10‰ for D/H ratios and 1‰ for $^{18}O/^{16}O$ ratios. However, evaporation processes strongly affect the isotopic composition, because they cause a preferential depletion in the lighter isotopes, which become enriched in the vapor phase. Consequently, the remaining water will be isotopically enriched so that highly saline waters generally have the highest D and ^{18}O contents. This effect is well illustrated by water from the Red Sea (Craig 1966) shown in Fig. 39. Low salinities, which are caused by fresh water and melt water dilution, correlate with low D and ^{18}O concentrations (Epstein and Mayeda 1953; Redfield and Friedman 1965).

One very important fact concerns the circulation of deep water masses in the oceans. At least half of all the water currently entering the deep ocean is generated in the Norwegian Sea at the northern end of the Atlantic Ocean. This water flows down the Atlantic basins around Africa, through the Indian Ocean, and finally up into the Pacific Ocean. Joining this North Atlantic Deep Water flow toward the Deep Pacific is water which has been recooled in the Antarctic Ocean.

A $\delta^{18}O$ versus salinity diagram for North Atlantic ocean water samples is shown in Fig. 40 (after Broecker 1974). Those samples with salinities of about 36‰ have $\delta^{18}O$-values about 1‰ higher than samples of SMOW. Waters taken close to Greenland with a salinity of 16‰ have an isotopic composition of −11‰. As can be seen in Fig. 40, all the North Atlantic surface waters (NASW) fall along one line, suggesting that these waters are mixtures of normal ocean water with fresh water. The same holds true for North Atlantic deep water samples (NADW), taken from 3000 m depth. However, deep Pacific and Antarctic waters

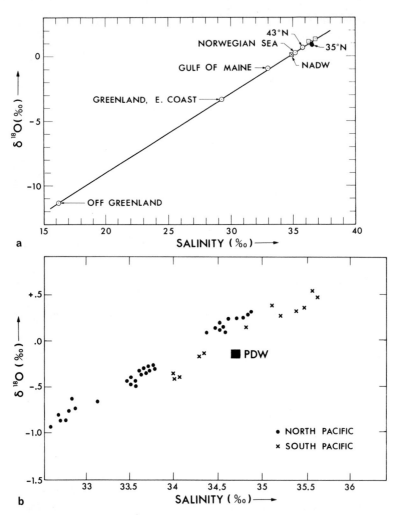

Fig. 40a,b. Plots of $\delta^{18}O$ versus salinity. **a** Samples from the North Atlantic, suggesting that North Atlantic Deep Water (NADW) consists dominantly of water sinking at the northern end of the Atlantic. **b** Samples throughout the Pacific Ocean (deep water in the Pacific cannot be generated from any mixture of these waters) (Broecker 1974)

differ from NADW (see Fig. 41). Thus, one may conclude that NADW consists almost entirely of surface water from the North Atlantic, because, if more than 30% of Antarctic water or Mediterranean water were mixed with it, NADW would not fall on the Atlantic ^{18}O-salinity line.

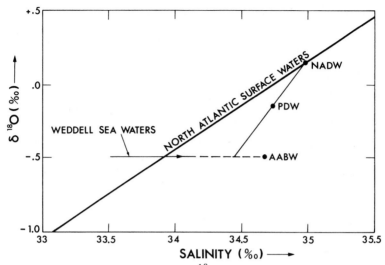

Fig. 41. Relationship between the salinity and ^{18}O of major deep waters: NADW, AABW, and PDW. Pacific Deep Water (PDW) can only be generated by mixing roughly equal amounts of water originating in the Northern Atlantic (NADW) and water originating along the edge of the Antarctic continent (AABW) (Broecker 1974)

Figure 40b shows the relationship for Pacific surface waters. Note that the Pacific Deep Water point (PDW) does not fall on the North or South Pacific water surface water line. Actually, the deep ocean points of PDW, NADW, and AABW (Antarctic Bottom Water) in Fig. 41 are almost on a line, suggesting that PDW could be produced by mixing NADW with water sinking in the Antarctic.

Two surface sources account for the bulk of deep water in the ocean:
1. NADW falls along a line that defines the water found at the surface of the North Atlantic, and
2. Antarctic Bottom Water (AABW) falls along an extension of the Weddell Sea Line.

Each of the known sources of deep water, the water sinking in the Norwegian Sea and the water sinking along the Antarctic coast, have isotopic composition-salinity relationships consistent with that in the Deep Pacific. A 50/50 mixture of these two would be identical to PDW.

One important question concerning the isotopic composition of ocean water is how constant its isotopic composition has been throughout geologic history. If all the ice sheets in the world were melted, the

δ^{18}O-value of the ocean would be lowered by about 1‰ and the δD-value by about 10‰. Estimates for the maximum enrichment of the ocean in ^{18}O during the Pleistocene glaciation range from 0.5‰ (Emiliani 1966) to 1.8‰ (Craig 1965; Shackleton 1968). At least throughout Phanerozoic time, the isotopic composition of ocean water has probably fluctuated within those limits. There are major uncertainties about the Precambrian. Whereas Becker and Clayton (1976) argued that the Precambrian ocean had a δ^{18}O-value at least as light as -3.5‰, Knauth and Epstein (1976) and Kolodny and Epstein (1976) believed that these oceans were roughly similar in isotopic composition to the modern ocean.

As has already been demonstrated magmatic waters have δ^{18}O-values between +6 to +8‰ and δD-values between -80‰ and -50‰. If the oceans originate from such waters, why are they now at 0‰? The apparent shift in isotopic compositions is probably due to the sedimentation of authigenic minerals rich in ^{18}O and poor in D relative to ocean water. For instance, much oxygen enriched in ^{18}O is bound in the form of cherts, carbonates, and clay minerals. Savin and Epstein (1970b) estimated on the basis of material balance calculations that the volume of ^{18}O-rich sediments can account for an ^{18}O-depletion of about 6‰ in the oceans. If the oceans have grown progressively with time, they could have remained more or less constant in ^{18}O. If the volume of the oceans has remained constant for the last 3 billion years, this model would require a progressive shift in the δ^{18}O-value of about 1‰ every 500 million years. Kokubu et al. (1961) suggested that the apparent enrichment of ocean water in deuterium relative to the magmatic water is due to preferential loss of dissociated H-atoms or ions to the exosphere over geologic time. Murozumi (1961) calculated the amounts of water photochemically decomposed in the upper atmosphere in order to account for this isotope enrichment. He estimated more than 10^{23} g water, representing some 10% of the present ocean, to have been lost by this effect.

3.6.4 Pore Water

Knowledge of the chemical composition of sedimentary pore waters has increased considerably since the beginning of the Deep-Sea Drilling Program. From numerous drill sites it has been found that significant variations exist in the chemical and isotopic composition of pore waters relative to ocean water.

Lawrence et al. (1975) and Perry et al. (1976) noted a decrease in $\delta^{18}O$ of the pore waters from an initial value very near 0‰ (Atlantic Deep Water) to about −2‰ at depths around 200 m. This decrease in ^{18}O of the pore water is due to the formation of authigenic clay minerals such as smectite and sepiolite from complete alteration of basaltic material and volcanic ash. In all cases where significant isotopic gradients have been found in the pore waters, they have been accompanied by increases in the Ca^{2+} concentration and decreases in the Mg^{2+} concentration of the pore water. The alteration reactions least effective in causing ^{18}O-depletion in the pore waters are the recrystallization of fossil carbonate to limestones and of biogenic silica to chert. In fact, at temperatures above 20 °C, these reactions will act to increase the $\delta^{18}O$ of the pore water (Lawrence et al. 1975).

3.6.5 Formation Waters and Oil Field Brines

Oil field brines are the best-known examples of subsurface saline waters. The processes involved in the development of saline formation waters are complicated by the extensive changes that have taken place in the brines after sediment deposition. Clayton et al. (1966), Hitchon and Friedman (1969), and Kharaka et al. (1974) have shown convincingly that the water now present in the formation brines is largely meteoric in origin. Formation waters show a wide range in δD, $\delta^{18}O$, and salinity, but the waters within a sedimentary basin are usually isotopically distinct. Just as with the surface meteoric waters, there is a general decrease in isotopic composition as one moves to higher latitudes (see Fig. 42).

The oxygen isotopic composition of the water is affected by exchange with carbonate minerals and by isotope fractionation across membranes. It is well known that shales and compacted clays can act as semipermeable membranes which prevent passage of ions in solution while allowing passage of water (ultrafiltration or salt filtration). Coplen and Hanshaw (1973) have shown experimentally that ultrafiltration may be accompanied by hydrogen and oxygen isotope fractionation. The ultrafiltrates were depleted in D and in ^{18}O relative to the residual solution.

Somewhat unusual isotopic compositions have been observed in highly saline deep waters from Precambrian crystalline rocks, which plot above or to the left of the Meteoric Water Line (Frape et al. 1984;

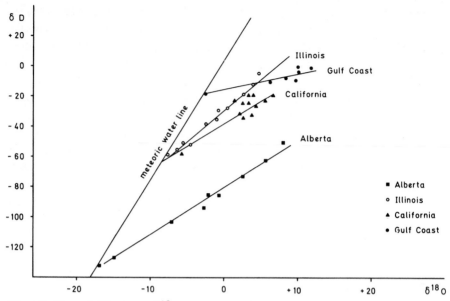

Fig. 42. Plot of δD versus $\delta^{18}O$ for oil field brines (formation waters) from the midcontinent region of the United States. (After Taylor 1974a)

Kelly et al. 1986). How to explain the unusual isotopic composition of these brines is not at all clear. Kelly et al. (1986) advocated a multi-step model, in which the formation waters evolve in Paleozoic sediments, then deeply infiltrate the crystalline basement under high temperature conditions and water/rock ratios. Later on, as the sedimentary cover was removed, the waters "back-react" under the new conditions of low temperatures and low water/rock ratios. In a final stage these waters have been contaminated to varying degrees by fresh meteoric waters.

3.6.6 Water in Hydrated Salt Minerals

Very little data exist on the isotopic composition of water in hydrated minerals (Matsuo et al. 1972; Matsubaya and Sakai 1973; Stewart 1974; Sofer 1978). To interpret such isotopic data it is necessary to know the equilibrium fractionation factors between the hydration water and the solution from which they are deposited. If the isotopic compositions of the parent water and the crystalline water are known, the temperature of precipitation may be obtained. From the fractionation factor and a temperature estimate, the isotopic composition of the water from which the hydrated salt minerals were precipitated can be determined. This may yield information about the environmental conditions of depositions such as whether deposition occurred from fresh or marine water and whether such water had undergone evaporation. Sofer (1978) showed that the isotopic composition of the hydration water of gypsum may be used to specify the mechanisms of its formation, either from an evaporating brine, by hydration of anhydrite, or through oxidation of sulfides in groundwaters. Since the isotopic record in primary gypsum is destroyed both by dehydration and exchange processes with water, one cannot expect to find ancient marine gypsum samples which have retained their original water composition. Only under arid conditions may the primary isotopic record be preserved (Sofer 1978).

3.7 The Isotopic Composition of Dissolved and Suspended Compounds in Ocean and Fresh Waters

3.7.1 Nitrogen

The major sources of nitrogen in the ocean are river runoff, rain, and fixation of molecular nitrogen by organisms. Sinks originate by burial on the sediments and especially by denitrification, which is the dominant process producing large N-isotope fractionations. Denitrification seems to be the principal mechanism that keeps marine nitrogen at higher δ^{15}N-values than atmospheric nitrogen.

Due to the transient nature of marine nitrate, δ^{15}N-values of dissolved nitrate vary considerably. Cline and Kaplan (1975) observed in the North Pacific Ocean a δ^{15}N-range from +6.5‰ in the surface region to +18.8‰ in the active denitrification zone. Saino and Hattori (1980) presented a vertical ^{15}N-profile on particulate organic matter in the Indian Ocean. δ^{15}N-values decrease near the surface and reach a minimum near 30-m water depth and then increase through the 30–500 m depth by 10‰. Below 500 m the δ^{15}N-value remained constant around 13‰. The ^{15}N-enrichment with depth may result from isotope fractionations during oxidative degradation of particulate organic matter. The minimum δ^{15}N-values near the surface may be due to preferential ^{14}N-incorporation during nitrate uptake by phytoplankton.

Altabet and Deuser (1985) observed seasonal ^{15}N-variations in particles sinking to the ocean bottom. They suggested that the δ^{15}N-values of sinking particles represent a monitor for nitrate flux in the euphotic zone. If the sediments preserve this $\delta^{15}N$-record it may represent a means by which primary production in the oceans of the past can be studied.

Sweeney et al. (1978) have shown that nitrogen in suspended marine organic matter is significantly heavier than nitrogen in suspended terrestrial organic matter. In near-shore environments $\delta^{15}N$-values can, therefore, act as a tracer to determine the source of nitrogen (Mariotti et al. 1984).

3.7.2 Oxygen

As early as 1951 Rakestraw et al. demonstrated that dissolved oxygen in the oceans is enriched in ^{18}O relative to atmospheric oxygen. Extreme enrichments, up to 14‰ (Kroopnick and Craig 1976), occur in the oxygen minimum region of the deep ocean due to preferential consumption of ^{16}O by bacteria in abyssal ocean waters, which is evidence for a "deep metabolism" (see also Fig. 43).

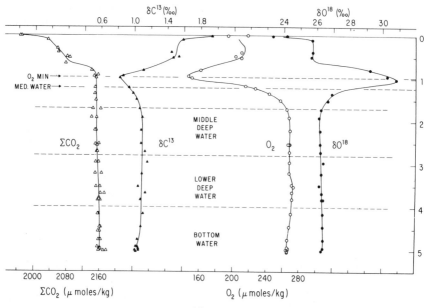

Fig. 43. Vertical profiles of CO_2, $\delta^{13}C$, dissolved O_2, and $\delta^{18}O$ in the dissolved O_2 in the North Atlantic (Kroopnick et al. 1972)

3.7.3 Carbon Species in Water

3.7.3.1 Bicarbonate in Ocean Water

In addition to organic carbon, four other carbon species exist in natural water: dissolved CO_2, H_2CO_3, HCO_3^-, and CO_3^{2-}, all of which tend to equilibrate as a function of temperature. As previously mentioned, the concentration and the isotopic composition of the individual species also vary with pH.

HCO_3^- is the dominant C-bearing species in ocean water. The C-isotope composition in a vertical profile is shown in Fig. 43. Most surface waters in the central ocean basins have $\delta^{13}C$-values of about +2.2‰ (Deuser and Hunt 1969; Kroopnick 1985). However, this value changes downward into deeper water masses due to continuous flux of organic and skeletal detritus into the deep water (see Fig. 43). In a very detailed study Kroopnick (1985) analyzed 2252 samples from 107 hydrographic stations and demonstrated that the distribution of $\delta^{13}C$ is controlled mainly by the input of organically produced material and its subsequent oxidations as it falls through the water column. Other factors which influence the $\delta^{13}C$-value are the dissolution of inorganic carbonate and the addition of anthropogenic CO_2. It is estimated that from a preindustrial value of 2.5‰ the $\delta^{13}C$-value of the total dissolved CO_2 in the ocean has decreased by 0.5‰.

3.7.3.2 POM

Particulate organic matter (POM) in the ocean originates in large parts from the detrital remains of plankton in the euphotic zone and reflects living plankton populations. As POM sinks, biological reworking changes its chemical composition, with labile compounds such as amino acids and sugars being degraded in preference to the more refractory lipid components. The extent of biological degradation depends on residence time in the water column. Most reported POM profiles exhibit a general trend of surface isotopic values comparable to those for living plankton towards increasingly lower $\delta^{13}C$-values with depth. Eadie and Jeffrey (1973) and Jeffrey et al. (1983) interpreted this trend as the loss of labile, isotopically enriched amino acids and sugars through biological reworking, leaving the more refractory, isotopically light lipid components.

C/N ratios of POM increase with depth of the water column. This implies that nitrogen is more rapidly lost than carbon during degradation of POM. This is the reason for the much greater variation in $\delta^{15}N$-values than in $\delta^{13}C$-values (Saino and Hattori 1980; Altabet and McCarthy 1985).

3.7.3.3 Carbon Isotope Composition of Pore Waters

Initially, the pore water at the sediment/water interface has a $\delta^{13}C$-value near that of seawater. In oxic environments with little organic matter no major change in $\delta^{13}C$ with depth should occur. In more organic carbon-rich sediments a decrease in the pore water $\delta^{13}C$ is observed. The decomposition of organic matter in sediments consumes oxygen and releases isotopically light CO_2 to the pore water, while the dissolution of $CaCO_3$ adds CO_2 which is isotopically heavy. The carbon isotope composition of pore waters at a given locality reflects modification by these two processes. The net result of these two sources is to make pore waters isotopically lighter than the overlying bottom water. Nissenbaum et al. (1972) and Grossman (1984a) have documented a strong negative $\delta^{13}C$-signal in several reducing marine environments. McCorkle et al. (1985) have shown that steep gradients in pore water $\delta^{13}C$ exist in the first few centimeters below the sediment-water inter-

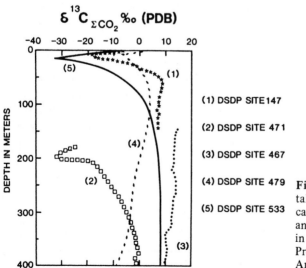

(1) DSDP SITE 147

(2) DSDP SITE 471

(3) DSDP SITE 467

(4) DSDP SITE 479

(5) DSDP SITE 533

Fig. 44. $\delta^{13}C$-profiles of total CO_2 (dissolved inorganic carbon) from pore waters of anoxic sediments recovered in various Deep Sea Drilling Project Sites (Anderson and Arthur 1983)

face. The observed δ^{13}C-profiles vary systematically with the rain of organic matter to the seafloor, with higher carbon rain rates resulting in isotopically lighter δ^{13}C-values.

One would expect that pore waters have δ^{13}C-values no more negative than organic matter. However, a more complex situation is actually observed due to bacterial methanogenesis (see Fig. 44). Bacterial methane production generally follows sulfate reduction in anaerobic carbon-rich sediments, the two microbiological environments being distinct from one another. Since methane-producing bacteria produce very ^{12}C-rich methane, the residual pore water becomes enriched in ^{13}C. As bacterial methane production continues, the pore waters evolve to higher δ^{13}C-values (see Fig. 44). The observed trends cannot simply be interpreted in terms of amounts of sulfate reduction and methane formation, rather carbon losses and gains from the pore water systems must also be taken into account.

3.7.3.4 Bicarbonate in Fresh Waters

Dissolved carbonate in fresh water exhibits an extremely variable isotopic composition, because it represents varying mixtures of carbonate species derived from weathering of carbonates and that originating from biogenic sources like freshwater plankton or CO_2 from organic matter in soils (Hitchon and Krouse 1972; Longinelli and Edmond 1983). Comparison of the data from the Mackenzie River (Hitchon and Krouse 1972) with those from the Amazon basin (Longinelli and Edmond 1983) reveals an interesting difference. The Mackenzie River data have a major δ^{13}C-peak at about $-9‰$ with a "tail" to lower values. The Amazon River data are displaced to about $-20‰$ with a broad distribution range. These differences are consistent with a dominance of carbonate weathering in the Mackenzie River drainage system, whereas in the tropical environment of the Amazon River biological CO_2 predominates.

3.7.4 Sulfate in Ocean and Fresh Water

3.7.4.1 Sulfur Isotope Composition of Ocean Water

Modern ocean water with its large sulfate reservoir has a fairly constant isotopic sulfur composition of $+21‰$ (Rees et al. 1978). An interesting question is whether an isotope fractionation occurs in the sulfate during evaporation of seawater. Nielsen and Ricke (1964) showed that later evaporites within the different evaporation cycles are depleted in ^{34}S by about 2‰ relative to earlier precipitates. Nevertheless, the difference that might occur in the late stages may be neglected if we consider the gypsum-brine relationship. Assuming that calcium sulfates preserve the δ^{34}S-value of the ancient oceans, then it may be concluded that gypsum, anhydrite, and other sulfate-containing evaporite minerals provide information about the isotopic composition of oceanic sulfate during the geologic past. This topic will be discussed in more detail in the following chapter.

3.7.4.2 Sulfur Isotope Composition of Fresh Water

A compilation of river water δ^{34}S-values is shown in Fig. 45. The data of Hitchon and Krouse (1972) for water samples from the Mackenzie

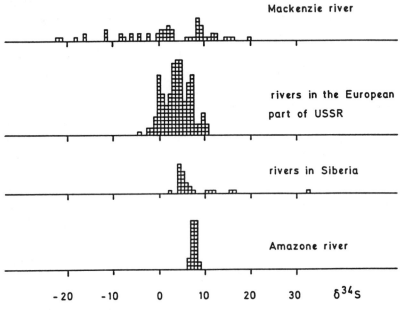

Fig. 45. Frequency distribution of δ^{34}S-values in river sulfate. (Data source: Amazon, Longinelli and Edmond 1983; Mackenzie River, Hitchon and Krouse 1972; rivers in Siberia and in the European part of the USSR, Rabinovich and Grinenko 1979)

River drainage system exhibit a wide range of δ^{34}S-values reflecting contributions from marine evaporites and shales. Surprisingly, Longinelli and Edmond (1983) found a very narrow range for the Amazon River which was interpreted as representing a dominant Andean source of Permian evaporites with a lesser admixture of sulfide sulfur.

Rabinowitch and Grinenko (1979) reported time-series measurements for the large European and Asian rivers in the Soviet Union. The European river systems are approximately normally distributed with a mean value of +6‰, while in the Asian systems δ^{34}S-values below 4‰ are, with one exception, lacking. Rabinowitch and Grinenko (1979) found a mean value of 8.2‰ for the 22 largest rivers in the USSR, which coincides with the average isotopic composition of sediments.

3.7.4.3 Oxygen Isotope Composition of Ocean Water Sulfate

Oceanic sulfate has a very constant oxygen isotope composition of 9.6‰ (Lloyd 1967, 1968; Longinelli and Craig 1967). From the theoretical calculations of Urey (1947) it is quite clear that this value does not represent equilibrium with the δ^{18}O-value of the ocean water, but how this value has been achieved in the worlds ocean is still controversial. Lloyd (1967, 1968) proposed a model in which the fast bacterial turnover of sulfate at the sea bottom determines the oxygen isotope composition of dissolved sulfate. This conclusion was questioned by Holser et al. (1979), who argued that the oxygen isotope composition of seawater sulfate should be controlled by a dynamic balance of sulfate inputs (mainly from weathering of sulfides and sulfates) and sulfate outputs (mainly through evaporite formation and sulfate reduction). When considering the oxygen isotopic variations of evaporite minerals through geologic time, one has to consider the fractionation of around 3.5‰ (Lloyd 1968) between the crystallized evaporite mineral and ocean water. Cortecci and Longinelli (1971, 1973) have observed that the ^{18}O-content of sulfate in living shells is close to that of the dissolved oceanic sulfate. Fossil shells, however, exhibit a wide range in sulfate ^{18}O-content, which they interpret as the product of postdepositional changes. Thus, the ^{18}O-content of fossil shells apparently gives no information on the past ocean sulfate δ^{18}O-values.

3.7.4.4 Oxygen Isotope Composition of Fresh Water Sulfate

The oxygen isotope composition of rainfall sulfate is highly variable with $\delta^{18}O$-values ranging from +5 to +17‰ (Cortecci and Longinelli 1970). From the $\delta^{34}S$-values of the samples analyzed by these authors, it appears that most of their rainfall sulfate is not oceanic, but is produced by oxidation of sulfur produced during the burning of fossil fuels. Because the isotopic composition of sulfate in rivers reflects the isotopic composition of the sulfate sources, it is not surprising that sulfates from nonmarine environments range over a wide spectrum (Longinelli and Cortecci 1970; Longinelli and Edmond 1983). Longinelli and Edmond (1983) argued that the variations in the sulfate oxygen of the Amazon River require exchange with the water and dissolved oxygen via partial redox processes.

3.8 Changes in the Isotopic Composition of the Ocean During Geologic History

The question of whether or not the chemical and isotopic composition of the ocean has remained constant throughout geologic history, has been discussed quite frequently in the literature. Hoefs (1981) has summarized current thinking on this controversial topic.

3.8.1 Sulfur

Perhaps the best documented trend of isotope variation is that for the sulfur isotope distribution in marine sulfate. In 1964 Nielsen and Ricke and Thode and Monster published independently the first two "age curves" which showed that the isotopic composition of gypsum and anhydrite in marine evaporites was different at different times in the geologic past. Since then, this curve has been updated with many more analyses (see Holser and Kaplan 1966; Nielsen 1972; Holser 1977; Claypool et al. 1980), demonstrating that especially during the Phanerozoic several pronounced maxima and minima in $\delta^{34}S$ exist, the extremes lying close to +10‰ (Permian) and +30‰ (Cambrian) (see Fig. 46). Because the isotope fractionation between the sulfate-containing evaporite and the sulfate in ocean water is almost negligible, the observed trend in evaporite sulfate should closely reflect fluctu-

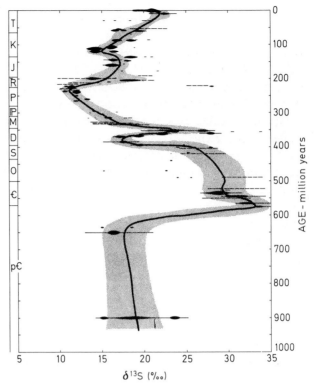

Fig. 46. δ^{34}S age curve of oceanic sulfate (Holser 1977)

ations in the sulfur isotope composition of marine sulfate through geologic time. The present δ^{34}S-value of marine sulfate is primarily the result of the activity of sulfate-reducing microorganisms.

Changes in the δ^{34}S of marine sulfate during the geologic past may be caused by major changes in the budget between the individual reservoirs: during periods of high biological sulfate reduction, which should take place under favorable paleogeographic conditions, the δ^{34}S of ocean water should increase. In contrast, periods of extended weathering introduce additional light continental sulfur into the ocean which decreases the δ^{34}S-value of ocean sulfate. Such periods of extended weathering are geologically plausible in periods of high tectonic, mountain-building activity.

Early models of Nielsen (1965) and Holser and Kaplan (1966) discussed quantitatively the consequences of such changes in the weathering rate and bacterial reduction rate. These earlier models did not, however, consider the possible influence of varying rates of evaporite formation on the δ^{34}S-values of marine sulfate. Rees (1970) was the first to point to the importance of sulfur extraction by evaporite formation and postulated that the δ^{34}S of the ocean "should have tended to high values in periods when evaporite formation was of minor importance and to low values in periods of major evaporite formation. This is qualitatively the case for the Cambrian and the Permian ..."

Recently, a further mechanism of sulfate extraction became quite evident, i.e., the annual cycling of large quantities of seawater through midocean ridges which can have a remarkable effect on the chemistry of ocean water. Claypool et al. (1980) investigated the oxygen isotope together with the sulfur isotope composition of evaporites. Their model approximates each of the flux pairs of sulfide in and sulfide out of the ocean and sulfate in and sulfate out of the ocean by a single net flux of sulfide or sulfate.

Whatever the actual causes may be for the fluctuations of the δ^{34}S-values of oceanic sulfate during the geologic past, it is obvious that this behavior of oceanic sulfate deviates strongly from that expected from an extreme concept of a steady-state ocean. In such a steady-state view the partition into reduced and oxidized reservoirs would be at a fixed ratio. According to Garrels and Perry (1974) the range of δ^{34}S from +30‰ to +10‰ corresponds to a variation of ±30% in the average total amount of sulfate stored in sedimentary rocks and in the ocean. In addition, it is obvious that fluctuations in the three major mechanisms for sulfate removal (1) evaporite formation; (2) bacterial reduction; and (3) cycling through midocean ridges; all have occurred largely in response to changes in the geography of the ocean basins and/or of the adjacent seas.

While the partial cycle between ocean and evaporites only involves sulfate transfer from one reservoir to the other, bacterial sulfate reduction, as well as the weathering of sulfides from argillaceous sediments, change the valence state of the sulfur. Therefore, during a period with increased rate of one of these two processes, appreciable amounts either of organic compounds or of free atmospheric oxygen are needed. Especially in the latter case, oxygen consumption during weathering is appreciable.

3.8.2 Carbon

In the system ocean-atmosphere-biosphere the ocean contains approximately 90% exchangeable carbon, whereas the biosphere contains only 8% and the atmosphere 2%. The ocean is thus the major long-term regulator for the atmospheric CO_2 concentration. Secular variations in the oceanic, dissolved bicarbonate reservoir are generally interpreted as representing changes in the output of carbon from the oceans. At present, a $C_{org}/C_{carbonate}$ output ratio of about $1:4$ maintains a steady state $\delta^{13}C$ of about 0% in the oceanic, total dissolved carbon reservoir. Shifts towards higher $\delta^{13}C$-values in limestones of a given age may be due to an increase of organic carbon burial relative to carbonate carbon burial. Negative $\delta^{13}C$-shifts may accordingly indicate a decrease in the rate of carbon burial. Unusually heavy $\delta^{13}C$-values in Cretaceous carbonates have been thus related to unusually high burial rates of organic carbon (Scholle and Arthur 1980; Hilbrecht and Hoefs 1986; Jenkyns and Clayton 1986).

Such periods have been called "anoxic oceanic events" and the best documented example is the Cenomanian/Turonian boundary. The present stratigraphic resolution of the different sections is not sufficient to establish whether or not the "$\delta^{13}C$-anomaly" is isochronous worldwide.

Similar drastic ^{13}C-changes of carbonates have been reported for other geologic time periods (for the Permian/Triassic boundary by Magaritz et al. 1983 and for the Upper Proterozoic by Knoll et al. 1986). It is still debatable whether these changes represent worldwide phenomena which characterize the ocean water at those specific time intervals or represent more local phenomena. Furthermore, it is also possible that in some cases the observed ^{13}C-patterns are partially due to secondary diagenetic processes. However, in thick beds of more or less pure limestones with minor amounts of organic carbon diagenetic reactions should have only a negligible effect on the primary isotopic composition.

Because of the $1:4$ ratio of organic carbon to carbonate carbon withdrawal from the ocean, the organic carbon reservoir is much more sensitive to such secular variations. However, diagenetic reactions during burial and maturation of organic matter may change the isotopic composition by several ‰ (see p. 158) which complicates the use of organic carbon in the search for secular isotope variations. Nevertheless, as Knoll et al. (1986) suggested measurement of both organic and carbonate carbon greatly reduces the chance of misinterpretations.

Due to the relatively rapid exchange in the ocean-atmosphere-biosphere system, an increase in photosynthetic activity may also lead to secular changes in the system. The larger the amount of carbon accumulated in the biosphere, the more positive is the $\delta^{13}C$ of the oceanic carbon reservoir. A decrease in photosynthetic activity should, of course, lead to a shift in the opposite direction. Attempts to relate differences in photosynthetic activity with $\delta^{13}C$-values have been made by Welte et al. (1975) and Arneth et al. (1985).

Such $\delta^{13}C$-excursions can be modelled quantitatively in order to estimate changes of C_{org} both forward and backward in time (Garrels and Lerman 1984). Finally, it should be mentioned that because the geochemical cycles of carbon and sulfur are coupled, the observed changes in the sulfur isotope composition of evaporites should lead to a concomitant change in the carbon isotope composition. And, indeed, careful inspection of the literature data led Veizer et al. (1980) to conclude that a negative correlation between the $\delta^{13}C$-values of sedimentary carbonates and $\delta^{34}S$-values of the sulfates does exist.

3.8.3 Oxygen

To alter the oxygen isotope composition of ocean water demands huge amounts of water being isotopically very different from the ocean water composition. Short-term variations of $\delta^{18}O$ occur during the ice ages. When continental glaciers grow, ^{16}O is preferentially removed from the ocean, the reverse occurs when glaciers melt. Thus, the advance and retreat of glaciation during the ice ages must have changed the $\delta^{18}O$ of seawater.

Besides these short-term fluctuations, long-term unidirectional processes also have to be considered. As is known for many years the $\delta^{18}O$-values of marine cherts and limestones tend to decrease with increasing geologic age (Knauth and Lowe 1978; Veizer and Hoefs 1976). The significance of these trends is still not settled; continuous postdepositional exchange with interstitial solutions was the first explanation for these trends and undoubtedly many of the samples which define trends have been subjected to diagenesis. Nevertheless, this hypothesis appears to be contradictory to several important observations. The $\delta^{18}O$ versus age trends for cherts and carbonates are nearly parallel despite the fact that the susceptibility of quartz and calcite to isotope exchange with fluids is quite different. Shemesh et al. (1983)

reported a similar trend for phosphorites and concluded that the stability of the phosphate ion and its inertness of phosphate-water isotope exchange excludes meteoric water diagenesis as an explanation for the trend. These phosphorite data suggest that the trend towards lower $\delta^{18}O$-values reflect changes in temperatures and/or changes in the isotopic composition of the ocean.

This interpretation again is in conflict with data from altered ocean crust. Muehlenbachs and Clayton (1976) presented a model in which the isotopic composition of ocean water is held constant by two different processes: (1) low-temperature weathering of oceanic crust which depletes ocean water in ^{18}O because ^{18}O is preferentially bound in the weathering products, whereas (2) high-temperature hydrothermal alteration of MORB basalts enriches ocean water in ^{18}O because ^{16}O is consumed by the hydrothermal alteration reactions. These two processes, being opposite in sense and roughly equal in magnitude, thus buffer the isotopic composition of ocean water. Similar conclusions have been drawn by Gregory and Taylor (1981) who postulated that ocean water had a constant $\delta^{18}O$-value during almost all of earth history. In summary, various strong arguments contradict each other, which leaves the issue far from being resolved.

3.9 Atmosphere

The basic chemical composition of the atmosphere is quite simple, being made up almost entirely of three elements: nitrogen, oxygen, and argon. Other elements and compounds are present in amounts that, although small, are nevertheless significant in terms of important properties of the atmosphere, such as the ozone content. The atmosphere is moderately homogeneous in season as well as at elevation except for water and ozone; the former is continually derived from and returned to the hydrosphere and the latter is largely concentrated in the stratosphere. Other interchanges between the ocean and the atmosphere take place within the carbon, nitrogen, and sulfur cycles. The average abundances of some important atmospheric constituents are shown in Table 22. There is increasing awareness that isotope techniques can be very useful in evaluating the sources of anthropogenic pollution of the atmosphere, such as CO_2, SO_2, and nitrogen oxides.

The constituents of the atmosphere have been derived largely from degassing of the earth's mantle. From observations of volcanic and cos-

Table 22. Average contents of some important constituents in the atmosphere (on a water-free basis)

Gas	Abundance by volume (%)
N_2	78.09
O_2	20.95
Ar	0.93
CO_2	0.032
CH_4	0.00002
H_2	0.00005
N_2O	0.00005
O_3	0.000007 (Summer)
	0.000002 (Winter)

mic gases it has been concluded that the primeval atmosphere was free of oxygen and that reducing conditions prevailed. The first free oxygen was probably produced through photochemical dissociation of water vapor in the upper atmosphere, which would produce oxygen and hydrogen, with the hydrogen escaping into outer space. However, this free oxygen probably did not initially accumulate in the atmosphere, but was used up in oxidizing the more reduced constituents of the atmosphere. This stage came to an end when oxygen production exceeded oxygen use, which probably occurred when photosynthesis reached a certain level of oxygen production.

The amount of argon in the atmosphere, which is 99.69% ^{40}Ar, is anomalously high when compared with that of the other inert gases. This is evidently due to the production of ^{40}Ar by the radioactive decay of ^{40}K throughout geologic time.

3.9.1 Nitrogen

Atmospheric nitrogen collected from many altitudes shows a constant isotopic composition (Dole et al. 1954; Sweeney et al. 1978). Air samples collected over a 6-month period at several locations had a constant ^{15}N/^{14}N ratio, within 0.2‰ (Hoering 1956).

Besides the overwhelming predominance of elemental nitrogen, there are various other nitrogen compounds in the atmosphere as trace compounds. Of these, nitrous oxide (N_2O) is an important greenhouse gas, which thus influences the energy budget of the atmosphere. Nitrous

oxide is mainly produced by bacterial processes of nitrification and denitrification in soils and oceans. It is destroyed photochemically in the
stratosphere. $\delta^{15}N$- and $\delta^{18}O$-measurements (Yoshida and Matsuo 1983;
Yoshida et al. 1984; Whalen and Yoshinara 1985) have shown that
N_2O is isotopically variable in the atmosphere depending upon its specific source.

Besides its natural occurrence, N_2O plays a special role in stable
isotope investigations, because it condenses when CO_2 is extracted
from air in a liquid nitrogen trap. Because the masses of the isotopic
N_2O molecules equal those of CO_2, N_2O interferes with the $^{13}C/^{12}C$
and $^{18}O/^{16}O$ ratios of CO_2 and thus a correction for the N_2O atmosphere concentration is required (Craig and Keeling 1963; Mook and
van der Hoek 1983).

3.9.2 Oxygen

Atmospheric oxygen has a rather constant isotopic composition with
a $\delta^{18}O$-value of +23‰ (Dole et al. 1954; Kroopnick and Craig 1972;
Horibe et al. 1973). Urey (1947) calculated that if equilibrium was obtained between atmospheric oxygen and water, then atmospheric oxygen should be enriched in ^{18}O by 6‰ at 25 °C. This means that atmospheric oxygen cannot be in equilibrium with the hydrosphere and thus,
enrichment of free O_2 in ^{18}O, the so-called Dole effect, must have
another explanation.

It was originally believed that photosynthesis of green plants might
control the $^{18}O/^{16}O$ ratio in the atmosphere. It is commonly accepted
that molecular oxygen produced during photosynthesis arises from the
splitting of H_2O molecules and not from the splitting of CO_2. Dole and
Jenks (1944) determined that the liberated oxygen was enriched in
^{18}O by about 5‰ relative to the water from which it was derived.
Thus, photosynthetic oxygen has an $^{18}O/^{16}O$ ratio approximately expected for isotopic equilibrium between water and free oxygen. Therefore, photosynthesis cannot account for the Dole effect. To solve the
problem, Rabinowitch (1945) suggested that the cause of the Dole effect might be isotope fractionation caused by the preferential uptake
of ^{16}O during respiration. This is supported by the observation of Lane
and Dole (1956), who found that oxygen isotope enrichment during
respiration in several plants, bacteria, and in man varied from 7‰ to
25‰. Kroopnick (1975) measured the oxygen isotope fractionation

during respiration on natural populations in ocean water and found that respiration can cause an enrichment of about 21‰. It is, therefore, reasonable to assume that the $\delta^{18}O$-value of atmospheric oxygen is balanced between input from photosynthesis and output by respiration.

The analysis of fossil air in ice cores yields information about the isotope composition during the past tens of thousands years. When snow transforms into ice, atmospheric air is trapped in the form of bubbles. Fireman and Norris (1982), Horibe et al. (1985), and Bender et al. (1985) demonstrated that the $\delta^{18}O$-value of this atmospheric oxygen was ^{18}O-enriched and varied along with that of seawater (during the ice ages, seawater is ^{18}O-enriched).

Ozone. In situ mass spectrometric measurements of stratospheric ozone by Mauersberger (1981) have shown large ^{18}O-enrichments, as much as 40% above tropospheric values. He explained this enrichment as being due to preferential dissociation of $^{18}O^{16}O$ leading to an overabundance of ^{18}O-atoms. However, as pointed out by Kaye and Strobel (1983) the problem is that isotope exchange of the exchange reaction:

$$^{18}O + {}^{16}O^{16}O \rightleftharpoons {}^{16}O + {}^{18}O^{16}O$$

is much faster than ozone formation, which should prevent any enhancement of heavy ozone in the stratosphere.

As has been shown recently by Thiemens and Heidenreich (1983), photochemical reactions may produce anomalous isotopic patterns. These authors reported equal enrichment of ^{17}O and ^{18}O in ozone produced by an electric discharge in oxygen and interpreted their results by self-shielding of $^{16}O_2$. Navon and Wasserburg (1985) analyzed the oxygen isotope shifts during photodissociation of oxygen and found that the remaining O_2 is enriched in ^{16}O. However, it is necessary to separate the anomalous oxygen from the gas reservoir in order to preserve the isotope shifts. Furthermore, Navon and Wasserburg (1985) demonstrated that the effects found by Thiemens and Heidenreich (1983) cannot be explained by self-shielding of UV radiation as the pressure is below the minimum needed for self-shielding to occur.

3.9.3 Carbon

3.9.3.1 Carbon Dioxide

$\delta^{13}C$. The CO_2 content of the atmosphere controls many processes of geologic importance (e.g., pH of ocean water and the "greenhouse effect" of shielding solar energy). Daily, seasonal, secular, local, and regional changes in atmospheric content have been observed as regular fluctuations. Careful determinations by Keeling (1958, 1960, 1961) have shown that daily variations, which depend on respiration, exist over continents. Respiration of plants reaches a distinct maximum around midnight or in the early morning hours. Respiratory plant CO_2 has a $\delta^{13}C$-value between $-26\%o$ and $-21\%o$. At night, when respiration of plants reaches maximum values, there is a measurable contribution of respiratory CO_2. This relation between CO_2 content and $\delta^{13}C$ is demonstrated in Fig. 47.

The burning of fossil fuel has significantly increased the CO_2 content of the atmosphere. Farmer and Baxter (1974) noted an increase from about 290 ppm CO_2 in 1900 to 320 ppm in 1970. Freyer and Wiesberg (1973) and Farmer and Baxter (1974) suggested that the carbon isotope composition of tree rings is, in fact, a record of atmospheric carbon isotopic variations. Besides the burning of fossil fuel, other factors such as increased oxidation of plant debris caused by increased

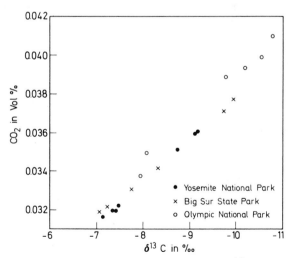

Fig. 47. Relationship between atmospheric CO_2 concentration and $\delta^{13}C_{(CO_2)}$. (After Keeling 1958)

cultivation of arable land could have influenced the carbon isotopic composition of atmospheric CO_2.

The annual combustion of 10^{15} g fossil fuel with an average $\delta^{13}C$-value of $-27\%o$ would change $\delta^{13}C$ of the atmospheric CO_2 by $-0.2\%o$ yr^{-1}. The observed change is about ten times as small (Mook et al. 1983). There are two reasons for this apparent discrepancy:

1. only about 60% of the injected CO_2 remains airborne, the other part is mainly dissolved in the oceans;
2. oceanic, dissolved carbon exchanges isotopically with atmospheric CO_2.

One of the most promising methods of reconstructing atmospheric CO_2 concentrations of earlier times are measurements on air trapped in polar ice. Such measurements have revealed that at the end of the last glaciation, the atmospheric CO_2 concentration was about 30% lower than during the Holocene. Friedli et al. (1984) made $\delta^{13}C$-measurements on CO_2 separated from air extracted from South Pole ice which yielded a 1.1%o higher $\delta^{13}C$-value than air-CO_2 in 1980. This is consistent with the measured CO_2 concentration in these samples and with model-based estimations.

By measuring the ^{13}C-content of planktonic and benthic foraminifera, Shackleton et al. (1983) reached very similar conclusions and postulated low CO_2 concentrations during the last glacial ages (see also discussion on p. 175).

$\delta^{18}O$. Atmospheric CO_2 has a $\delta^{18}O$-value of $+41\%o$, which means that atmospheric CO_2 is in approximate equilibrium with ocean water at 25 °C. Bottinga and Craig (1969) showed that exchange with ocean water regulates the average composition of atmospheric CO_2, although exchange with atmospheric water may cause small perturbations.

3.9.3.2 Other Carbon Compounds

Stevens et al. (1972) reported regular seasonal variations in the carbon and oxygen isotopic composition of atmospheric carbon monoxide. They estimated the worldwide average $\delta^{13}C$-value for engine CO to be $-27.4\%o \pm 0.3\%o$. Bainbridge et al. (1961) determined one $\delta^{13}C$-value of atmospheric methane of $-39\%o$.

3.9.4 Hydrogen

Free atmospheric hydrogen is present to approximately five parts in 10^7 in tropospheric air. The deuterium concentration is in the vicinity of $+70‰ \pm 30‰$ (Friedman and Scholz 1974). Gonsior et al. (1966) found δD-values from -600 to $-200‰$ for industrial hydrogen, which was mainly ascribed to hydrogen released from automobile exhausts. The δD-values of atmospheric hydrogen are higher than in any natural material found on earth. Because all equilibrium isotope exchange reactions that are known concentrate hydrogen in H_2 relative to the other reacting phase, kinetic effects seem to account for this heavy hydrogen.

3.9.5 Sulfur

Sulfur is found in trace compounds in the atmosphere where it occurs in aerosols as sulfate and in the gaseous state as H_2S and SO_2. The major contributions to atmospheric sulfur are: (1) industrial sulfur, (2) ocean spray sulfate, (3) bacterial sulfur, mostly from tidal flats, and (4) volcanic sulfur. Mizutani and Rafter (1969) concluded that seawater spray was one of the main sources of sulfate in the rain studies, the other main source being industrial activity, which produces SO_2. The contribution of each source varied widely, depending mainly upon meteorological conditions. However, about one-half of the samples studied showed a dominant contribution of seawater sulfate.

Anthropogenic vs Natural Sources. Sulfur compounds arising from anthropogenic and natural sources are mixed in the atmosphere and hydrosphere. The complexities involved in the isotopic composition of atmospheric sulfur have been discussed by Nielsen (1974). Grey and Jensen (1972) argued that, in the Salt Lake City Region of Utah, the atmospheric sulfur comes from automobile exhausts, from biological H_2S production, and from the plume of a large copper smelter. During their investigation, an extended strike brought the smelter to a standstill and thus enabled the determination of δ-values with and without the smelter exhaust. The results are shown graphically in Fig. 48. Normally, the premises are much more complicated, which limits the "fingerprint" character of the S-isotope composition of atmospheric sulfur to such rare cases as described above.

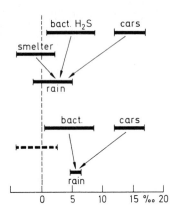

Fig. 48. δ^{34}S-values in rain water in the Salt Lake area (USA) with and without contribution from copper smelter (Nielsen 1974)

For another very unique situation in Alberta (Canada) where the industrial SO_2 had a δ^{34}S-value near 20‰, while the δ^{34}S of uncontaminated soil was near 0‰, Krouse (1980) was able to give semiquantitative estimates. However, it must be clearly stated that these data cannot be transferred to other localities. The isotopic composition of the industrial sources are generally so variable that the assessment of anthropogenic contributions to the atmosphere is extremely difficult.

3.10 Biosphere

As used here, the term "biosphere" includes the total sum of living matter, plants, animals, and microorganisms, and the residues of living matter in the geologic environment, such as coal and petroleum. A fairly close balance exists between photosynthesis and respiration, although over the whole of geologic time respiration has been exceeded by photosynthesis, and the energy thus derived was stored mostly in disseminated organic matter and, of course, in coal and petroleum.

Questions concerning the origin of coal and petroleum center around three topics: the nature and composition of the parent organisms, the mode of accumulation of the organic material, and the reactions whereby it was transformed into the end products.

Petroleum (frequently also called crude oil) is a naturally occurring complex mixture, composed mainly of hydrocarbons, but also with varying amounts of heterocompounds containing S, N, O, and metalloorganic molecules, such as vanadium and nickel porphyins. Although there are, without any doubt, numerous compounds that have been formed more or less directly from biologically produced molecules, the majority of petroleum components are of secondary origin, either decomposition products or products of condensation and polymerization reactions.

3.10.1 Living Organic Matter

3.10.1.1 Carbon

The complexities involved in the photosynthetic fixation of carbon have already been discussed briefly on p. 34. Wickman (1952) and Craig (1953) were the first to demonstrate that marine plants are about 10‰ enriched in ^{13}C relative to terrestrial plants. Since then, numerous studies have broadened this view and provided a much more detailed picture of isotope variations in the biosphere. The reasons for the large C-isotope differences found in plants were only satisfactory explained after the discovery of new photosynthetic pathways in the late 1960's. The bulk of the plant kingdom fixes CO_2 during the pathway described by Calvin (also called C_3-pathway). The two new pathways are known as Hatch-Slack (or C_4-pathway) and CAM (Crassulacean Acid Metabolism, diurnal process of acidification and deacidification). The differences in isotopic composition characteristic for each of the pathways are due to different enzymatic processes and the different sizes of the metabolic pools of carbon.

Figure 49 summarizes the variability of $\delta^{13}C$-values exhibited by some major groups of higher plants, algae, and microorganisms. Especially noteworthy is that the $\delta^{13}C$-ranges of C_3 and C_4 plants virtually do not overlap and that the methanogenic bacteria show an extremely large variation range.

One of the most important groups of all living matter is marine phytoplankton. Natural oceanic phytoplankton populations vary in $\delta^{13}C$-values by about 15‰ (Sackett et al. 1973; Wong and Sackett 1978). Rau et al. (1982) showed that latitudinal trends in the $^{13}C/^{12}C$ ratio of plankton differ significantly between the northern and the southern oceans: south of the equator the correlation between latitude and the plankton $\delta^{13}C$-value is significant, whereas a much weaker relationship exists in the northern oceans. This is strong evidence against a simple temperature-plankton $\delta^{13}C$-relationship as originally proposed by Sackett et al. (1973). Therefore, factors other than temperature must also play an important role in determining plankton $\delta^{13}C$.

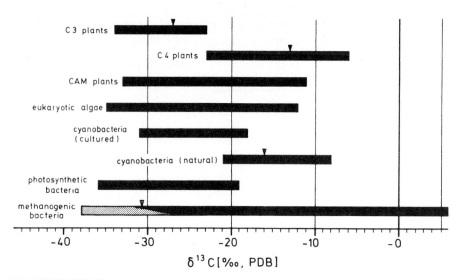

Fig. 49. Carbon isotope composition of extant higher plants, algae, and autotrophic prokaryotes. Means for some groups are indicated by *triangles*. Note the virtual absence of overlap in the ranges of C3 and C4 plants (Schidlowski et al. 1983)

Chemical Components of Plant Material. A number of investigators have studied the isotopic composition of the major biochemical constituents of plants (Park and Epstein 1960; Abelson and Hoering 1961; Parker 1964; Degens et al. 1968b; Smith and Epstein 1970; DeNiro and Epstein 1977). Figure 50 demonstrates that differences in ^{13}C-contents exist between different chemical plant components: sugar, cellulose, and hemicellulose exhibit values close to the mean plant carbon isotopic composition, whereas pectin appears to be enriched in ^{13}C and lignin and lipids are depleted in ^{13}C relative to the total plant. In the latter class the size of the isotopic difference is especially pronounced (see Fig. 50). This is not surprising because the lipid fraction includes a wide variety of organic compounds. For example, Degens et al. (1968b) found that the $CHCl_3$ extractable lipids had lighter δ^{13}C-values than the C_2H_5OH extractable lipids.

Amino acids as a whole exhibit a general ^{13}C-enrichment with respect to the total plant. By separating different amino acids Abelson and Hoering (1961) were able to demonstrate that there are large variations among individual amino acids. In the case of Chlorella, for example, some typical values are: glutamic acid – 18.7‰, aspartic acid

Fig. 50. Differences in ^{13}C-contents of chemical constituents and the bulk plant material (Deines 1980b)

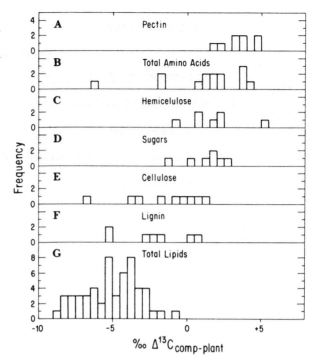

$-6.6‰$, serine $-5.7‰$, alanine $-10.3‰$, leucine $-22.7‰$, and tyrosine $-19.8‰$. Thus, any approach to isotope fractionations of amino acids as a whole must fail. However, the situation is even more complex. By using a similar method to that of Abelson and Hoering (1961), Macko et al. (1983) observed a wide range in δ^{13}C- and δ^{15}N-values of individual amino acids, many appear to be associated with kinetic fractionations that follow the metabolic pathways of amino acids. These results illustrate the diversity of isotope fractionations that occur within the cells of a single organism.

Respiratory Processes. Although plants fix CO_2 in the form of organic compounds, a certain amount of CO_2 is also released as the result of respiration. Measurements made by trapping the CO_2 in a CO_2-free atmosphere indicate that there is little difference between respired CO_2 and total plant for both C_3 and C_4 plants. However, not all of the CO_2, formed as a result of respiration and other CO_2-forming processes is actually released to the environment. Some of this CO_2 is re-fixed with some isotope fractionation (O'Leary 1981). Consequently, the isotopic composition measured for respired carbon may differ from that of total carbon formed by respiratory processes.

Soil CO_2. Soil CO_2 originates from the decomposition of organic material and from plant root respiration. Available δ^{13}C-measurements of soil CO_2 have been compiled by Deines (1980b) and reveal the same bimodality observed for C_3 and C_4 plants. Long-term observations on the isotope variation in soil CO_2 by Parada et al. (1983) have shown seasonal variations of $\sim\pm 5‰$. Some variations might be due to mixing with atmospheric CO_2, which is more noticeable in winter, when soil CO_2 concentrations are lowered. Other variations could be caused by decomposition of organic matter with different isotopic composition and different decay rates.

Animals. Already Craig (1953) noted that δ^{13}C-values for animal tissues fall in the same range as their food supply. DeNiro and Epstein (1978) demonstrated clearly that the carbon isotope composition of an animal greatly depends on its diet. Studies by Haines (1976) and Minson et al. (1975) have shown that the large differences in the ^{13}C-value between plants possessing either the C_3 or C_4 photosynthetic pathways are reflected in animals which derive their carbon predominantly from C_3 or C_4 plants. Furthermore, animals feeding on marine organisms have different isotopic compositions from those feeding on terrestrial organisms (Schoeninger and DeNiro 1984).

3.10.1.2 Hydrogen

During photosynthesis, plants remove hydrogen from water and transfer it to organic compounds. The fixation of CO_2 and H_2O to organic matter leads to a deuterium depletion in the plants relative to the environmental water (Schiegl and Vogel 1970; Smith and Epstein 1970). After the formation of the organic matter the oxygen-bound hydrogen atoms are readily exchangeable, while the carbon-bound hydrogen seems to be nonexchangeable (Epstein et al. 1976). This has to be taken into account when correlations between the D/H ratio in plants and the environmental water are attempted (see also Sect. 3.10.2).

There are systematic differences in the hydrogen-isotope ratios among classes of compounds in plants. Lipids usually contain less deuterium than the protein and the carbohydrate of the extracted plant

(Hoering 1975; Estep and Hoering 1980). The lipids may be subdivided into two different groups having greatly different deuterium contents. The first contains the fatty acid-saturated hydrocarbons which have a common biosynthetic pathway involving synthesis from two carbon fragments. The second group contains the pythol, sterols, and carotenes which synthesized via the five-carbon isoprenoid pathway (Hoering 1975).

3.10.1.3 Oxygen

The experimental difficulties in determining the oxygen isotope composition of biological materials lie in the rapid exchange between organically bound oxygen, in particular the oxygen of carbonyl and carboxyl functional groups with water. Thus, it is not surprising that studies on the oxygen isotope fractionation within living systems have been limited to that associated with the biosynthesis of cellulose, the oxygen of which is only very slowly exchangeable at physiological pH (Epstein et al. 1977; DeNiro and Epstein 1979, 1981). Epstein et al. (1977) analyzed the ^{18}O-content of cellulose from aquatic and terrestrial plants and compared the $\delta^{18}O$-values obtained to those of the water used by the plants ($\delta^{18}O$-values range from +14 to +33‰). For aquatic plants the fractionation factor between the oxygen in the cellulose and that in the water medium is about 1.027. A model which accounts for this fractionation factor is that two-thirds of the cellulose oxygen comes from the dissolved CO_2 and one-third from the oxygen of the water.

The relationship between ^{18}O in cellulose and in water from terrestrial plants is more complicated because evaporative transpiration of water takes place through their leaves. In cases in which an aquatic and a terrestrial plant have similar δD-values, making a comparison possible, the $\delta^{18}O$-values of the terrestrial plants are higher by 4 to 16‰. DeNiro and Epstein (1979) investigated the relationship between the oxygen isotope ratios of plant cellulose, carbon dioxide, and water. They argued that the oxygen derived from CO_2 undergoes complete exchange with the water oxygen in the plant during the synthesis of cellulose. This equilibration implies that the $\delta^{18}O$-value of cellulose is primarily a function of the ^{18}O-content of the water in the plant.

3.10.1.4 Sulfur

Sulfur is a key element of life constituting on average between 0.5 and 1.5% (dry weight) of plant and animal matter. It occurs mainly in proteins that typically display a C/S ratio of about 50. The processes responsible for the direct primary production of organically combined sulfur are the direct assimilation of sulfate by living plants and microbiological assimilatory processes in which organic sulfur compounds are synthesized. During these processes, sulfate is first phosphorylated to give "activated" sulfate species which in turn are reduced via sulfite and other intermediates to the sulfide level. At present, only a limited number of measurements of $^{34}S/^{32}S$ ratios of biological material are available. Mekhtiyeva and Pankina (1968) and Mekhtiyeva et al. (1976) have demonstrated that sulfur of aquatic plants from a given water is slightly lighter than the sulfur of the dissolved sulfate. The same results have been obtained by Kaplan et al. (1963) for marine organisms, plants, and animals.

3.10.1.5 Nitrogen

Nitrogen uptake in terrestrial plants is primarily implemented by the fixation of atmospheric N_2 mediated by soil bacteria. Since this process is not related with an appreciable isotope effect (Hoering and Ford 1960; Delwiche and Steyn 1970), the organic matter in terrestrial plants should have a $\delta^{15}N$-value near that of atmospheric nitrogen.

3.10.2 Tree Rings

3.10.2.1 Deuterium and Oxygen

The approach to such studies is to compare the δD- and $\delta^{18}O$-values in organic material from growth rings which represent various ages and stages in the development of the tree. A depletion in D and ^{18}O in growth rings from numerous specimens of the same tree species growing in the same location could be an indication of lower temperature, while an enrichment in D and ^{18}O may then reflect a higher temperature.

Because of an intracellular heterogeneity in the δD-distribution of plant organic matter, Epstein et al. (1976) suggested that those studies which attempted to determine climatic changes based upon analyses of whole wood samples may be erroneous due to the presence of a mixture of cellulose, starch, lignins, and lipids within the wood tissues and that D/H ratios of only single, specified components should be compared. Epstein et al. (1976) chose cellulose as that component of wood which could be isolated in the purest form, by first replacing the exchangeable OH groups of the polymer with nitrate.

Burk and Stuiver (1981) have shown that the oxygen isotope composition of cellulose can be used as a temperature indicator in specific West Coast areas of the United States where humidity values are fairly constant. These authors suggested that the $\delta^{18}O$-values of the source water, humidity, leaf boundary-layer dynamics and the $\delta^{18}O$-composition of atmospheric water vapor must be considered when evaluating the temperature dependence of oxygen isotope ratios in tree rings.

3.10.2.2 Carbon

There is debate about the meaning of $^{13}C/^{12}C$ variations in tree rings. Two different interpretations are frequently discussed: one relates the decrease in $^{13}C/^{12}C$ ratios over the period 1850 to 1950 to an increase in fossil fuel combustion (Freyer 1979), whereas the other explains the $^{13}C/^{12}C$ variations with climatic temperature changes (Mazany et al. 1980). Further work must be done to identify and eliminate sources of "noise" in the tree ring record. The recent model of Francey and Farquhar (1982), in which carbon isotope variations are related to physiological properties of a leaf, is an important step forward in understanding the meaning of $^{13}C/^{12}C$ variations in tree rings.

3.10.3 Organic Matter in Sediments

Immediately after burial of the biological organic material into the sediments, complex diagenetic changes occur in the organic matter. The biopolymers, e.g., polysaccharides and proteins, are attacked by microorganisms and are partly broken down to soluble components, while other parts polymerize and react to high molecular weight polycondensation products, i.e., humic substances. With these diagenetic changes carbon isotope shifts of a few or several per mil are connected. They include isotope effects during bacterial degradation of the biopolymers which preferentially eliminate ^{13}C-enriched carbohydrates and proteins and preserve ^{12}C-enriched lipids. Decarboxylation reactions remove ^{13}C-enriched carboxyl groups leading to ^{13}C-depletion in the residue. As has been mentioned above, humic substances are considered to represent the first transformation products of the organic matter. They are defined as dark-brown polymers that are divided into alkali soluble, but acid insoluble, humic acids and into alkali- and acid-soluble fulvic acids. Nissenbaum and Kaplan (1972) demonstrated that humic acids are generally depleted in ^{13}C relative to fulvic acids. Thus, fulvic acids are closer in $\delta^{13}C$ to plant carbon and are considered by Nissenbaum and Schallinger (1974) to be an intermediate in the huminification process.

Considered as a whole, recent marine sediments show a mean $\delta^{13}C$-value of -25% (Deines 1980b). With transformation to kerogen some ^{13}C-loss occurs, leading to an average $\delta^{13}C$-value of -27.5% (Hayes et al. 1983). This ^{13}C-depletion might be best explained by the large losses of CO_2 that occur during the transformation to kerogen and which are especially pronounced during the decarboxylation of some ^{13}C-rich carboxyl groups.

With further thermal maturation the opposite effect of an ^{13}C-enrichment is observed. Experimental studies of Chung and Sackett (1979), Peters et al. (1981), and Lewan (1983) indicate that thermal alteration produces a maximum ^{13}C-change of about $+2\%$ in kerogens. Changes of more than $2-3\%$ are most probably not due to isotope fractionation during normal thermal degradation of kerogen, but due to isotope exchange reactions between kerogen and carbonates.

Recent marine organic carbon is ^{13}C-enriched relative to terrestrial organic carbon (Deines 1980b). This distinction has been used to differentiate between these two sources in sediments (Brown et al. 1972). However, Dean et al. (1986) and Arthur et al. (1985) failed to observe a consistent relationship between $\delta^{13}C$-values of organic carbon and independent geochemical indicators of marine and terrestrial organic matter for samples older than Miocene in age. On the contrary, just the oposite is apparently observed: those samples with the highest $\delta^{13}C$-values had chemical compositions indicating the greatest contributions of marine organic matter. Arthur et al. (1985) concluded that marine organic carbon in sediments that are Cretaceous or older have $\delta^{13}C$-values that are 5 to 7% more negative than marine organic carbon in Holocene sediments. The reasons for this inverse relationship are not fully understood. Arthur et al. (1985) and Dean et al. (1986)

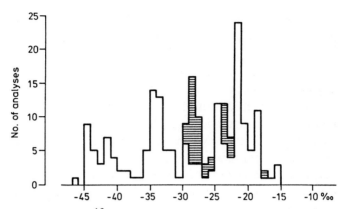

Fig. 51. Frequency distribution of δ^{13}C-values of Precambrian graphites. □ Archean, ▤ Proterozoic (Strauss 1986)

postulated that marine photosynthesis in Mid-Cretaceous or earlier oceans resulted in larger fractionation producing organic carbon with lighter δ^{13}C-values.

Extremely low δ^{13}C-values have been observed in Archean organic matter. About 25% of around 250 samples analyzed are anomalously depleted in ^{13}C with δ^{13}C-values between -47 and $-35‰$ (Strauss 1986), while the remaining 75% cover a "normal" spread between -35 and $-15‰$ (see Fig. 51). As far as it is known, such low δ^{13}C-values have to be attributed to methanogenic bacteria and their metabolic processes. Such extremely ^{13}C-depleted organic matter seems to be restricted to the Archean and may represent special conditions during the evolution of life.

3.10.4 Oil

In recent years the combination of stable isotope data (δ^{13}C, δD, δ^{34}S, δ^{15}N) on crude oils and natural gas has become a powerful tool in petroleum exploration. The papers of Fuex (1977), Stahl (1977), Schoell (1984a,b), and Sofer (1984) summarize recent work in this area.

The isotopic composition of crude oil is mainly determined by the isotopic composition of its source material, more specifically the type of kerogen and the sedimentary environment in which it has been formed. Secondary effects like biodegradation and water washing have only little effect on its isotopic composition. Because the isotopically lightest fractions are preferentially consumed during secondary altera- tion processes, biodegradation and water washing lead to a small ^{13}C- enrichment. In laboratory experiments of bacterial oxidation Stahl (1980) was able to demonstrate a small ^{13}C-enrichment in the remain- ing saturated hydrocarbons, while the asphaltene fraction showed a tendency to become isotopically lighter.

As far as it is known today, very small changes in the ^{13}C/^{12}C ratio do occur during migration. Silverman (1965) observed a shift of 0.4‰ during secondary migration of 6 km in the Quriquiru field in Venezuela. Heterocompounds, being the most polar petroleum constituents, tend to be absorbed on mineral surfaces. Aromatics are more polar and water soluble than saturates and, therefore, preferentially removed during migration. This leads to a small ^{13}C-decrease in the crude oil with in- creasing migration paths.

Sofer (1984) could not support earlier reports stating that marine oils are isotopically heavier than terrigenous oils and that the difference can be utilized to distinguish between them. However, isotopic dif- ferences between oils derived from terrigenous and marine organic mat- ter manifest in the isotopic relationship between the saturate and aro- matic hydrocarbon fractions.

The various classes of chemical compounds in crude oils show small, but characteristic, differences in their carbon isotope composition. With increasing polarity the ^{13}C-content increases from the saturated hydrocarbons to the aromatic hydrocarbons, to the heterocomponents (N, S, O compounds) to the asphaltene fractions. As recently demon- strated by Schoell (1984a,b), the same relationship is also true for hydrogen. Thus, the combination of carbon and hydrogen isotope ratio determinations effectively increases the importance of this method for oil-oil and oil-source rock correlations.

Figure 52 schematically demonstrates the changes in isotopic com- positions of extracts in relation to kerogens for various maturities (after Schoell 1984b). From comparison of the type patterns it becomes evident that an isotopic relationship between extract and kerogen can only be expected in the mature stage of kerogens. This has been sup- ported by Johns and Hoefs (1985) on immature extracts from the Vienna Basin. How these "petroleum-type curves" can be successfully applied for exploration questions is shown in Fig. 53, which shows a positive oil-oil correlation and a negative crude-oil-source rock correla- tion. The Tertiary and Jurassic oils are isotopically more or less identical,

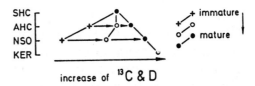

Fig. 52. Changes in isotopic composition of extracts in relation to kerogen. From comparison of the type patterns of immature and mature stages, it is evident that an isotopic relationship between extract and kerogen can only be expected in the mature stage of kerogen (*SHC* saturated hydrocarbons; *AHC* aromatic hydrocarbons; *NSO* components; *KER* kerogen) (Schoell 1984b)

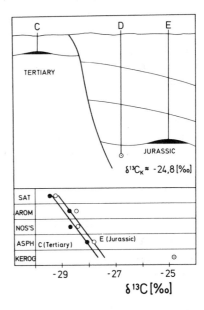

Fig. 53. "Petroleum-type curves" of different oil components from the North Sea. The diagram shows a positive oil-oil correlation and a negative source rock-oil correlation (*SAT* saturated hydrocarbons; *AROM* aromatic hydrocarbons; *NSO's* heterocomponents; *ASPH* asphaltenes) (Stahl et al. 1977)

which points to an origin from the same source rocks. The δ^{13}C-value of the kerogen is not in line with the type line, thus indicating a migration from elsewhere into both reservoirs.

Yeh and Epstein (1981) and Schoell (1984a,b) investigated the hydrogen isotope composition of crude oils. Because there are considerable variations in deuterium concentrations in the biological precursor materials and in kerogens, it is not surprising that there is a considerable range in δD-values (between -200 and $-80‰$, Yeh and Epstein 1981). As far as it is known, D/H exchange processes with pore waters have no major effect on the D-composition of crude oils, at least at temperatures below 160 °C.

Since the pioneering studies of Silverman and Epstein (1958) and Silverman (1964, 1967) it is generally agreed that crude oil is isotopically lighter than the kerogen from which it is derived, but it is similar in isotopic composition to the lipid fraction. However, comparison of the deuterium variations in oils and kerogens (Schoell 1984a,b) shows that many oils have similar deuterium concentrations to the kerogens. This suggests that possibly not only lipidic compounds are responsible for the formation of oils, but the restructuring of the kerogen as a whole leads to the formation of extractable compounds.

Sulfur in Oil. Recently, Krouse (1977) and Thode (1981) reviewed the use of sulfur isotopes in petroleum exploration. Sulfur isotope studies in crude oil may be useful in several respects: (1) to identify crude oil from specific source beds, (2) to group oils into genetic families, (3) to follow their migration, and (4) under favorable conditions to identify oil alteration processes. Oils from widely distributed pools in the same reservoir rocks have similar $\delta^{34}S$-values despite marked differences in sulfur contents, whereas oils from different source beds have variable sulfur isotope compositions because of different environmental conditions during source rock deposition. In the Williston Basin Thode (1981) was able to correlate crude oils with their source rocks and to distinguish three major types of crude oil on the basis of their sulfur isotope composition.

Although the reduced sulfur in crude oil is present in a large number of aliphatic and aromatic compounds with various degrees of complexity and stability, Monster (1972) demonstrated that the sulfur isotope composition in a particular compound class is principally the same as that of the bulk oil. In oil-source rock correlation studies Monster and Thode (unpubl. results) demonstrated that the sulfur in the crude oil is slightly more ^{34}S-rich than the kerogen sulfur, but nearly identical in $\delta^{34}S$-value to that of the solvent-extractable organic sulfur in the source rock.

During thermal maturation, oils maintain their characteristic $\delta^{34}S$-values (Harrison and Thode 1958; Thode et al. 1958), even though sulfur is lost. However, during the very mature stages of thermal alteration Orr (1974) found that $\delta^{34}S$-values in the light fractions of the crude oil change considerably, whereas the heavy asphaltenes tend to retain their original $\delta^{34}S$-values. According to Thode (1981), long distance secondary migration over some 150 km results in little or no change in $\delta^{34}S$-value. However, crude oil alteration, such as water washing and biodegradation, may change sulfur isotope ratios of the bulk oil.

3.10.5 Coal

Carbon and hydrogen isotope compositions of coals are rather variable (Schiegl and Vogel 1970; Redding et al. 1980; Smith et al. 1982). Different plant communities and climates may account for these variations. Several studies, summarized by Maass et al. (1978), have indicated that with increasing grade of coalification very little change in the carbon isotope composition occurs. This may be due to the fact that during coalification the amount of methane and other higher hydrocarbons liberated is small compared to the total carbon reservoir in coals. With respect to hydrogen the reservoir is smaller, which may explain why δD-differences up to 50‰ have been observed by Redding et al. (1980). Schwarzkopf (cited in Schoell 1984b) found systematic δD-differences among different coal macerals, which they explained to represent primary differences within the plant constituents. On the other hand Smith et al. (1982) argued that the complex reactions taking place during the conversion of land plant debris into coals lead to a homogenization of the isotopic differences that initially characterized contributing plant materials.

Because of the problems associated with the combustion of coals, the origin and distribution of sulfur in coals is of special significance. Sulfur in coals usually occurs in different forms, as organic sulfur, as pyrite, sulfates, and elemental sulfur. Pyrite and organic sulfur are the most abundant forms. Organic sulfur is primarily derived from two sources: the original organically bound plant sulfur preserved during the coalification process and biogenic sulfides which reacted with organic compounds during the biochemical alteration of plant debris.

Studies by Smith and Batts (1974), Smith et al. (1982), Price and Shieh (1979), and Hackley and Anderson (1986) have shown that organic sulfur in coal exhibits rather characteristic isotope variations which correlate with sulfur contents. In low-sulfur coals $\delta^{34}S$-values of organic sulfur are rather homogeneous and reflect the primary plant sulfur. In contrast, high-sulfur coals are more variable and typically more negative in $\delta^{34}S$-values, suggesting a significant contribution from bacteriogenic sulfides. The range of $\delta^{34}S$-values in massive pyrite is even more variable and shows no systematic correlation with organic sulfur. This is possibly due to the occurrence of several pyrite generations (Price and Shieh 1979; Hackley and Anderson 1986).

3.10.6 Natural Gas

Natural gases have been found in a wide variety of environments. While methane is always a major constituent of the gas, other components are higher hydrocarbons (ethane, propane, butane), CO_2, H_2S, N_2, H_2, and rare gases. Two different processes are responsible for the formation of the major methane occurrences. The most useful parameters in distinguishing the two different types are their $^{13}C/^{12}C$ and D/H ratios.

Biogenic Gas. According to Rice and Claypool (1981), over 20% of the world's natural gas accumulations are of biogenic origin. Biogenic methane commonly occurs in recent anoxic sediments and is well documented in freshwater environments, such as lakes and swamps and in marine environments, such as estuaries and shelf regions. Two primary metabolic pathways are generally recognized for methanogenesis: fermentation of acetate and reduction of CO_2. Although both pathways may occur in both marine and freshwater environments, CO_2 reduction is dominant in the sulfate-free zone of marine sediments, while acetate fermentation is dominant in freshwater sediments.

During the microbial action kinetic isotope fractionations on the organic material by methanogenic bacteria result in methane very much enriched in ^{12}C, typically with $\delta^{13}C$-values between -110 to $-50‰$ (Schoell 1980, 1984b; Rice and Claypool 1981; Whiticar et al. 1986). In marine sediments the methane formed by CO_2 reduction is often more depleted in ^{13}C than methane formed by acetate fermentation in freshwater sediments. Thus, typical ranges for marine sediments are between -110 and $-60‰$, while methane from freshwater sediments ranges from -65 to $-50‰$ (Whiticar et al. 1986).

The distribution between methane of freshwater and of marine origin is even more pronounced on the basis of hydrogen isotopes. Marine bacterial methane has δD-values between -250 to $-170‰$, while biogenic methane in freshwater sediments is strongly depleted in D with δD-values between -400 to $-250‰$ (Whiticar et al. 1986).

Different sources of hydrogen in biogenic methanes can be accounted for these large differences in hydrogen isotope composition. Formation water supplies the hydrogen during CO_2 reduction, whereas during fermentation three-quarters of the hydrogen come directly from the methyl group, being extremely depleted in D.

Thermogenic Gas. Thermogenic gas is produced when organic matter is buried to greater depths. Increasing temperatures modify the organic matter due to various chemical variations, such as cracking and hydrogen disproportion in the kerogen. $^{12}C-^{12}C$ bonds are preferentially broken during the first stages of organic matter maturation. As this results in an ^{13}C-enrichment of the residue, more $^{13}C-^{12}C$ bonds are broken with increasing temperatures, producing higher $\delta^{13}C$-values. Thermal cracking experiments carried out by Sackett (1978) have confirmed this view and showed that the resulting methane is 4 to 25‰ lower than the parent material. Thus, thermogenic gas typically has $\delta^{13}C$-values between -50 and -20‰ (Rice 1983; Schoell 1980, 1984b). Gases generated from nonmarine (humic) source rocks are isotopically heavier than those generated from marine (sapropelic) source rocks at equivalent levels of maturity.

In contrast to the $\delta^{13}C$-values, δD-values are independent of the composition of the precursor material, but solely depend on the maturity of kerogen. The combination of $\delta^{13}C$- and δD-determinations on natural gases is one of the most promising tools for gas-gas correlations. In a $\delta^{13}C$- versus δD-diagram (see Fig. 54) not only can a clear distinction of biogenic and thermogenic gases from different environ-

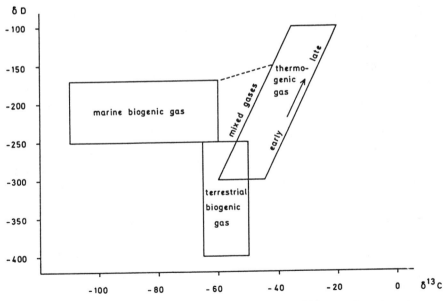

Fig. 54. Natural gas genetic classification diagram using $\delta^{13}C$ and δD of methane. (After Whiticar et al. 1986)

ments be made, but it is also possible to delineate mixtures between the different types.

A well-defined correlation is observed between the δ^{13}C-values of natural gases and the vitrinite reflectance of the sedimentary organic matter from which the gas is derived. The vitrinite reflectance is a measure of the maturity of the organic matter, it changes from about R_0 $\approx 0.3\%$ to $R_0 \approx 3.5\%$ in mature kerogen.

This relationship can be successfully applied for gas-source rock correlations (see Fig. 55). Figure 55 gives an example from the Arctic. A gas was discovered in a Jurassic sand, which could have originated in the relatively immature Lower Cretaceous-Jurassic section (R_0 of 0.5 to 0.65%) or from thermally overcooked Triassic black shale (R_0 of 1.0 to 2.0%). The δ^{13}C-value of the gas was $-37.2\%_0$ and the vitrinite reflectance R_0 1.9%, and thus the gas could only have originated in the Triassic black shales at a considerable depth below the horizon. It was apparent to the exploration geologists that any search for similar-type deposits was dependent on the existence of the Triassic black shale beneath the reservoir rock.

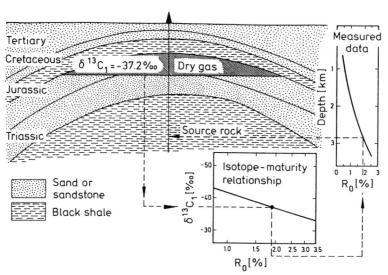

Fig. 55. Correlation between the isotopic composition of methane and the maturity of kerogen as indicated by the vitrinite reflectance. The diagram shows a positive gas-source rock correlation from the Canadian Arctic (Stahl 1979)

3.11 Sedimentary Rocks

Sediments are the weathering products and residues of magmatic, metamorphic, and sedimentary rocks after transport and accumulation in
water and air. Classification of sedimentary rocks is based on easily recognizable characteristics that reflect something of the mode of transport and the environment of deposition. It is customary to consider
sedimentary rocks in two categories: clastic and chemical. Transported
fragmental debris of all kinds — sands, gravel, shell fragments — make
up the clastic component of the rock. Inorganic precipitates from water
obviously belong to the chemical category. But the chemical components also include biogenic material extracted from waters and secret-
ed as skeletons of living organisms. According to their very different
mode of formation, sedimentary rocks may be quite variable in isotopic
composition. Thus, the $\delta^{18}O$-values of sedimentary rocks span a large
range from about +10 (certain sandstones) to about +44‰ (cherts).

3.11.1 Clay Minerals

The major processes that produce clays are the weathering of rocks in
contact with ocean and fresh water, the diagenesis of sediments at low
temperatures, and the alteration of country rocks by hydrothermal
fluids at elevated temperatures. Clays from these sources are usually
distinguishable isotopically (Savin and Epstein 1970a; Lawrence and
Taylor 1971; Sheppard et al. 1971).

Because weathering normally involves large amounts of water relative to the amount of parent rock, the isotopic composition of the
parent rock should have little influence on the isotopic composition
of the weathering products. Lawrence and Taylor (1971) confirmed
that the isotopic composition of weathered rocks mainly reflects the
isotopic variations of meteoric waters. It follows that clay minerals
which originated from fresh waters or have undergone diagenetic, isotopic exchange have more negative δ-values than those for minerals
whose isotopic compositions have been established in a marine environment. Thus, clay minerals formed in contact with meteoric waters
should have δD- and $\delta^{18}O$-values which depend on the meteoric water
relationship $\delta D = 8\ \delta^{18}O + 10$. Therefore, on a δD- versus $\delta^{18}O$-dia-
gram (shown in Fig. 56) sedimentary clay minerals plot on lines which
are parallel to the meteoric water line. Many clay-rich soils analyzed

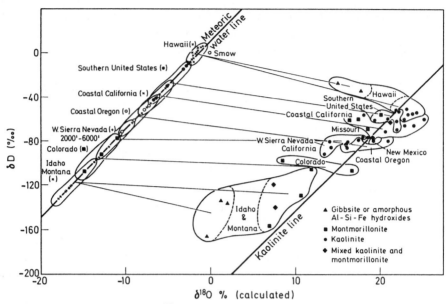

Fig. 56. Plot of δD versus $\delta^{18}O$ of clay minerals and hydroxides from Quaternary soil zones (Lawrence and Taylor 1971)

by Lawrence and Taylor (1971, 1972) lie close to the kaolinite line and obviously reflect the climatic conditions at the time of formation.

Generally, authigenic components may be distinguished from detrital components. $^{18}O/^{16}O$ ratios of detrital minerals appear to reflect the provenance and mode of origin. Detrital quartz, for instance, seems to be resistant to weathering, and it will retain its original ^{18}O-content as established in the parent rock. Rex et al. (1969) found that the $^{18}O/^{16}O$ ratio of quartz isolated from Hawaiian soils, Pacific sediments, and tropospheric dusts are remarkably uniform. The authors suggested a common eolian origin of this quartz from continental land masses. The question remains: If authigenic minerals form under equilibrium conditions, do they retain their original composition or is there a subsequent isotope exchange with pore fluids? Clay minerals can exchange isotopes with water depending upon the temperature, and the chemistry and grain size of the mineral. While interlayer water isotopically equilibrates with water vapor at room temperatures within a few days, structural water normally does not at temperatures typical of sedimentary environments (James and Baker 1976; O'Neil and Kharaka 1976; Yeh and Savin 1976; Eslinger and Yeh 1981). As temperatures rise, the rate of isotope exchange increases with notable exchange of hydrogen at temperatures near 100 °C and of oxygen near 300 °C (O'Neil and Kharaka 1976). These experimental findings have gained support from natural systems, where Yeh and Savin (1976) and Eslinger and Yeh (1981) were able to show that oxygen isotope exchange is significant for only the very finest size fractions ($< 0.2 \mu m$).

Whole-Rock Composition of Sediments. The isotopic composition of whole-rock ocean sediments cannot be interpreted without knowledge of their chemical and mineralogical composition. Ocean sediments are complex mixtures of many minerals and generally consist of illite, smectite, mixed-layer clays, chlorite, kaolinite, quartz, and feldspar. Commonly, the relative proportions of these minerals change as a function of particle size.

Knowing the isotopic composition of the pure phases, the average composition of marine shales relatively similar in mineralogy and chemistry to recent ocean sediments can be calculated. Savin and Epstein (1970b) found a variation range of +14 to 19‰ for the oxygen isotope composition of shales. Yeh and Savin (1977) determined $\delta^{18}O$-values for shales from wells drilled through argillaceous sediments in the Gulf of Mexico. The $\delta^{18}O$-variations observed indicate that the rocks are not isotopically equilibrated systems. In comparison with the coarser fractions, the finer fractions of clay minerals are always richer in ^{18}O. The disequilibrium among clay fractions becomes less pronounced as the temperatures of diagenesis increase.

δD-values of shales range from -73 to -33‰ (Yeh 1980). The δD-variation among different size fractions of a shale is about 15 to 20‰ in the upper parts of the sedimentary column and decreases with depth of burial. The large range is mostly due to the variations in proportion of clays of different origins. The differences in δD-values among different size fractions indicate isotopic disequilibrium between clays of different sizes. Yeh (1980) concluded that significant fractionations occurred between residual and expelled pore water and that the conversion of montmorillonite to illite during burial diagenesis of shales is the most important mechanism of late-stage dehydration.

3.11.2 Cherts

Cherts have the highest $^{18}O/^{16}O$ ratios found in rocks. This is due to the large oxygen isotope fractionation factor between quartz and water at low temperatures. Cherts are very similar in chemical and mineralogical composition, however, their oxygen isotope composition may vary by as much as 25‰.

Cherts, like carbonates, show temporal isotopic variations, the older cherts having lower $\delta^{18}O$-values (Degens and Epstein 1962). Cherts of different geologic ages may contain a record of temperature, isotopic composition of ocean water, and the diagenetic history. There has been some discussion as to which factor is more important, the temperature (Knauth and Epstein 1976), the isotopic composition of ocean water (Perry 1967; Perry and Tan 1972), or the diagenetic history (Kolodny and Epstein 1976).

Knauth and Epstein (1975) and Murata et al. (1977) analyzed the various forms of silica with increasing sedimentary burial. Murata et al. (1977) found that the oxygen isotope ratios in different silica phases from three different diagenetic zones decrease abruptly at the transition from biogenic opal into disordered cristobalite and again at the transition from ordered cristobalite into microquartz (quartzose chert). These stepwise changes indicate that each phase retains its original composition during progressive burial until some limiting depth where it is transformed into another phase. This decrease in $\delta^{18}O$ with increasing burial seems to reflect a rise in temperature or an isotope exchange with some kind of isotopically light water.

There is strong evidence that most cherts of Paleozoic and younger age originate from biogenic amorphous silica. However, a purely inorganic origin has been found for cherts in sodium carbonate lakes of East Africa. O'Neil and Hay (1973) concluded from oxygen isotope analyses that such East-African cherts formed from their precursors in lake waters of widely varying salinity.

3.11.3 Carbonates

3.11.3.1 Marine Organisms and "Paleotemperatures"

In 1946 Urey presented a paper concerning the thermodynamics of isotopic systems and suggested that variations in the temperature of precipitation of calcium carbonate from water should lead to measurable variations in the $^{18}O/^{16}O$ ratio of the calcium carbonate. He postulated that the determination of temperatures of the ancient oceans should be possible, in principle, by measuring the ^{18}O-content of fossil calcite shells. The first paleotemperature scale was introduced by

McCrea (1950) and refined by Epstein et al. (1953), who obtained the following empirical relationship, slightly modified by Craig (1965):

$$T\,(°C) = 16.9 - 4.2\,\Delta + 0.13\,\Delta^2 \ ,$$

where Δ is the per mil difference between CO_2 derived from carbonate by reaction with H_3PO_4 at 25 °C and CO_2 equilibrated at 25 °C with the water from which the carbonate was deposited.

Three problems make the interpretation of paleotemperature determinations rather complicated:

1. the unknown ^{18}O-content of the ancient oceans;
2. metabolic effects on carbonate precipitation;
3. the isotopic preservation of primary oxygen in the carbonates.

1. We have to assume that ancient ocean water has had a more or less constant isotopic composition, similar to that at present. However, a crucial point is the question of "paleosalinities". We must know if the organism to be analyzed has lived in ocean water of 35‰ salinity. Ocean water of higher salinities has a higher ^{18}O-content, because ^{16}O is preferentially concentrated in the vapor phase during evaporation. Ocean water of lower salinity has a lower ^{18}O-content, because it is diluted by fresh waters. Epstein and Mayeda (1953) estimated that a variation in salinity of 1‰ would be accompanied by 1 °C error in temperature determinations in a nonglacial period of the history of the Earth.

2. Some organisms, such as many foraminifera species, which apparently deposit calcite or aragonite in isotope equilibrium with ocean water and other organisms (e.g., echinoderms, asteriodea, ophiuroidea, and crinoidea) do not precipitate their carbonates in equilibrium with their environment (Weber and Raup 1966a,b; Weber 1968). These so-called vital effects are accounted for by an isotope exchange reaction between respiratory CO_2 and dissolved bicarbonate at or near the site of skeletal deposition.

Knowledge of the ecologic behavior of shell-secreting organisms is also essential. If nonextinct species are used for thermometry the assumption must be made that their depth habitats have not changed with time. Another important point is the question of whether the $CaCO_3$-secreting organisms grow shells only during a portion of the local temperature range or throughout the entire range. Epstein and Lowenstam (1953) have shown that growth of skeletons of most species does not take place during the entire year. The majority of the pelecypods, for instance, seem to grow primarily in the warm temperature range, whereas the gastropods show winter as well as summer growth. Some forms retain shell growth at low temperatures.

3. The isotopic composition of oxygen in an aragonite or calcite shell will remain unchanged until the shell material dissolves and recrystallizes during diagenesis. Some of the criteria by which unaltered samples might be recognized have been discussed by Lowenstam (1961). But the problem of how to prove the preservation is still unsolved.

Mineralogy can also play a role in the isotopic composition of carbonates (Sharma and Clayton 1965). For instance, the $\delta^{18}O$-value of aragonite at 25 °C is 0.6‰ higher than in coexisting calcite and the ^{13}C-content of aragonite is enriched by 1.8‰ relative to calcite (Rubinson and Clayton 1969).

In recent years most "paleoclimate" studies have concentrated on foraminifera. Since the first pioneering paper of Emiliani (1955) numerous cores from the Atlantic, Caribbean, and equatorial Pacific have been analyzed and, when correlated accurately, produced a well established oxygen isotopic curve for the past hundred thousands of years (Emiliani 1972; Shackleton and Opdyke 1973; Emiliani and Shackleton 1974; Emiliani 1978).

$\delta^{18}O$-values exist from both planktonic and benthic species. From these core studies it is quite obvious that similar $\delta^{18}O$-variations are observed in all areas. With independently dated time scales on hand, these $\delta^{18}O$-variations result in synchronous isotope signals in the sedimentary record because the mixing time of the oceans is relatively short ($\sim 10^3$ years). These synchronous signals provide stratigraphic markers enabling correlations between cores which may be thousands of kilometers apart.

Differences in oxygen isotope composition of foraminifera can be caused by both glacially controlled changes in the isotopic composition of ocean water and variations in the temperature of the ocean. There has been some controversy concerning the extent of the "temperature factor" as opposed to the "ice volume factor". While Emiliani (1955, 1966) originally favored the temperature factor, Shackleton and Opdyke (1973), and later many others, favored the ice volume factor. Although the resolution of the $\delta^{18}O$-values into these two effects cannot yet be adequately done for all times and all ocean areas, the problem can be partly resolved by separately analyzing planktonic and benthic foraminifera. Recall that bottom water in the oceans (see p. 125f.) is produced at high latitudes. It could be expected that the temperature of this water is more or less constant, as long as ice caps exist at the Poles. Thus, the oxygen isotope composition of benthic dwelling organisms should preferentially reflect the change in the isotopic composition of the water, while the $\delta^{18}O$-values of planktonic foraminifera should be affected by both temperature and isotopic water composition.

Most species of benthic foraminifera have been shown to precipitate carbonate slightly out of isotopic equilibrium with ambient seawater. Adjustments of these species to equilibrium have been proposed by Shackleton and Opdyke (1973), Shackleton (1977a), Belanger et al. (1981), and Graham et al. (1980). However, for several species, there is considerable range in the estimates of the adjustments. As suggested by Vincent et al. (1981) the degree of disequilibrium might be even variable in space and time.

Oxygen isotope variations in foraminifera have been very successful when applied in the Pleistocene regardless of the exact proportion of the temperature, ice-volume effect, and species-specific factors. One such example is a very detailed record of sea level variation, another the global synchroneity of a biostratigraphic marker (Thierstein et al. 1977). As shown in Fig. 57 there are several striking features of the Pleistocene record: the most obvious one is the cyclicity, secondly the fluctuations never go beyond a certain maximum value on either side of the range. This seems to imply that very effective feedback mechanisms are at work stopping the cooling and warming trends at some maximum level. Furthermore, the curve shown in Fig. 57 is characterized by a "sawtooth" slope, resulting from maximum warm periods followed immediately by maximum cold periods. This may mean the maximum ice cover is abruptly melted by rapid warming when a certain critical maximum ice cover is reached.

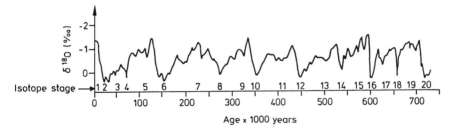

Fig. 57. Composite $\delta^{18}O$-fluctuations in the foraminifera species *G. sacculifer* from Caribbean cores showing constancy of ^{18}O-maxima and minima (Emiliani 1978)

Broecker (1982) summarized the Pleistocene $\delta^{18}O$-records of planktonic and benthic foraminifera from deep-sea cores and showed that
there is no significant difference between the average amplitudes for
both records. This similarity demands that the glacial to interglacial
change in surface ocean temperature be quite small. He further demonstrated that only a small portion of the observed ^{18}O-change from glacial
to interglacial conditions for benthic shells can be attributed to temperature change, by far the larger portion must be due to the change in
ice volume.

Savin (1977) tried to trace the oxygen isotope record back throughout the Tertiary. In addition to the problems mentioned above, other
questions such as the importance of diagenetic recrystallization complicate the record (Killingley 1983). Nevertheless, the evidence for a
global cooling throughout the Tertiary is well established. Figure 58
presents the oxygen isotope record from foraminifera at the DSDP
Site 167 in the Pacific Ocean (Savin 1977). A gradual, although several
steplike events are clearly evident, deep water temperature decrease
from nearly 12 °C during Late Cretaceous to present values of 1°–2 °C
is observed.

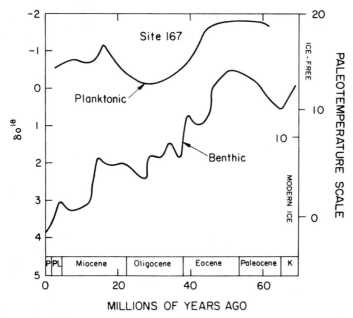

Fig. 58. Oxygen isotopic values from benthic and planktonic foraminifera from
DSDP Site 167 from the Pacific Ocean. Interpreted paleotemperatures with two
scales, with polar ice and without polar ice. (After Savin 1977)

3.11.3.2 Carbon in Foraminifera

In recent years, a large number of investigations have been undertaken to use the $\delta^{13}C$-values of foraminifera as a paleooceanographic tracer (Shackleton and Kennett 1975; Williams et al. 1977, 1981; Bender and Keigwin 1979; Broecker 1982; Shackleton et al. 1983). The carbon isotope composition in foraminifera is, however, influenced by many factors and, therefore, more difficult to interpret than the oxygen isotope record. The first good record of carbon isotope variations in Cenozoic deep-sea carbonates was given by Shackleton and Kennett (1975). They clearly demonstrated that planktonic and benthic foraminifera yield consistent differences in $\delta^{13}C$-values, the former being enriched in ^{13}C by about 1‰ relative to the latter.

This ^{13}C-enrichment in planktonic foraminifera is due to photosynthesis which removes ^{12}C preferentially from the surficial layer. A portion of this precipitated organic matter settles into deeper water where it is reoxidized, which causes a slight ^{12}C-enrichment in the deeper water masses. Besides these internal oceanographic processes, external factors may also influence the carbon isotope composition, which act on a much slower time scale than the internal processes. One very important external factor is obviously the sea level change: periods of high sea level coincide with times of ^{13}C-enrichment. During times of transgression, higher rates of organic carbon are buried into marginal sediments. For instance, the excess removal of 1% of the ocean's carbon into marine sediments results in an ^{13}C-increase of 0.2‰ (assuming a $\delta^{13}C$-value of – 20‰ for the organic carbon).

A further complication results from the fact that, in general, the $\delta^{13}C$-value of foraminifera is not equal to dissolved bicarbonate, which is interpreted as indicating disequilibrium due to vital effects. Grossman (1984b), however, by analyzing live benthic foraminifera, demonstrated that when mineralogy, temperature, and dissolved inorganic carbon are considered, foraminiferal carbonate-HCO_3^- fractionation may not be very different from inorganic precipitated carbonate. Assuming that vital effects are either nonexistent or on the average invariant with time, then systematic variations in C-isotope composition may reflect variations in bottom water $\delta^{13}C$. With these prerequisites Bender and Keigwin (1979) attempted to trace the age and movement of deep water masses. Shackleton et al. (1983) used $\delta^{13}C$-data to confirm reduced CO_2 concentrations in the ice age atmosphere, which has also been verified by directly measuring the CO_2 content of air bubbles in ice cores.

3.11.3.3 Fresh Water Carbonates

Since fresh water is, in general, depleted in ^{18}O relative to ocean water and more variable in the $^{13}C/^{12}C$ ratio due to a relatively high contribution of soil-derived organic CO_2, fresh water carbonates are generally lighter and show a much broader variation range in carbon and oxygen isotope composition than marine carbonates (Clayton and Degens 1959; Keith and Weber 1964; Keith et al. 1964). This very general distinction has been used to determine paleoenvironmental conditions. However, decisive in this connection is the degree of evaporation of the fresh water body.

Carbonates deposited from highly evaporated fresh water lakes may not only have higher $\delta^{18}O$-values than marine carbonates, but also have even more pronounced higher $\delta^{13}C$-values (Rothe and Hoefs 1977). The ^{13}C-enrichment may be interpreted as reflecting an increased CO_2 exchange between the atmosphere and the shallowing water body.

A further complication may arise from stratification of fresh water bodies either intermittently (i.e., seasonally) or permanently (thermal or salinity). During stratification, ^{12}C is transferred from the surface to deeper waters by sinking of dead organic matter. Degradation in deeper water masses leads to relatively ^{12}C-enriched dissolved carbon dioxide, while the surface waters exhibit a ^{13}C-enrichment. McKenzie (1982) observed a maximum gradient of 5 to 6‰ during summer thermal stratification, while during winter-spring mixing there is no $\delta^{13}C$-gradient.

There have been several attempts to use the isotopic composition of fresh water shells as a paleoclimatic indicator. Objects of studies have been mollusks (Fritz and Poplawski 1974), gastropods (Abell 1986), and land snails (Yapp 1979; Magaritz and Heller 1980). However, since fresh waters are highly variable in isotopic composition and can be easily altered by evaporative processes, quantitative interpretations of climatic changes are nearly impossible. Nevertheless, qualitative changes of environmental conditions are clearly indicated. Thus, Magaritz and Heller (1980) found that snails from an arid zone are enriched by 2‰ in ^{18}O compared to the same species from a moderate climate zone.

3.11.3.4 Dolomites

The "dolomite problem", i.e., the origin of dolomite and the conditions promoting the dolomitization of limestones, is still being debated.

Although many claims of primary dolomite have been made, the preferred view now is that most dolomites are of replacement origin. Land (1980) concluded that there is no unique environment of dolomitization. Aside from the basic chemical constraint that a solution must be oversaturated with dolomite in order to crystallize it, dolomite may form in a variety of chemical environments. Within the last few thousand years dolomite has formed from hypersaline, subtidal waters of marine derivation. McKenzie (1984), studying dolomitization in coastal sabkhas from the Persian Gulf, postulated that aragonite and perhaps high Mg-calcite serve as intermediates in the formation of dolomite via a dissolution-precipitation process. At the other end of the salinity spectrum, dolomite forms in the mixing zone between meteoric and ocean waters (Land 1980). Such a model has also been favored for very old Precambrian deposits (Tucker 1983).

Two problems complicate the interpretation of isotope data to delineate the origin and diagenesis of dolomites. First, it has not been possible to determine directly the equilibrium oxygen isotope fractionations between dolomite and water at sedimentary temperatures, because the laboratory synthesis of dolomite at these low temperatures is still problematic. Furthermore, the fractionation may depend partly on the crystal structure, more specifically on the composition and the degree of crystalline order and, in this respect, it is well known that dolomite is a very complex mineral. Secondly, dolomitization on a massive scale appears to occur under open-system conditions, simply because large quantities of magnesium have to be supplied. Extrapolations of high-temperature, experimental, dolomite-water fractionations to low temperatures suggest that at 25 $^{\circ}$C dolomite should be enriched in ^{18}O relative to calcite by 4 to 7‰. In contrast, the oxygen isotope fractionation observed between Holocene calcite and dolomite is somewhat lower, namely in the range between 2 and 4‰ (Land 1980; McKenzie 1981).

A very important site of dolomite formation is the deep-sea environment (Pisciotti and Mahoney 1981; Kelts and McKenzie 1982). Along continental margins and in small ocean basins dolomite forms as cement, layers, and concretionary zones in associations with rapidly deposited, fine-grained, organic-rich sediments. As shown by Deuser (1970), among others, these dolomites can be extremely variable in C-isotope composition with $\delta^{13}C$-values ranging from -60 to $+20$‰. In contrast to this very large variation for deep-sea dolomites, most platform dolomites fall in the relatively small $\delta^{13}C$-range between -2 to $+4$‰ (Land

1980). The low δ^{13}C-values of deep-sea dolomites are characteristic of formation at shallow depths from organic matter via microbial reduction of sulfate, whereas the high δ^{13}C-values are typical of dolomites formed below this zone where dissolved $H^{12}CO_3^-$ is preferentially removed by reduction of CO_2 to methane during methanogenesis. Thus, in the sequence of diagenetic alteration of organic matter, carbonate is continuously precipitated and each alteration reaction carries a distinct C-isotope signature which is preserved in the diagenetic carbonate. Differences in the sedimentation rate, the amount of organic matter available, and the geothermal gradient will affect these reactions and the extent of δ^{13}C-variation (Pisciotti and Mahoney 1981; Kelts and McKenzie 1982).

3.11.4 Diagenesis of Limestones

Isotope data on several thousand limestone samples have been reported in the literature to date. The tendency toward lower $^{18}O/^{16}O$ ratios with their increasing age is a well-documented fact (Keith and Weber 1964; Veizer and Hoefs 1976), although the reasons for this isotope shift are still under debate (see discussion on the isotopic evolution of ocean water). The majority of isotope analyses of limestones have involved whole-rock samples, but in recent years individual components such as different generations of cements have been analyzed (Hudson 1977; Dickson and Coleman 1980; Moldovanyi and Lohmann 1984; Given and Lohmann 1985).

The original marine carbonate assemblage of aragonite, Mg-calcite, and low-Mg-calcite is converted into stable diagenetic low-Mg-calcite through a process of dissolution and reprecipitation during which isotopes from the dissolving phases mix with the intervening water. This diagenetic, mineralogical stabilization usually proceeds in discrete microenvironments which often preserve original textures. Once stabilized, carbonates are normally not subject to perpetual dissolution-precipitation and isotopic reequilibrium with younger diagenetic fluids. This is supported by results of Given and Lohmann (1985), who found that two distinct secondary calcite phases, representing a fine-scale intermixture, have maintained isotopic integrity despite intimate association.

The diagenetic process can proceed in waters of meteoric or marine derivation, the former are typical for shallow marine sequences, while the latter are common for deep-sea carbonates. Several studies have reported a general decrease of the ^{18}O-content in pelagic carbonate sediments with increasing age and depth of burial (cf. Matter et al. 1975; McKenzie et al. 1978). The progressive decrease of $\delta^{18}O$ appears to reflect precipitation of cement at progressively higher temperatures. In contrast, the $\delta^{13}C$-values are little altered and obviously reflect the composition in the original sediment. A special case is carbonate diagenesis in carbonates closely associated with basalts. Many oceanic limestones lying above basalt or interbedded with them show alteration which is commonly attributed to contact thermal metamorphism. However, isotope studies show that the alteration frequently occurred at relatively low temperatures (Bernoulli et al. 1978; McKenzie and Kelts 1979).

Clear trends in the isotopic composition of cements can be established for near-surface diagenesis (Allan and Matthews (1982) and burial diagenesis (Milliken et al. 1981). Allan and Matthews (1973, 1982) determined the effects of subaerial diagenesis on the isotopic composition of carbonates. These studies showed that subaerial carbonate sediments can be recognized from characteristic isotope patterns preserved in vertical stratigraphic sections. In particular, they observed $\delta^{13}C$-depleted carbonates at the exposure surface which are interpreted as representing soil-derived CO_2 from the vegetation on the exposure surface. Beeunas and Knauth (1985) observed equivalent isotope trends in the 1.2 Ga Mescal Limestone of central Arizona and suggested that a vegetative land cover existed on the Precambrian exposure surface.

Studies of sequential cement generations by Dickson and Coleman (1980) suggest that early cements exhibit higher $\delta^{18}O$- and $\delta^{13}C$-values with successive cements becoming more depleted in both ^{13}C and ^{18}O. This ^{18}O-trend is attributed to increasing temperatures and to isotopic evolution of pore waters during burial. The $\delta^{13}C$-trend is interpreted as an increase of organic-derived CO_2 during burial. A more unusual effect of diagenesis is the formation of carbonate concretions in essentially uniform argillaceous sediments. Isotope studies of Hoefs (1970), Sass and Kolodny (1972), Irwin et al. (1977), Hudson (1977), and Gautier (1982) suggest that microbiological activity created localized supersaturation of calcite in which dissolved carbonate species were produced more rapidly than they could be dispersed by diffusion. Extremely variable $\delta^{13}C$-values in these concretions indicate that different microbiological processes participated in concretionary growth. Irwin et al. (1977) presented a model in which organic matter is diagenetically modified by (1) sulfate reduction, (2) fermentation, and (3) thermally-induced, abiotic CO_2 formation which can be distinguished by their $\delta^{13}C$-values; (1) $\sim -25\%$; (2) $\sim +15\%$; and (3) $\sim -20\%$.

3.11.5 Diagenesis of Clastic Rocks

A very detailed study of Milliken et al. (1981) constrained the diagenetic history for the Frio sandstone, Gulf Coast area. Quartz is most commonly the first cement and constitutes around 2.5% of the average sandstone volume. The average $\delta^{18}O$ of quartz cement is 31‰ ± 1.5, indicating precipitation at considerably cooler temperatures than that at which most clay mineral transformations take place. Calcite is the dominant cement in Frio sandstone, constituting about 5% of the total sandstone volume and generally postdates quartz precipitation. Its isotopic composition is relatively constant around 23 ± 2‰. The volume of water required to precipitate quartz and calcite cements far exceeds the volume of pore water deposited with, near, or beneath the sands. Thus, an external water source is required.

Longstaffe (1983) summarized case studies of the application of stable isotope research in clastic diagenesis. One set of case studies estimates the crystallization temperature of authigenic minerals from experimentally determined oxygen isotope mineral-water fractionations (such as kaolinite-water, Land and Dutton 1978; and smectite-water, Yeh and Savin 1977). The calculated temperatures can be used to determine the sequence of authigenic mineral formation and to estimate geothermal gradients and maximum depths of burial. Another group of examples concerns studies of shale diagenesis. Oxygen isotope geothermometry using quartz-illite/smectite can provide estimates of the maximum temperature to which the shale has been heated. Examples have been reported from the Precambrian Belt Supergroup, the Texas Gulf Coast, and the Great Valley Sequence (Eslinger and Savin 1973a,b; Yeh and Savin 1977; Suchecki and Land 1983). The latter authors demonstrated that illite/smectite reactions during burial can control the geochemical evolution of formation fluids. A final example presented by Longstaffe (1983) discusses the control that meteoric water can exert upon diagenesis. He concluded that isotope signatures of authigenic minerals precipitated from an isotopically characteristic fluid may provide a method by which the paleohydrology of a sandstone can be inferred.

3.11.6 Phosphates

As was pointed out by Urey et al. (1951), the development of a temperature scale using calcium carbonate and another oxygen compound precipitated by marine organisms (e.g., phosphates) would permit a temperature scale independent of the oxygen isotope composition of ocean waters. This was originally presented by Longinelli (1965, 1966). However, Longinelli and Nuti (1973) gave a revised phosphate-water isotope temperature scale and showed that the slope of the calculated equation is practically identical to that of the carbonate equation. This means that the difference in the δ-values of carbonate and phosphate in equilibrium with ocean water is constant and independent of the temperature and of the $\delta^{18}O$-value of the water.

The major advantage of the phosphate thermometer is that it is a system which is insensitive towards diagenetic reactions (Shemesh et al. 1983). As shown by Kolodny et al. (1983) only in enzyme-catalyzed reactions is phosphate oxygen readily exchangeable with environmental water. This inertness towards postdepositional recrystallization is illustrated in Fig. 59, where the isotopic composition of cherts, limestones, and phosphorites from a single area in the Negev, Israel is compared. The $\delta^{18}O$-values of carbonates and cherts vary widely, with probably only the highest values indicative of marine conditions. In contrast, the phosphorites are very uniform in isotopic composition. The small spread in phosphorite $\delta^{18}O$-values is strong evidence for the resistance of PO_4 to dissolution-reprecipitation processes.

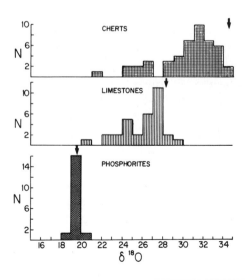

Fig. 59. Histogram of $\delta^{18}O$-values for cherts, limestones, and phosphorites of the Mishash Formation of Campanian age in Israel. *Arrows* mark expected composition of cherts and carbonates in equilibrium with phosphate of $\delta^{18}O$ = 19.5‰ (Shemesh et al. 1983)

Another interesting observation made by Shemesh et al. (1983) is that the $\delta^{18}O$-values of phosphorites decrease with increasing geologic age, a similar trend is also observed for limestones and cherts. Because this trend cannot be explained by postdepositional exchange with ^{18}O-depleted meteoric water, the $\delta^{18}O$-change implies either higher temperatures of phosphorite formation in the past or the progressive depletion of the ocean in ^{18}O through time.

3.11.7 Sedimentary Sulfides

Due to the activity of sulfate-reducing bacteria (producing isotopically light sulfide and isotopically heavy residual sulfate), most sulfur isotope fractionation takes place in the uppermost layers of the muds of shallow sea basins and tidal flats. Thode et al. (1960), Vinogradov et al. (1962), Kaplan et al. (1963), Nakai and Jensen (1964), Hartmann and Nielsen (1969), and others have observed that depletions occurred in ^{34}S from 15 to 62‰ for sulfides relative to associated sulfates in various natural environments. Normally, sedimentary sulfides should have a $\delta^{34}S$-value between -30 and $-10‰$, although numerous examples exist where the sedimentary sulfides show a ^{34}S-enrichment.

Unconsolidated mud remains slightly permeable to dissolved marine sulfate and, therefore, the availability of sulfate within the uppermost centimeters of the sediment is unlimited. With increasing sediment depths, a gradual transition from the open system of the free water column toward an almost completely closed system at depth takes place. If the bottom water above a sediment is poorly aerated, then the sediment may be reduced almost to the sediment-water interface and very large amounts of isotopically light sulfide can accumulate within this transition zone. Such an example, shown in Fig. 60, has been described by Hartmann and Nielsen (1969) in the Baltic Sea. Here, the sulfide concentration increases progressively from zero at the sediment surface to about ten times the original sulfate concentration in the interstitial water at a depth of 5 cm. Because the open-system condition is restricted to the uppermost few centimeters of the sediment, the production rate of isotopically light biogenic sulfide is inversely correlated with the rate of sediment accumulation.

$\delta^{34}S$ of pyrite may depend on its texture (Raiswell 1982). Predominantly framboidal pyrites have lighter $\delta^{34}S$-values than euhedral pyrites. The former have evolved in an open system relative to seawater sulfate, the latter at greater depths under closed-system conditions.

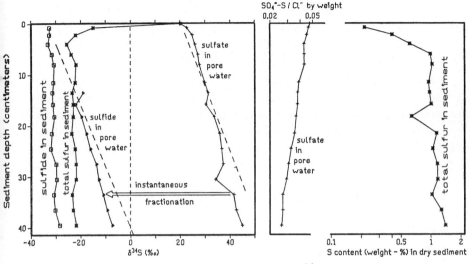

Fig. 60. Variation in sulfur content (*right*) and in $\delta^{34}S$ (*left*) from Kiel Bay, Baltic Sea. *Dotted lines* indicate theoretical Rayleigh fractionation curves (Nielsen 1978)

Berner (1972, 1984) has discussed the factors governing the formation of sedimentary pyrite in detail. Pyrite is formed in those marine sediments where organic matter accumulates faster than it can be destroyed, pore waters become anaerobic, and the process of bacterial sulfate reduction begins. As H_2S is formed, some of it reacts with detrital iron minerals to form black iron sulfide. The primary limitations upon how much sulfate can be transformed and fixed as pyrite are:

1. the availability of bacterially metabolizable organic matter;
2. the concentration and rate of deposition of detrital iron compounds which can react with H_2S;
3. the rate of replenishment of sulfate in the sediment via diffusion from the overlying water.

Laboratory studies (Harrison and Thode 1958; Kemp and Thode 1968) have demonstrated that the rate of sulfate reduction is much more strongly dependent upon the concentration of bacterially metabolizable organic compounds than on the concentration of sulfate. The population of sulfate reducers in marine sediments decreases rapidly with depth, most likely as a result of the loss of organic matter. The readily decomposable material rapidly disappears so that the sulfate reducers become dependent upon fermentative microorganisms to break down long-chain polymers and other macromolecules which are otherwise not available to the sulfate reducers. No matter how much H_2S is produced in a sediment, no more pyrite can form than the amount of iron available for reaction with H_2S.

Accepting that a difference in $\delta^{34}S$-values of -30 to $-50\permil$ between bacteriogenic sulfide and marine sulfate exists in present-day sedimentary environments, similar fractionations in ancient sedimentary rocks may be interpreted as evidence for the activity of sulfate-reducing bacteria. The presence or absence of such fractionations in sedimentary sulfur may thus constrain the time of emergence of sulfate-reducing microorganisms.

In early Archaean sedimentary rocks (> 3.0 Ga) most sulfides and the rare sulfates have isotopic compositions near $0\permil$ (Monster et al. 1979; Cameron 1982), indicating an absence of bacterial reduction in oceanic sulfate. According to Cameron (1982) and Hattori et al. (1983), the onset of bacterial reduction was at around 2.3 Ga. Pyrites in strata older than 2.3 Ga have $\delta^{34}S$-values of $\sim 0\permil$, while younger pyrites are depleted in ^{34}S and have a wide range in sulfur isotope composition. Thode and Goodwin (1983) and Goodwin et al. (1985) presented evidence for an even earlier development of bacterial sulfate reduction. However, the criteria which are used to distinguish sulfide of biogenic and of hydrothermal origin are not unequivocal, therefore, the exact beginning of bacterial sulfate reduction is still debatable.

3.12 Metamorphic Rocks

The application of stable isotopes to metamorphic rocks has concentrated primarily (1) on the nature of fluid-rock interactions and (2) on the determination of metamorphic or equilibration temperatures.

3.12.1 Metamorphic Fluids: Their Flow, Sources, and Water/Rock Ratios

The problem addressed here is the extent to which the isotopic characteristic of the metamorphic system was modified by a fluid phase (see also the recent review by Valley 1986). Two end-member situations can be postulated in which coexisting minerals would reequilibrate during metamorphism.

1. In the case of equilibration of minerals with a *pervasive* fluid of uniform composition a pervasive fluid moves independently of structural and lithologic control and each mineral becomes isotopically homogeneous despite whatever differences in isotopic composition may have existed prior to metamorphism.

2. Local equilibration between adjacent mineral grains by *channelized* fluids, which move along vein systems, shear zones, or other channel-ways, such as rock contacts or more permeable lithologic units. This fluid flow leads to equilibration on the scale of individual beds or units, but will not result in isotopic homogenization of different rock types. Channelized flow favors chemical heterogeneity, allowing some rocks to remain unaffected. Most C–O–H studies have demonstrated that fluid flow is channelized along structural weaknesses or more permeable lithologies (Rumble et al. 1982; Graham et al. 1983; Rumble and Spear 1983; Tracy et al. 1983; Valley and O'Neil 1984; Nabelek et al. 1984; Bebout and Carlson 1986).

Fluid involvement in fault and shear zones is an established phenomenon. Kerrick et al. (1984) demonstrated that, in general, flow regimes follow a sequence of events: during initiation of the structures, locally-derived fluids at low water/rock ratios predominate which, as the structures propagate, change to metamorphic fluids with high water/rock ratios along conduits. Later in the tectonic evolution and at shallower crustal levels there is often invasion of surface waters into the faults.

Prograde metamorphism of sediments causes the liberation of volatiles. Dehydration is most common, decarbonation occurs in carbonate-bearing rocks, desulfidation can be locally important. The liberation of volatiles can be described by two end-member processes (Rumble 1982; Valley 1986):

1. Rayleigh volatilization during which rocks interact only with fluids generated internally by devolatilization reactions between the rocks' minerals. The conditions of Rayleigh distillation require that once fluid is generated it is isolated immediately from the rock.
2. Batch volatilization, where all fluid is evolved before any is permitted to escape. Most natural processes actually fall between these extremes, but these two end-member situations provide useful limits.

Dehydration is the best known and most common example of metamorphic volatilization. The effect of dehydration reactions on the $\delta^{18}O$ value of a rock will always be small, less than a 1‰. The temperature effect is the most important due to the crossover in the sign of fractionations. At temperatures below 400°– 500 °C, H_2O is isotopically lighter than an average rock and dehydration will cause ^{18}O-enrichment. At temperatures above 500 °C oxygen in H_2O is heavier and dehydration causes ^{18}O-depletion. Thus, reactions tend to cancel each other, depending on the details of the reaction path. Even with conservatively made assumptions to maximize the isotope effect, it will be under all circumstances less than 1‰ (Valley 1986).

In contrast to $\delta^{18}O$, the effect of dehydration on δD may be much larger because the amount of hydrogen in the rock is much smaller. Because the isotopic composition of the fluids will be buffered at high temperatures by the isotopic composition of the surrounding rocks from which they were derived or passed through, oxygen isotope ratios often do not provide clear constraints on the origin of the fluids, whereas H-isotope ratios in many cases may do so. Possible sources for metamorphic fluids are:

1. magmatic water derived from deep levels in the crust or even in the upper mantle;
2. water liberated during metamorphic dehydration reactions;
3. meteoric water derived from the earth's surface;
4. connate formation waters (brines) trapped at deep levels in the sedimentary pile;
5. seawater derived from the ocean, possibly in a rift zone (Wickham and Taylor 1985).

On the basis of their δD-value, fluids (1) and (2) are not necessarily distinguishable from each other, but fluids (3) and (5) are. In order to calculate the isotopic composition of the fluid it is necessary to know the fractionation factors between the minerals and water as a function of temperature and composition.

Studies of prograde regional metamorphic mineral assemblages have suggested the preservation of hydrogen isotope equilibrium amongst coexisting hydrous minerals (Rye et al. 1976; Hoernes and Friedrichsen 1978, 1980). Much of the available literature pertains to muscovite and biotite δD-values, the latter being always more negative than the former. However, Graham (1981) has argued that this apparent preservation of hydrogen isotope equilibrium may be a consequence of the similarity of grain size and diffusion parameters. Describing methods of calculating diffusion coefficients Graham (1981) presented experimental evidence that in the absence of a fluid phase, H-isotope exchange is slower by at least two orders of magnitude than in the presence of a fluid phase. Rapid diffusion of hydrogen implies that in slowly cooling regional metamorphic terranes closure temperatures for cessation of H-isotope exchange may be far below the temperature of formation. Thus, it is often difficult to establish whether hydrogen isotope equilibrium is commonly preserved between hydrous minerals in metamorphic rocks.

Decarbonation is a very important process that effects the isotopic composition of carbonate rocks during metamorphism (e.g., Shieh and Taylor 1969b; Taylor and O'Neil 1977; Matthews and Kolodny 1978; Rumble 1982; Nabelek et al. 1984). Both equilibrium and kinetic fractionations may be operative during the process, but in any event, the CO_2 leaving the system is relatively enriched in both ^{18}O and ^{13}C. Consequently, decarbonation results in a lowering of both $\delta^{13}C$- and $\delta^{18}O$-values of the metamorphosed carbonates. Significant shifts in both ^{13}C and ^{18}O may occur at isograd boundaries, where large amounts of volatile-producing reactions take place (Lattanzi et al. 1980). The magnitude of ^{18}O- and ^{13}C-depletions are directly linked and can be calculated if the reaction stoichiometry is known.

Since prograde reactions in carbonate rocks are dependent on H_2O/CO_2 ratios in the fluid phase and the carbonates themselves generate only CO_2, it is very likely that the observed reactions occur because of H_2O infiltration. Rumble et al. (1982), Graham et al. (1983) and others have demonstrated from both phase equilibrium and isotopic data that greenschist and amphibolite facies metacarbonates were infiltrated by between 1 and 5 rock volumes of H_2O. These are the volumes of H_2O that have reacted with the rock and therefore represent lower limits of the water that actually passed through the rocks. Such high fluid-rock ratios have only been well documented in metacarbonate units. Their implication for the more common pelitic and psammitic units is difficult to establish, because these rocks produce a water-rich fluid, and the effects of externally derived waters are thus difficult to document.

Nitrogen in metamorphic rocks is mainly fixed as ammonium within the crystal lattice of micas and other silicate minerals and, to a lesser amount, in fluid inclusions as molecular nitrogen. Haendel et al. (1986) determined the content and isotopic composition of ammonium-nitrogen in metasedimentary rocks with increasing metamorphic grade and found a decrease in nitrogen content and an increase in $\delta^{15}N$-values (see Fig. 61).

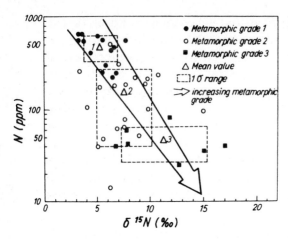

Fig. 61. Correlation between δ^{15}N-values and nitrogen concentrations in regional metamorphic rocks (Haendel et al. 1986)

From model experiments they concluded that isotope exchange mainly takes place between ammonium and molecular nitrogen, which is fast enough to reach isotopic equilibrium even at temperatures around 400 °C.

3.12.2 Temperature Determination in Metamorphic Rocks

Although most attempts to apply stable isotope geothermometers to metamorphic rocks have utilized oxygen isotope fractionations, in recent years carbon isotope fractionations between carbonates and graphite have also been successfully applied (see below). In this context the problems of choosing suitable temperature calibrations have been already discussed on p. 16f. and will not be elucidated here.

It is common belief that the metamorphic assemblage observed at the earth's surface represents the peak grade of metamorphism. However, with respect to oxygen isotope temperatures this is not necessarily the case. The detailed summary of O-isotope relationships in mineral triplets by Deines (1977) demonstrated clearly that concordant temperatures indicative of maximum metamorphic temperatures do occur, but not as a rule. Oxygen isotope temperatures comparable to maximum temperatures of metamorphism will only be preserved if water is lost from the assemblage at this temperature or if oxygen diffusion in a mineral pair in the presence of water is sufficiently slow. In addition to diffusion coefficients, the degree of isotope discordancy also depends upon grain size, role and nature of the fluid phase, rock and grain permeabilities, rate of cooling, and other factors. For example, Hoefs et al. (1982) demonstrated on iron ores from the Iron Quadrangle, Brazil, that temperatures obtained from quartz-iron oxide fractionations depend upon the deformation history. Iron ores which have been overprinted by later deformation events are selectively reset to lower "isotopic" temperatures. In these rocks the qualitative relationship exists such that the more closely spaced the schistosity planes, the larger the extent of temperature lowering.

Inspection of the literature data (Deines 1977; Javoy 1977; Hoernes and Friedrichsen 1978, 1980; Hoernes and Hoffer 1979; Matthews and Schliestedt 1984) indicates the following general pattern of apparent temperatures:

1. "high" temperatures: quartz-rutile, quartz-garnet, quartz-iron oxides, quartz-pyroxene, and
2. "low" temperatures: quartz-muscovite, quartz-biotite, quartz-feldspar, quartz-calcite.

The preferred interpretation of this subdivision is that, due to different diffusion rates, some minerals exchange their oxygen isotopes with a fluid phase down to relatively low temperatures during retrograde cooling of a metamorphic event. Some retrograde temperatures may even represent a distinctly younger metamorphic event in a polyphase metamorphic rock. However, this cannot be differentiated by oxygen isotope determinations alone. A different interpretation has been proposed by Hoernes and Hoffer (1978), who argued that the $^{18}O/^{16}O$ ratio of biotite, for instance, is fixed at the time of crystallization and its initial composition is preserved when the temperature increases. In a later paper Hoernes and Hoffer (1985) argued that at relatively high temperatures the initiation of specific dehydration reactions, their production, and their escape rates from the metamorphic system are so fast as to inhibit a complete reequilibration of the solids with the fluid. The oxygen isotope disequilibria frequently discussed in the literature may result from this kinetic effect rather than from retrograde exchange.

Very informative are temperature studies of low-grade metamorphic rocks, because isotope fractionations are large at these low temperatures and temperature information hard to receive. However, because of the fine-grained nature of such rocks, mineral separations are much more difficult. Furthermore, oxygen isotope geothermometry relies on the assumption that fine-grained diagenetic quartz and mixed-layer illite/smectite continuously equilibrate with one another during burial. Eslinger et al. (1979) demonstrated that there is no well-established example of mineral pair thermometry at temperatures below 100 °C and this may well be true up to 200 °C. However, studies on such very low-grade metamorphic rocks can be used to provide a measure of the extent of isotope exchange toward the equilibrium value. Analyzing the Precambrian Belt Series Eslinger and Savin (1973b) showed that extensive oxygen isotope exchange can occur at relatively low temperatures (225° to 310 °C) in rocks that appear lithologically to be ordinary shale and carbonate. Their data indicate some degree of disequilibrium between carbonate and quartz, because carbonate may have been more affected by retrograde effects than silicate. Eslinger and Yeh (1986) estimated maximum burial temperatures for carbonates and shales to be 160° to 250 °C, which agreed with other petrologic estimates.

Becker and Clayton (1976) studied the banded iron formation from the Hamersley Range, Western Australia and concluded that chert and iron oxide had undergone burial metamorphism at a temperature of 270° to 310 °C. Similar temperatures have been obtained by Hoefs et al. (in press) on the much younger Urucum deposits in Brazil.

Other low-temperature metamorphic reactions are serpentinization processes which have been investigated by Barnes and O'Neil (1969), Wenner and Taylor (1974), Magaritz and Taylor (1974), and Ikin and Harmon (1983). Magaritz and Taylor (1974) demonstrated that antigorites typically have very uniform δD- and $\delta^{18}O$-values identical to those of most metamorphic chlorites. Thus, the antigorites seem to form by reaction with metamorphic fluids. The δD-values of lizardite-chrysotile serpentines, on the other hand, show a latitudinal dependence as a result of interactions with meteoric groundwaters at relatively low temperatures (≈ 100 °C or less). Wenner and Taylor (1973) demonstrated that oceanic serpentines have higher δD- and lower $\delta^{18}O$-values than the serpentines of the continental ophiolite complexes, and concluded that heated ocean water was involved in submarine serpentinization.

Carbon Isotope Fractionation Between Calcite and Graphite. Recently, carbon isotope fractionations between calcite and graphite in marble have been applied as geothermometers by Valley and O'Neil (1981), Wada and Suzuki (1983), and Morikiyo (1984). Valley and O'Neil (1981) calibrated their temperature scale empirically against the potassium feldspar-plagioclase and magnetite-ilmenite thermometers, while Wada and Suzuki (1983) calibrated it against the calcite-dolomite solvus thermometer. Although the slope of all temperature scales is slightly different, geologically meaningful temperatures are only obtained in the high temperature range at temperatures above 500 °C. At temperatures below 500 °C kinetics of the isotope exchange reaction may become so important that a general application of the thermometer at lower temperatures is hazardous.

In terranes where organic carbonaceous matter can be traced from low to high grade, a progressive increase in $\delta^{13}C$-values is commonly observed (Hoefs and Frey 1976; Barker and Friedman 1969). This increase is partly due to progressive loss of isotopically light methane and partly due to exchange with isotopically heavy carbonates. At high metamorphic temperatures, the low $\delta^{13}C$-values that are characteristic of sedimentary organic matter should not be preserved in the presence of high $\delta^{13}C$-carbonates. Thus, high $\delta^{13}C$-values in graphite are not a sufficient criterion to infer an abiogenic origin in high-grade marbles.

3.12.3 Contact Metamorphism

Because the oxygen isotope composition of igneous rocks is quite different from that of sedimentary and low-grade metamorphic rocks, studies of the variation of oxygen isotopes in the vicinity of an intrusive contact offers the possibility of investigating the extent of isotope exchange between the intrusive and its country rock. Typically, exchange between the igneous intrusion and the adjacent pelitic country rock takes place within a short distance of the intrusive contact. The width of the exchanged zone correlates well with the size of the intrusions, the presumed intrusive temperatures, the length of heating time, and the availability of fluids. Figure 62 shows the percent oxygen isotope exchange between intrusive and country rock as a function of distance from the contact for several contact zones (Shieh and Taylor 1969a). The narrowness and steepness of an isotope gradient (see Fig. 62) in the exchanged zones suggest that such small-scale isotope exchange occurred essentially in the solid state by a diffusion-controlled recrystallization process.

In many contact metamorphic aureoles values of $\delta^{18}O$ and $\delta^{13}C$ vary systematically. Table 23 summarizes 16 studies, mostly of contact aureoles, that show coupled O–C trends. In all cases $\delta^{18}O$-

Table 23. Studies demonstrating coupled O–C depletion trends in metamorphosed carbonates (Valley 1986)

	Width of aureole or traverse	Pressure/ depth	Maximum temperature (°C)	X(CO_2)	Range in $\delta^{18}O$	Range in $\delta^{13}C$	Pluton	Rock types, comments
1. Trenton limestone Mount Royal, Quebec	100 m				14	6	Essexite, nepheline-syenite	Limestone, marble, calcite from intrusion
2. Marysville, Montana	1–3 km	1 kbar	525	0.95	18	6	Grano-diorite	Marble, hydrothermal assemblages Depletion attributed to Rayleigh volatilization Largest isotope shift at diopside isograd
3. Pine Creek, California W-skarn	Roof pendant	<2 kbar	600	<0.25	22	14	Quartz-monzonite	Marble, calc-silicate, skarn in pendant Gradients in $\delta^{18}O$ up to 10‰ per 10 cm across skarn Depletion attributed to mixing with magmatic O + C
4. Osgood Mts., Nevada	<1 km	<2 kbar	>550	<0.15	9	9	Grano-diorite	Marble, calc-silicate hornfels, skarn 3 Stages of skarn formation
5. Skye, Scotland					20	7	Granite	Marble, skarn Some depletion due to meteoric water
6. Elkhorn, Montana	Marble 240 m Skarn 21 m	1 kbar	525	<0.25	15	7	Quartz-diorite	Dolostone, marble, skarn
7. Notch Peak Stock, Utah	3 km	1.5 kbar	600	<0.2	12	12	Quartz-monzonite	Metamorphosed calcareous argillite, marble
8. Weolag W-Mo Deposit, Korea	2.5 km	1.0–2.4 km	>400		15	11	Granite	Limestone, calc-silicate, skarn

9. Tauern Area, Austria	3 m	>4 kbar	450–600		8	2	–	Traverse across a 3 m-thick marble layer
10. Birch Creek, California	750 m		540		13	4	Granite	Marble, skarn
11. McArthur R. Pb–Zn Deposits, Australia			350		7	2.5	–	Dolomite, values change in relation to distance to the Emu Fault. Range in values attributed to variable T=350°–150 °C and constant fluid composition
12. Providencia Pb–Zn Deposits, Mexico			365		10	10		Limestone, late-stage hydro-thermal calcites. Variable T=365°–200 °C
13. Gaspé Cu Deposits, Quebec			350		17	14		13a = Drill hole GMS4, limestone, marble, vein calcite, skarn. 13b = Drill hole GMS2, limestone, marble. Variable T=350°–150 °C
14. CanTung W-skarn, N.W. Territories			500		9	9	Monzo-granite	Limestone, marble, calc-silicate, skarn. Variable skarn T=400°–270 °C
15. Mottled Zone, Israel		<25 atm	1300		13	25	None	Natural combustion of bituminous marl
16. Bergell Aureole, Italy	<20 cm	2 kbar	400	0.1–0.25	15.3	8	Tonalite	Metasomatic zone around veins in marble. Sharp isotopic gradients, 5–14 per mil cm^{-1} at infiltration fronts

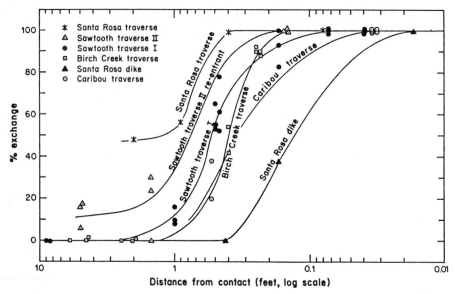

Fig. 62. Percent oxygen isotope exchange between intrusive and country rock as a function of distance from the intrusive contacts (Shieh and Taylor 1969a)

scales from 3 m to 3 km. Important questions to be answered are to what extent these depletion trends are affected by volatilization, by infiltration, and mixing or by changing p-T-X conditions. In general, the effects of volatilization, infiltration, and disequilibrium all tend towards lower δ^{13}C- and δ^{18}O-values in calc-silicates and marbles with increasing metamorphic grade. However, detailed analysis shows that volatilization, although always leading to a depletion in heavy isotopes, is not the dominant cause of large shifts. Most contact aureoles studied have been infiltrated by fluids and O−C−H isotope ratios frequently enable identification of fluid sources. For example, Taylor and O'Neil (1977) presented evidence that magmatic water moved out from the pluton during the first stage of development of a skarn, then, at a later stage, meteoric water constituted between 20 and 50‰ of the fluid volume and in the final stage, meteoric water was the predominant component.

Many of the large isotopic changes documented in Table 23 have previously been misinterpreted to result largely from volatilization. However, as already mentioned, volatilization is not a sufficient process to cause the large isotope shifts seen in Table 23; these must result largely from exchange with infiltrating fluids. The effect of infiltration

is greatest when sufficient fluids are available, when contrasts in isotopic composition are great, and when permeability is high (Valley
1986).

One important factor in determining the permeability of a metamorphic rock is the transient effect of volatilization itself (Rumble
and Spear 1983; Nabelek et al. 1984; Valley and O'Neil 1984; Valley
1986). During volatilization reactions, small fluid overpressures are
created. At the same time, the volume of the solid rock is reduced by
the removal of material. Both of these effects may lead to an enhancement of permeability that will be short-lived in the ductile environment.

The infiltration of surface derived fluids into a contact aureole requires that fluid pressures be approximately hydrostatic or less. Thus,
if stable isotope ratios indicate exchange with large amounts of surface
waters, then p_{H_2O} must have been much less than $p_{lithostatic}$. Perhaps
the deepest known penetration of surface-derived fluids is the Trois
Seigneurs Massif, Pyrenees where seawater infiltration may have contributed to partial melting of migmatites at a depth of around 11 to
12 km (Wickham and Taylor 1985).

3.12.4 Regional Metamorphism

There have been extensive debates whether metamorphic fluids and
the rocks with which they interact form "open" or "closed" systems.
If metamorphism occurs under closed-system conditions with respect
to externally derived fluids, then evolved fluids must leave their rock
of origin without interacting with other rocks which are also closed
systems. This would represent a perfectly channelized system. If metamorphism occurs under open-system conditions, then fluid flow
must be pervasive such that all rocks interact with the homogenizing
fluid. Because in many rocks steep isotope gradients have been observed, indicating that homogenization has not occurred, most rocks
fall somewhere between these end-member conditions.

The processes that determine isotopic compositions in systems with
large amounts of pervasive fluids are very different from those in the
presence of small amounts of fluids or highly channelized fluid movement. In the former case, some studies propose isotopic homogenization which would require convective circulation of fluids during regional metamorphism. This is supported by the general observation
that low-grade metamorphic pelites have $\delta^{18}O$-values between 15 and

18‰, whereas high-grade gneisses often have $\delta^{18}O$-values between 6 and 10‰ (Shieh and Schwarcz 1974; Longstaffe and Schwarcz 1977). In the latter case, other studies report fine-scale gradients which would suggest that isotope exchange was very limited and took place without large amounts of fluid (Rumble et al. 1982; Graham et al. 1983; Valley and O'Neil 1984; Bebout and Carlson 1986).

The results of these studies and many others indicate that the exact nature of fluid migration is highly variable among metamorphic rocks. For example, Hoernes and Hoffer (1985) observed a 5‰ difference between quartzes from very low-grade and quartzes formed under greenschist conditions, but a further increase in metamorphic grade did not result in any further systematic decrease in $\delta^{18}O$-values. These authors argued that a free-fluid phase is only present during the low-grade transformations, while at high grades the fluid phase had already escaped from the system and left behind a "dry" system, which had no effect on the overall oxygen isotope composition with increasing temperatures.

3.12.5 Granulite Facies Metamorphism

Since the pioneering work of Touret (1971), it is well known that lower crustal rocks are characterized by the presence of an H_2O-poor, CO_2-rich fluid phase. Although the source of this CO_2 is still a matter of debate, many workers have suggested mantle outgassing as the most feasible CO_2 source (Newton et al. 1980). One of the main arguments for this assumption is that the volume of the lower continental crust is so great and the required CO_2/H_2O ratio so high that only a mantle source can deliver sufficient quantities of CO_2.

Three possible sources must be considered for the CO_2 present in granulite facies rocks:

1. CO_2 may be generated by decarbonation reactions in siliceous dolomite and other impure carbonate rocks. Both equilibrium and kinetic fractionation effects may be operative during this process, but in any event the liberated CO_2 should be slightly enriched in ^{13}C relative to the carbonate.
2. CO_2 may be produced by oxidation of carbonaceous matter (graphite) during metamorphism. Since most of this graphite is depleted in ^{13}C, the resulting CO_2 should, therefore, have low $\delta^{13}C$-values.

3. CO_2 may be of deep-seated origin, possibly derived from degassing of the upper mantle. This deep-seated CO_2 should have δ^{13}C-values between -8 and $-5\%_o$.

Preliminary δ^{13}C-values from fluid inclusions in quartzes (Hoefs and Touret 1975) and from scapolite-rich granulites (Hoefs et al. 1981) show variable, but rather low δ^{13}C-values between -20 and $-10\%_o$. Such low δ^{13}C-values do not support a simple, mantle-degassing process, but rather favor the idea that a substantial portion of the CO_2 may be derived from a graphitic source within the crust.

Many granulites have rather low δ^{18}O-values between 6 and 8‰ (Wilson et al. 1970; Fourcade and Javoy 1973; Longstaffe 1979). Even more depleted δ^{18}O-values (down to 0.1‰) have been reported by Wilson and Baksi (1983). The normal range of δ^{18}O-values between 6–8‰ requires that the precursor rocks have similar low δ^{18}O-values and/or that extensive oxygen isotope exchange with a low ^{18}O-reservoir occurred. It is likely that the unusually low δ^{18}O-values below 6 record pre-granulitic facies events. The Australian rocks analyzed by Wilson and Baksi (1983) may be explained by pre-granulitic exchange with heated seawater. Similarly, the low ^{18}O-eclogites reported by Vogel and Garlick (1970) may have preserved their premetamorphic values and may be interpreted as representing altered oceanic crust.

References

Abell PI (1986) Oxygen isotope ratios in modern African gastropod shells: a data base for paleoclimatology. Chem Geol 58:183−193

Abelson PH, Hoering TC (1961) Carbon isotope fractionation in formation of amino acids by photosynthetic organism. Proc Natl Acad Sci USA 47:623

Allan JR, Matthews RK (1973) Carbon and oxygen isotopes as diagenetic and stratigraphic tools: Surface and subsurface data, Barbados, West Indies. Geology 5:16−20

Allan JR, Matthews RK (1982) Isotope signature associated with early meteoric diagenesis. Sedimentology 29:797−817

Allard P (1979) $^{13}C/^{12}C$ and $^{34}S/^{32}S$ ratios in magmatic gases from ridge volcanism in Afar. Nature (London) 282:56−58

Altabet MA, Deuser WG (1985) Seasonal variations in natural abundance of ^{15}N in particles sinking to the deep Sargossa Sea. Nature (London) 315:218−219

Altabet MA, McCarthy JJ (1985) Temporal and spatial variations in the natural abundance of ^{15}N in POM from a warm-core ring. Deep Sea Res 32:755−772

Anderson AT (1967) The dimensions of oxygen isotope equilibrium attainment during prograde metamorphism. J Geol 75:323−332

Anderson AT (1975) Some basaltic and andesitic gases. Rev Geophys Space Phys 13:37−55

Anderson AT, Clayton RN, Mayeda TK (1971) Oxygen isotope thermometry of mafic igneous rocks. J Geol 79:715−729

Anderson TF, Arthur MA (1983) Stable isotopes of oxygen and carbon and their application to sedimentologic and paleoenvironmental problems. In: Stable isotopes in sedimentary geology. SEPM Short Course No 10, Dallas, 1983, p 1-1−1-151

Arnason B (1977) The hydrogen and water isotope thermometer applied to geothermal areas in Iceland. Geothermics 5:75−80

Arnason B, Sigurgeirsson T (1968) Deuterium content of water vapour and hydrogen in volcanic gas at Surtsey, Iceland. Geochim Cosmochim Acta 32:807−813

Arneth JD, Matzigkeit U, Boos A (1985) Carbon isotope geochemistry of the Cretaceous-Tertiary section of the Wasserfallgraben, Lattengebirge, Southeast Germany. Earth Planet Sci Lett 75:50−58

Arnold M, Sheppard SMF (1981) East Pacific Rise at 21°N: isotopic composition and origin of the hydrothermal sulfur. Earth Planet Sci Lett 56:148−156

Arthur MA (1984) Carbon isotope anomalies? Nature (London) 310:450−451

Arthur MA, Dean WE, Claypool GE (1985) Anomalous ^{13}C enrichment in modern marine organic carbon. Nature (London) 315:216−218

Bachinski DJ (1969) Bond strength and sulfur isotope fractionation in coexisting sulfides. Econ Geol 64:56−65

Baertschi P (1976) Absolute ^{18}O content of standard mean ocean water. Earth Planet Sci Lett 31:341−344

Bainbridge KT, Nier AO (1950) Relative isotopic abundances of the elements. Preliminary Rep No 9. Nuclear Sciences Ser, Natl Res Council USA, Washington DC

Bainbridge AE, Suess HE, Friedman I (1961) Isotopic composition of atmospheric hydrogen and methane. Nature (London) 192:648–649

Barker F, Friedman I (1969) Carbon isotopes and pelites of the Precambrian Uncompahgre formation, Needle, Colorado. Bull Geol Soc Am 80:1403–1408

Barnes I, O'Neil JR (1969) The relationship between fluids in some fresh Alpine type ultramafics and possible modern serpentinization, western United States, Bull Geol Soc Am 80:1947–1960

Barnes IL, Garner EL, Gramlich JW, Machlan LA, Moody JR, Moore JR, Murphy TJ, Shields WR (1973) Isotopic abundance ratios and concentrations of selected elements in some Apollo 15 and 16 samples. Proc 4th Lunar Sci Conf, pp 1197–1207

Barrer RM, Denny AF (1964) Water in hydrates. I Fractionation of hydrogen isotopes by crystallization of salt hydrates. J Chem Soc 1964:4677–4684

Beaty DW, Taylor HP (1982) Some petrologic and oxygen isotopic relationships in the Amulet Mine, Noranda, Quebec, and their bearing on the origin of Archaean massive sulfide deposits. Econ Geol 77:95–108

Bebout GE, Carlson WD (1986) Fluid evolution and transport during metamorphism: evidence from the Llano Uplift, Texas. Contrib Mineral Petrol 92:518–529

Becker RH (1982) Nitrogen isotopic ratios of individual diamond samples. In: 5th Int Conf Geochronology, Cosmochronology, Isotope Geology, Jpn 1982, Abstr pp 21–22

Becker RH, Clayton RN (1975) Nitrogen abundances and isotopic compositions in lunar samples. Proc 6th Lunar Sci Conf 2:2131–2149

Becker RH, Clayton RN (1976) Oxygen isotope study of a Precambrian banded-iron-formation, Hamersley Range, Western Australia. Geochim Cosmochim Acta 40:1153–1165

Becker RH, Clayton RN (1977) Nitrogen isotopes in igneous rocks. EOS Trans Am Geophys Union 58:536

Becker RH, Epstein S (1982) Carbon, hydrogen and nitrogen isotopes in solvent-extractable organic matter from carbonaceous chondrites. Geochim Cosmochim Acta 46:97–103

Becker RH, Pepin RO (1984) The case for a martian origin of the shergottites: nitrogen and noble gases in EETA 79001. Earth Planet Sci Lett 69:225–242

Beeunas MA, Knauth LP (1985) Preserved stable isotopic signature of subaerial diagenesis in the 1.2 b.y. Mescal Limestone, central Arizona. Implications for the timing and development of a terrestrial plant cover. Geol Soc Am Bull 96:737–745

Belanger PE, Curry WB, Matthews RK (1981) Core-top evaluation of benthic foraminiferal isotopic ratios for paleo-oceanographic interpretation. Palaeogeogr-climatol-ecol 33:205–220

Bender ML, Keigwin LD (1979) Speculations about upper Miocene changes in abyssal Pacific dissolved bicarbonate $\delta^{13}C$. Earth Planet Sci Lett 45:383–393

Bender ML, Labeyrie LD, Raynaud D, Lorius C (1985) Isotopic composition of atmospheric O_2 in ice linked with deglaciation and global primary productivity. Nature (London) 318:344–352

Bentor YK (1986) A new approach to the problem of tektite genesis. Earth Planet Sci Lett 77:1–13

Berger WH (1979) Stable isotopes in foraminifera. In: SEPM short course No. 6, Houston, 1979, pp 156–198

Berger WH, Vincent E (1986) Deep-sea carbonates: reading the carbon isotope signal. Geol Rundsch 75:249−269

Berner RA (1972) Sulfate reduction, pyrite formation and the oceanic sulfur budget. In: Dryssen D, Jagner D (eds) The changing chemistry of the oceans. Interscience, New York, pp 347−361

Berner RA (1984) Sedimentary pyrite formation: an update. Geochim Cosmochim Acta 48:605−615

Bernoulli D, Garrison RE, McKenzie J (1978) Petrology, isotope geochemistry and origin of limestone and dolomite associated with basaltic breccia, Hole 373A, Tyrrhenian Basin. Initial Rep DSDP 42:541−553

Berthenrath R, Friedrichsen H, Hellner E (1973) Die Fraktionierung der Sauerstoffisotope $^{18}O/^{16}O$ im System Eisenoxid-Wasser. Fortschr Mineral 50:32−33

Biemann K, Owen T, Rushneck DR, Lafleur AL, Howarth DW (1976) The atmosphere of Mars near the surface: Isotope ratios and upper limits on noble gases. Science 194:76−78

Bigeleisen J (1965) Chemistry of isotopes. Science 147:463−471

Bigeleisen J, Mayer MG (1947) Calculation of equilibrium constants for isotopic exchange reactions. J Chem Phys 15:261−267

Bigeleisen J, Wolfsberg M (1958) Theoretical and experimental aspects of isotope effects in chemical kinetics. Adv Chem Phys 1:15−76

Bigeleisen J, Perlman ML, Prosser HC (1952) Conversion of hydrogenic materials for isotopic analysis. Anal Chem 24:1356

Blattner P, Bird GW (1974) Oxygen isotope fractionation between quartz and K-feldspar at 600 °C. Earth Planet Sci Lett 23:21−27

Blattner P, Hulston JR (1978) Proportional variations of geochemical $\delta^{18}O$ scales − an interlaboratory comparison. Geochim Cosmochim Acta 42:59−62

Blattner P, Dietrich V, Gansser A (1983) Contrasting ^{18}O enrichment and origins of High Himalayan and Transhimalayan intrusives. Earth Planet Sci Lett 65:276−286

Boettcher AL, O'Neil JR (1980) Stable isotope, chemical and petrographic studies of high-pressure amphiboles and micas: evidence for metasomatism in the mantle source regions of alkali basalts and kimberlites. Am J Sci 280A (Jackson Vol): 594−621

Böhlke JK, Kistler RW (1986) Rb−Sr, K−Ar and stable isotope evidence for the ages and sources of fluid components of gold-bearing quartz veins in the Northern Sierra Nevada Foothills Metamorphic Belt, California. Econ Geol 81:296−322

Borthwick J, Harmon RS (1982) A note regarding ClF₃ as an alternative to BrF₅ for oxygen isotope analysis. Geochim Cosmochim Acta 46:1665−1668

Bottinga Y (1969a) Calculated fractionation factors for carbon and hydrogen isotope exchange in the system calcite-carbon dioxide-graphite-methane-hydrogen-water-vapor. Geochim Cosmochim Acta 33:49−64

Bottinga Y (1969b) Carbon isotope fractionation between graphite, diamond and carbon dioxide. Earth Planet Sci Lett 5:301−307

Bottinga Y, Craig H (1969) Oxygen isotope fractionation between CO_2 and water and the isotopic composition of marine atmospheric CO_2. Earth Planet Sci Lett 5:285−295

Bottinga Y, Javoy M (1973) Comments on oxygen isotope geothermometry. Earth Planet Sci Lett 20:250−265

Bottinga Y, Javoy M (1975) Oxygen isotope partitioning among the minerals in igneous and metamorphic rocks. Rev Geophys Space Phys 13:401−418

Bowers TS, Taylor HP (1985) An integrated chemical and isotope model of the origin of mid-ocean ridge hot spring systems. J Geophys Res 90:12583–12606

Bowman JR, O'Neil JR, Essene EJ (1985) Contact skarn formation at Elkhorn, Montana. II. Origin and evolution of C–O–H skarn fluids. Am J Sci 285:621–660

Boyd AW, Brown F, Lounsbury M (1955) Mass spectrometric study of natural and neutron-irradiated chlorine. Can J Phys 33:35

Bremner JM, Keeney DR (1966) Determination and isotope ratio analysis of different forms of nitrogen in soils, III. Soil Sci Soc Am Proc 30:577–582

Brenninkmeijer CAM, Kraft P, Mook WG (1983) Oxygen isotope fractionation between CO_2 and H_2O. Isotope Geosci 1:181–190

Broecker WS (1974) Chemical oceanography. Harcourt Brace Jovanovich, New York

Broecker WS (1982) Ocean chemistry during glacial time. Geochim Cosmochim Acta 46:1689–1705

Brown FS, Baedecker MJ, Nissenbaum A, Kaplan IR (1972) Early diagenesis in a reducing fjord, Saanich Inlet, British Columbia. III. Changes in organic constituents of a sediment. Geochim Cosmochim Acta 36:1185–1203

Brown PE, Bowman JR, Kelly WC (1985) Petrologic and stable isotope constraints on the source and evolution of skarn-forming fluids at Pine Creek, California. Econ Geol 80:72–95

Burk RL, Stuiver M (1981) Oxygen isotope ratios in trees reflect mean annual temperature and humidity. Science 211:1414–1419

Cameron EM (1982) Sulphate and sulphate reduction in early Precambrian oceans. Nature (London) 296:145–148

Carr LP, Grady MM, Wright IP, Fallick AE, Pillinger CT (1983) Carbon, nitrogen and hydrogen in enstatite chondrites. Meteoritics 18:270

Cerling TE, Brown FH, Bowman JR (1985) Low-temperature alteration of volcanic glass: hydration, Na, K, ^{18}O and Ar mobility. Chem Geol Isotope Geosci Sect 52:281–293

Chambers LA, Trudinger PA (1979) Microbiological fractionation of stable sulfur isotopes. Geomicrobiology J 1:249–293

Charef A, Sheppard SMF (1986) Pb–Zn mineralization associated with diapirism: fluid inclusion and stable isotope evidence (H, C, O) for the origin of the fluids at Fedj-El-Adoun. Chem Geol (in press)

Chivas AR, Andrew AS, Sinha AK, O'Neil JR (1982) Geochemistry of Pliocene-Pleistocene oceanic arc plutonic complex, Guadalcanal. Nature (London) 300:139–143

Chung HM, Sackett WM (1979) Use of stable carbon isotope compositions of pyrolytically derived methane as maturity indices for carbonaceous materials. Geochim Cosmochim Acta 43:1979–1988

Claypool GE, Holser WT, Kaplan IR, Sakai H, Zak I (1980) The age curves of sulfur and oxygen isotopes in marine sulfate and their mutual interpretation. Chem Geol 28:199–260

Clayton RN (1959) Oxygen isotope fractionation in the system calcium carbonate-water. J Chem Phys 30:1246–1250

Clayton RN (1981) Isotopic thermometry. In: Newton RC, Navrotsky A, Wood BJ (eds) Thermodynamics of minerals and melts. Springer, Berlin Heidelberg New York, pp 85–109

Clayton RN, Degens ET (1959) Use of carbon isotope analyses of carbonates for differentiating freshwater and marine sediments. Bull Am Assoc Petrol Geol 43:890–897

Clayton RN, Epstein S (1958) The relationship between O^{18}/O^{16} ratios in coexisting quartz, carbonate and iron oxides from various geological deposits. J Geol 66:352−373

Clayton RN, Mayeda TK (1963) The use of bromine pentafluoride in the extraction of oxygen from oxides and silicates for isotopic analysis. Geochim Cosmochim Acta 27:43−52

Clayton RN, Mayeda TK (1978) Genetic relations between iron and stony meteorites. Earth Planet Sci Lett 40:168−174

Clayton RN, Mayeda TK (1983) Oxygen isotopes in eucrites, shergottites, nakhlites and chassignites. Earth Planet Sci Lett 62:1−6

Clayton RN, Mayeda TK (1984) The oxygen isotope record in Murchison and other carbonaceous chondrites. Earth Planet Sci Lett 67:151−161

Clayton RN, Steiner A (1975) Oxygen isotope studies of the geothermal system at Warakei, New Zealand. Geochim Cosmochim Acta 39:1179−1186

Clayton RN, Friedman I, Graf DL, Mayeda TK, Meents WF, Shimp NF (1966) The origin of saline formation waters. 1. Isotopic composition. J Geophys Res 71:3869−3882

Clayton RN, Muffler LJP, White (1968) Oxygen isotope study of calcite and silicates of the River Branch No. 1 well, Salton Sea Geothermal Field, California. Am J Sci 266:968−979

Clayton RN, O'Neil JR, Mayeda TK (1972) Oxygen isotope exchange between quartz and water. J Geophys Res 77:3057−3067

Clayton RN, Grossman L, Mayeda TK (1973a) A component of primitive nuclear composition in carbonaceous meteorites. Science 182:485−488

Clayton RN, Hurd JM, Mayeda TK (1973b) Oxygen isotopic compositions of Apollo 15, 16 and 17 samples and their bearing on lunar origin and petrogenesis. Proc 4th Lunar Sci Conf Geochim Cosmochim Acta Suppl 2:1535−1542

Clayton RN, Goldsmith JR, Karel KJ, Mayeda TK, Newton RC (1975) Limits on the effect of pressure in isotopic fractionation. Geochim Cosmochim Acta 39:1197−1201

Clayton RN, Onuma N, Mayeda TK (1976) A classification of meteorites based on oxygen isotopes. Earth Planet Sci Lett 30:10−18

Clayton RN, Onuma N, Grossman C, Mayeda TK (1977) Distribution of the presolar component in Allende and other carbonaceous chondrites. Earth Planet Sci Lett 34:209−224

Cline JD, Kaplan IR (1975) Isotopic fractionation of dissolved nitrate during denitrification in the eastern tropical North Pacific Ocean. Mar Chem 3:271−299

Cloud PE, Friedman I, Sisler FD (1958) Microbiological fractionation of the hydrogen isotopes. Science 127:1394−1395

Cole DR, Ohmoto H, Lasaga AC (1983) Isotopic exchange in mineral fluid systems. I. Theoretical evaluation of oxygen isotopic reactions and diffusion. Geochim Cosmochim Acta 47:1681−1693

Coplen TB, Hanshaw BB (1973) Ultrafiltration by a compacted clay membrane. I. Oxygen and hydrogen isotopic fractionation. Geochim Cosmochim Acta 37:2295−2310

Coplen TB, Kendall C (1982) Preparation and stable isotope determination of NBS-16 and NBS-17 carbon dioxide reference samples. Anal Chem 1982:2611−2612

Coplen TB, Kendall C, Hopple J (1983) Comparison of stable isotope reference samples. Nature (London) 302:236−238

Correns CW (1950) Zur Geochemie der Diagenese. I. Das Verhalten von $CaCO_3$ and SiO_2. Geochim Cosmochim Acta 1:49−54

Cortecci G, Longinelli A (1970) Isotopic composition of sulfate in rain water, Pisa, Italy. Earth Planet Sci Lett 8:36–40

Cortecci G, Longinelli A (1971) $^{18}O/^{16}O$ ratios in sulfate from living marine organisms. Earth Planet Sci Lett 11:273–276

Cortecci G, Longinelli A (1973) $^{18}O/^{16}O$ ratios in sulfate from fossil shells. Earth Planet Sci Lett 19:410–412

Craig H (1953) The geochemistry of the stable carbon isotopes. Geochim Cosmochim Acta 3:53–92

Craig H (1957) Isotopic standards for carbon and oxygen and correction factors for mass-spectrometric analysis of carbon dioxide. Geochim Cosmochim Acta 12:133–149

Craig H (1961a) Isotopic variations in meteoric waters. Science 133:1702–1703

Craig H (1961b) Standard for reporting concentration of deuterium and oxygen-18 in natural waters. Science 133:1833–1834

Craig H (1963) The isotopic geochemistry of water and carbon in geothermal areas. In: Nuclear geology of geothermal areas, p 17–53. Spoleto, Sept 9–13, 1963

Craig H (1965) The measurement of oxygen isotope paleotemperatures. Proc Spoleto Conf Stable Isotopes Oceanogr Stud Paleotemp 3

Craig H (1966) Isotopic composition and origin of the Red Sea and Salton Sea geothermal brines. Science 154:1544–1548

Craig H, Gordon L (1965) Deuterium and oxygen-18 variations in the ocean and the marine atmosphere. In: Symposium on marine geochemistry. Graduate School of Oceanography, Univ Rhode Island, Occ Publ No 3:277

Craig H, Keeling CD (1963) The effects of atmospheric N_2O on the measured isotopic composition of atmospheric CO_2. Geochim Cosmochim Acta 27:549–551

Craig H, Lupton JE (1976) Primordial neon, helium and hydrogen in oceanic basalts. Earth Planet Sci Lett 31:369–385

Craig H, Boato G, White DE (1956) Isotopic geochemistry of thermal waters. Proc 2nd Conf Nucl Process Geol Settings, p 29

Criss RE, Taylor HP (1983) An $^{18}O/^{16}O$ and D/H study of Tertiary hydrothermal systems in the Southern half of the Idaho batholith. Geol Soc Am Bull 94:640–653

Criss RE, Ekren EB, Hardyman RF (1984) Casto Ring Zone: A 4500 km^2 fossil hydrothermal system in the Challis Volcanic Field, Central Idaho. Geology 12:331–334

Criss RE, Champion DE, McIntyre DH (1985) Oxygen isotope, aeromagnetic and gravity anomalies associated with hydrothermally altered zones in the Yankee Fork Mining District, Custer County, Idaho. Econ Geol 80:1277–1296

Curry WB, Lohmann GP (1982) Carbon isotopic changes in benthic foraminifera from the western South Atlantic: reconstruction of glacial abyssal circulation patterns.Quat Res 18:218–235

Czamanske GK, Rye RO (1974) Experimentally determined sulfur isotope fractionations between sphalerite and galena in the temperature range 600 $^{\circ}$C to 275 $^{\circ}$C. Econ Geol 69:17–25

Dansgaard W (1964) Stable isotopes in precipitation. Tellus 16:436–468

Dansgaard W, Johnsen SJ, Moller J, Langway CC (1969) One thousand centuries of climatic record from Camp Century on the Greenland ice sheet. Science 166:377–381

Dansgaard W, Johnsen SF, Reeh N, Gundestrup N, Clausen HB, Hammer CU (1975) Climatic changes, Norsemen and modern man. Nature (London) 255:24–28

Dansgaard W, Clausen HB, Gundestrup N, Hammer CU, Johnsen SF, Kristindottir PM, Reeh N (1982) A new Greenland deep ice core. Science 218:1273–1277

Davidson J (1985) Mechanisms of contamination in Lesser Antilles island arc magmas from radiogenic and oxygen isotope relationships. Earth Planet Sci Lett 72:163−174

Dean WE, Arthur MA, Claypool GE (1986) Depletion of ^{13}C in Cretaceous marine organic matter: source, diagenetic or environmental signal? Mar Geol 70:119−158

Degens ET, Epstein S (1962) Relationship between ^{18}O/^{16}O ratios in coexisting carbonates, cherts and diatomites. Bull Am Assoc Pet Geol 46:534−542

Degens ET, Guillard RRL, Sackett WM, Hellebust JA (1968a) Metabolic fractionation of carbon isotopes in marine plankton. I. Temperature and respiration experiments. Deep Sea Res 15:1−9

Degens ET, Behrendt M, Gotthardt B, Reppmann E (1968b) Metabolic fractionation of carbon isotopes in marine plankton. II. Data on samples collected off the coasts of Peru and Ecuador. Deep Sea Res 15:11−20

Deines P (1977) On the oxygen isotope distribution among mineral triplets in igneous and metamorphic rocks. Geochim Cosmochim Acta 41:1709−1730

Deines P (1980a) The carbon isotopic composition of diamonds: relationship to diamond shape, color, occurrence and vapor composition. Geochim Cosmochim Acta 44:943−962

Deines P (1980b) The isotopic composition of reduced organic carbon. In: Fritz P, Fontes JCh (eds) Handbook of environmental geochemistry, vol I. Elsevier, New York Amsterdam, pp 239−406

Deines P, Gold DP (1973) The isotopic composition of carbonatite and kimberlite carbonates and their bearing on the isotopic composition of deep-seated carbon. Geochim Cosmochim Acta 37:1709−1733

Deines P, Gurney JJ, Harris JW (1984) Associated chemical and carbon isotopic composition variations in diamonds from Finsch and Premier Kimberlite, South Africa. Geochim Cosmochim Acta 48:325−342

Delwiche CC, Steyn PL (1970) Nitrogen isotope fractionation in soils and microbial reactions. Environ Sci Technol 4:929

DeNiro MJ, Epstein S (1977) Mechanism of carbon isotope fractionation associated with lipid synthesis. Science 197:261−263

DeNiro MJ, Epstein S (1978) Influence of diet on the distribution of carbon isotopes in animals. Geochim Cosmochim Acta 42:495−506

DeNiro MJ, Epstein S (1979) Relationship between the oxygen isotope ratios of terrestrial plant cellulose, carbon dioxide and water. Science 204:51−53

DeNiro MJ, Epstein S (1981) Isotopic composition of cellulose from aquatic organisms. Geochim Cosmochim Acta 45:1885−1894

Dennis PF (1984) Oxygen self-diffusion in quartz under hydrothermal conditions. J Geophys Res 89:4047−4058

Desaulniers DE, Kaufmann RS, Cherry JO, Bentley HW (1986) ^{37}Cl−^{35}Cl variations in a diffusion-controlled groundwater system. Geochim Cosmochim Acta 50:1757−1764

Des Marais DJ (1983) Light element geochemistry and spallogenesis in lunar rocks. Geochim Cosmochim Acta 47:1769−1781

Des Marais DJ, Moore JG (1984) Carbon and its isotopes in mid-oceanic basaltic glasses. Earth Planet Sci Lett 69:43−57

Deuser WG (1970) Extreme ^{13}C/^{12}C variations in Quaternary dolomites from the continental shelf. Earth Planet. Sci Letters 8:118−124

Deuser WG, Hunt JM (1969) Stable isotope ratios of dissolved inorganic carbon in the Atlantic. Deep Sea Res 16:221−225

Deutsch S, Ambach W, Eisner H (1966) Oxygen isotope study of snow and firn of an Alpine glacier. Earth Planet Sci Lett 1:197−201

Dickson JAD, Coleman ML (1980) Changes in carbon and oxygen isotope composition during limestone diagenesis. Sedimentology 27:107–118

Dole M, Jenks G (1944) Isotopic composition of photosynthetic oxygen. Science 100:409

Dole M, Lange GA, Rudd DP, Zaukelies DA (1954) Isotopic composition of atmospheric oxygen and nitrogen. Geochim Cosmochim Acta 6:65–78

Donahue TM, Hoffman JH, Hodges RD, Watson AJ (1982) Venus was wet: a measurement of the ratio of deuterium to hydrogen. Science 216:630–633

Douthitt CB (1982) The geochemistry of the stable isotopes of silicon. Geochim Cosmochim Acta 46:1449–1458

Downs WF, Touysinhthiphonexay Y, Deines P (1981) A direct determination of the oxygen isotope fractionation between quartz and magnetite at 600° and 800 °C and 5 kbar. Geochim Cosmochim Acta 45:2065–2072

Dugan JP, Borthwick J, Harmon RS, Gagnier MA, Glahn JE, Kinsel EP, McLeod S, Viglino JA (1985) Guadinine hydrochloride method for determination of water oxygen isotope ratios and the oxygen-18 fractionation between carbon dioxide and water at 25 °C. Anal Chem 57:1734–1736

Dungan MA, Lindstrom MM, McMillan NJ, Moorbath S, Hoefs J, Haskin LA (1986) Open system magmatic evolution of the Taos Plateau Volcanic Field, Northern New Mexico. I. The petrology and geochemistry of the Servilleta basalts. J Geophys Res 91:5999–6028

Eadie BJ, Jeffrey LM (1973) $\delta^{13}C$ analyses of oceanic particulate organic matter. Mar Chem 1:199–209

Eggler DH, Wendlandt RF (1979) Experimental studies on the relationship between kimberlite magmas and partial melting of peridotites. In: Boyd FR, Meyer HOA (eds) Kimberlites, diatremes and diamonds: their geology, petrology and geochemistry. Am Geophys Union 1:330–338

Ellis AJ (1979) Chemical geothermometry in geothermal systems. Chem Geol 25:219–226

Elphick SC, Dennis PF, Graham CM (1986) An experimental study of the diffusion of oxygen in quartz and albite using an overgrowth technique. Contrib Min Petrol 92:322–330

Emiliani C (1955) Pleistocene temperatures. J Geol 63:538–578

Emiliani C (1966) Isotopic paleotemperatures. Science 154:851–857

Emiliani C (1972) Quaternary paleotemperatures and the duration of the high-temperature intervals. Science 178:398–401

Emiliani C (1978) The cause of the ice ages. Earth Planet Sci Lett 37:349–354

Emiliani C, Shackleton NJ (1974) The Brunhes epoch: Isotopic paleotemperatures and geochronology. Science 183:511–514

Epstein S, Lowenstam HA (1953) Temperature-shell-growth relations of recent and interglacial Pleistocene shoal-water biota from Bermuda. J Geol 61:424–438

Epstein S, Mayeda TK (1953) Variations of O^{18} content of waters from natural sources. Geochim Cosmochim Acta 4:213–224

Epstein S, Taylor HP (1970) $^{18}O/^{16}O$, $^{30}Si/^{28}Si$, D/H and $^{13}C/^{12}C$ studies of lunar rocks and minerals. Science 167:533–535

Epstein S, Taylor HP (1971) O^{18}/O^{16}, Si^{30}/Si^{28}, D/H and C^{13}/C^{12} ratios in lunar samples. Proc 2nd Lunar Sci Conf 2:1421–1441

Epstein S, Taylor HP (1972) O^{18}/O^{16}, Si^{30}/Si^{28}, C^{13}/C^{12} and D/H studies of Apollo 14 and 15 samples. Proc 3rd Lunar Sci Conf 2:1429–1454

Epstein S, Buchsbaum HA, Lowenstam HA, Urey HC (1953) Revised carbonate-water isotopic temperature scale. Bull Geol Soc Am 64:1315–1326

Epstein S, Sharp RP, Dow AJ (1965) Six-year record of oxygen and hydrogen isotope variations in South Pole firn. J Geophys Res 70:1809–1814

Epstein S, Yapp CJ, Hall JH (1976) The determination of the D/H ratio of non-exchangeable hydrogen in cellulose extracted from aquatic and land plants. Earth Planet Sci Lett 30:241–251

Epstein S, Thompson P, Yapp CJ (1977) Oxygen and hydrogen isotopic ratios in plant cellulose. Science 198:1209–1215

Eslinger EV, Savin SM (1973a) Mineralogy and oxygen isotope geochemistry of the hydrothermally altered rocks of the Ohaki-Broadlands, New Zealand geothermal area. Am J Sci 273:240

Eslinger EV, Savin SM (1973b) Oxygen isotope geothermometry of the burial metamorphic rocks of the Precambrian Belt Supergroup, Glacier National Park, Montana. Bull Geol Soc Am 84:2549–2560

Eslinger EV, Yeh HW (1981) Mineralogy, $^{18}O/^{16}O$ and D/H ratios of clay-rich sediments from Deep Sea Drilling Project site 180, Aleutian Trench: Clays Clay Minerals 29:309–315

Eslinger EV, Yeh HW (1986) Oxygen and hydrogen isotope geochemistry of Cretaceous bentonites and shales from the Disturbed Belt, Montana. Geochim Cosmochim Acta 50:59–68

Eslinger EV, Savin SM, Yeh HW (1979) Oxygen isotope geothermometry of diagenetically altered shales. Soc Econ Paleontol Mineral Spec Publ 26:113–124

Estep MF, Hoering TC (1980) Biogeochemistry of the stable hydrogen isotopes. Geochim Cosmochim Acta 44:1197–1206

Exley RA, Mattey DP, Clague DA, Pillinger CT (1986a) Carbon isotope systematics of a mantle "hot spot": a comparison of Loihi seamount and MORB glasses. Earth Planet Sci Lett 78:189–199

Exley RA, Mattey DP, Boyd SR, Pillinger CT (1986b) Nitrogen isotope geochemistry of basaltic glasses. Terra Cognita 6:191

Fallick AE, Hinton RW, McNaughton NJ, Pillinger CT (1983) D/H ratios in meteorites: some results and implications. Ann Geophys 1:129–134

Farmer JG, Baxter MS (1974) Atmospheric carbon dioxide levels as indicated by the stable isotope record in wood. Nature (London) 247:273–275

Faure G (1977) Principles of isotope geology. John Wiley & Son, New York

Faure G, Hoefs J, Mensing TM (1984) Effects of oxygen fugacity on sulfur isotope compositions and magnetite concentrations in the Kirkpatrick basalt, Mount Falla, Queen Alexandra Range, Antarctica. Isotope Geosci 2:301–311

Ferrara G, Laurenzi MA, Taylor HP, Tonarini S, Turi B (1985) Oxygen and strontium isotope studies of K-rich volcanic rocks from the Alban Hills, Italy. Earth Planet Sci Lett 75:13–28

Ferrara G, Preite-Martinez M, Taylor HP, Tonarini S, Turi B (1986) Evidence for crustal assimilation, mixing of magmas, and a ^{87}S-rich upper mantle. An oxygen and strontium isotope study of the M. Vulsini volcanic area, Central Italy. Contrib Mineral Petrol 92:269–280

Feder HM, Taube H (1952) Ionic hydration, an isotopic fractionation technique. J Chem Phys 20:1335

Ferry JM (1985a) Hydrothermal alteration of Tertiary igneous rocks from the Isle of Skye, northwest Scotland. I. Gabbros. Contr Mineral Petrol 91:264–282

Ferry JM (1985b) Hydrothermal alteration of Tertiary igneous rocks from the Isle of Skye, northwest Scotland. II. Granites. Contr Mineral Petrol 91:283–304

Field CW, Gustafson LB (1976) Sulfur isotopes in the porphyry copper deposit at El Salvador, Chile. Econ Geol 71:1533–1548

Fireman EL, Norris TL (1982) Ages and composition of gas trapped on Allan Hills and Byrd core ice. Earth Planet Sci Lett 60:339–350

Fontes JC, Gonfiantini R (1967) Comportement isotopique au cours de l'évaporation de deux bassins Sahariens. Earth Planet Sci Lett 3:258–266

Forester RW, Taylor HP (1977) $^{18}O/^{16}O$, D/H, and $^{13}C/^{12}C$ studies of the Tertiary igneous complex of Skye, Scotland. Am J Sci 277:136–177

Fourcade S, Javoy M (1973) Rapports $^{18}O/^{16}O$ dans les roches du vieux socle catazonal d'In Ouzzal (Sahara algérien). Contrib Mineral Petrol 42:235–244

Francey RJ, Farquhar GD (1982) An explanation of the $^{13}C/^{12}C$ variations in tree rings. Nature (London) 297:28–31

Frape SK, Fritz P, McNutt RH (1984) Water-rock interaction and chemistry of groundwaters from the Canadian Shield. Geochim Cosmochim Acta 48:1617–1627

Freer R, Dennis PF (1982) Oxygen diffusion studies. I: A preliminary ion microprobe investigation of oxygen diffusion in some rock-forming minerals. Min Mag 45:179–192

Freyer HD (1979) On the ^{13}C-record in tree rings. I. ^{13}C variations in northern hemisphere trees during the last 150 years. Tellus 31:124–137

Freyer HD, Wiesberg L (1973) ^{13}C-decrease in modern wood due to the large scale combustion of fossil fuels. Naturwissenschaften 60:517–518

Friedli H, Moor E, Oeschger H, Siegenthaler U, Stauffer B (1984) $^{13}C/^{12}C$ ratios in CO_2 extracted from Antarctic Ice. Geophys Res Lett 11:1145–1148

Friedman I (1953) Deuterium content of natural waters and other substances. Geochim Cosmochim Acta 4:89–103

Friedman I, O'Neil JR (1977) Compilation of stable isotope fractionation factors of geochemical interest. In: Data Geochem, 6th edn Geol Suvr Prof Pap 440KK

Friedman I, O'Neil JR (1978) Hydrogen. In: Wedepohl KH (ed) Handbook of geochemistry. Springer, Berlin Heidelberg New York

Friedman I, Scholz TG (1974) Isotopic composition of atmospheric hydrogen (1967–1969). J Geophys Res 79:785–788

Friedman I, O'Neil JR, Adami LH, Gleason JD, Hardcastle KG (1970) Water, hydrogen, deuterium, carbon-13, and oxygen-18 content of selected lunar material. Science 167:538–540

Friedman I, Hardcastle KG, Gleason JD (1974) Water and carbon in rusty lunar rock 66095. Science 185:346–349

Fritz P, Poplawski S (1974) ^{18}O and ^{13}C in the shells of freshwater molluscs and their environment. Earth Planet Sci Lett 24:91–98

Fry B, Cox J, Gest H, Hayes JM (1986) Discrimination between ^{34}S and ^{32}S during bacterial metabolism of inorganic sulfur compounds. J Bacteriol 165:328–330

Fuchs G, Thauer R, Ziegler H, Stiehler W (1979) Carbon isotope fractionation by Methanobacterium thermoautotrophicum. Arch Microbiol 120:135–139

Fuex AN (1977) The use of stable carbon isotopes in hydrocarbon exploration. J Geochem Explor 7:155–188

Fuex AN, Baker DR (1973) Stable carbon isotopes in selected granitic, mafic and ultramafic rocks. Geochim Cosmochim Acta 37:2509–2521

Galimov EM (1973) Carbon isotopes in oil-gas geology. Nedra, Moscow, p 384 (in Russian)

Galimov EM (1985) The relation between formation conditions and variations in isotope compositions of diamonds. Geochem Int 22, 1:118–141

Gardiner LR, Pillinger CT (1979) Static mass spectrometric analysis of active gases. Anal Chem 51:1230–1236

Garlick GD (1969) The stable isotopes of oxygen. In: Wedepohl KH (ed) Handbook of geochemistry, 8B. Springer, Berlin Heidelberg New York

Garlick GD, Dymond JR (1970) Oxygen isotope exchange between volcanic materials and ocean water. Geol Soc Am Bull 81:2137–2142

Garlick GD, MacGregor ID, Vogel DE (1971) Oxygen isotope ratios in eclogites from kimberlites. Science 172:1025–1027

Garrels RM, Lerman A (1984) Coupling of the sedimentary sulfur and carbon cycles – an improved model. Am J Sci 284:989–1007

Garrels RM, Perry EA (1974) Cycling of carbon, sulfur and oxygen through geologic time. In: Goldberg ED (ed) The sea, vol 5. Wiley and Son, New York, p 303

Gat JR (1971) Comments on the stable isotope method in regional groundwater investigation. Water Resource Res 7:980

Gat JR (1980) The isotopes of hydrogen and oxygen in precipitation. In: Handbook of environmental isotope geochemistry, vol 1. Elsevier, New York Amsterdam, pp 21–47

Gat JR, Dansgaard W (1972) Stable isotope survey of the fresh water occurrences in Israel and the northern Jordan Rift Valley. J Hydrol 16:177

Gat JR, Issar A (1974) Desert isotope hydrology: water sources of the Sinai desert. Geochim Cosmochim Acta 38:1117–1131

Gautier DL (1982) Siderite concretions: indicators of early diagenesis in the Ganimon Shale (Cretaceous). J Sediment Petrol 52:859–871

Gerlach TM, Thomas DM (1986) Carbon and sulphur isotopic composition of Kilauea parental magma. Nature (London) 319:480–483

Giggenbach W (1982) Carbon-13 exchange between CO_2 and CH_4 under geothermal conditions. Geochim Cosmochim Acta 46:159–165

Giggenbach W, Robinson BW (1976) Sulfur isotope geochemistry of White Island volcanic discharges. Abstr Int Conf Stable Isotopes, Lower Hutt, N Z, p 23

Giletti BJ (1986) Diffusion effect on oxygen isotope temperatures of slowly cooled igneous and metamorphic rocks. Earth Planet Sci Lett 77:218–228

Giletti BJ, Anderson TE (1975) Studies in diffusion. II. Oxygen in phlogopite mica. Earth Planet Sci Lett 28:225–233

Giletti BJ, Yund RA (1984) Oxygen diffusion in quartz. J Geophys Res 89:4039–4046

Giletti BJ, Semet MP, Yund RA (1978) Studies in diffusion. III. Oxygen in feldspars: an ion microprobe determination. Geochim Cosmochim Acta 42:45–57

Gilmour I, Pillinger CT (1983) The carbon isotopic composition of individual high molecular weight alkanes in the Murchison carbonaceous chondrite. Meteoritics 18:302

Given RK, Lohmann KC (1985) Derivation of the original isotopic composition of Permian marine cements. J Sediment Petrol 55:430–439

Godfrey JD (1962) The deuterium content of hydrous minerals from the East-Central Sierra Nevada and Yosemite National Park. Geochim Cosmochim Acta 26:1215–1245

Goldhaber MB, Kaplan IR (1974) The sedimentary sulfur cycle. In: Goldberg EB (ed) The sea, vol IV. Wiley and Son, New York

Gonfiantini R (1978) Standards for stable isotope measurements in natural compounds. Nature (London) 271:534–536

Gonfiantini R (1984) Advisory group meeting on stable isotope reference samples for geochemical and hydrological investigations. Rep Director General IAEA Vienna

Gonsior B, Friedman I, Lindenmayr G (1966) New tritium and deuterium measurements in atmospheric hydrogen. Tellus 18:256

Goodwin AM, Thode HG, Chou CL, Karkhansis SN (1985) Chemostratigraphy and origin of the late Archaen siderite-pyrite-rich Helen Iron Formation, Michipicoten belt, Canada. Can J Earth Sci 22:72–84

Grady MM, Wright IP, Fallick AE, Pillinger CT (1983) The stable isotope composition of carbon, nitrogen and hydrogen in some Yamato meteorites. Proc 8th Symp Antarctic Meteorites 1983, pp 289–302

Grady MM, Wright IP, Swart PK, Pillinger CT (1985) The carbon and nitrogen isotopic composition of ureilites: implications for their genesis. Geochim Cosmochim Acta 49:903–915

Graham AM, Graham CM, Harmon RS (1982) Origins of mantle waters: stable isotope evidence from amphibole-bearing plutonic cumulate blocks in calc-alkaline volcanics, Grenada, Lesser Antilles. Abstr 5th Int Conf Geochronology, Cosmochronology, Isotope Geology, pp 119–120

Graham CM (1981) Experimental hydrogen isotope studies. III. Diffusion of hydrogen in hydrous minerals and stable isotope exchange in metamorphic rocks. Contrib Mineral Petrol 76:216–228

Graham CM, Harmon RS (1983) Stable isotope evidence on the nature of crust-mantle interactions. In: Hawkesworth CJ, Norry MJ (eds) Continental basalts and mantle xenoliths. Shiva, pp 20–45

Graham CM, Sheppard SMF, Heaton THE (1980) Experimental hydrogen isotope studies. I. Systematics of hydrogen isotope fractionation in the systems epidote-H_2O, zoisite-H_2O and $AlO(OH)–H_2O$. Geochim Cosmochim Acta 44:353–364

Graham CM, Greig KM, Sheppard SMF, Turi B (1983) Genesis and mobility of the $H_2O–CO_2$ fluid phase during regional greenschist and epidote amphibolite facies metamorphism: a petrological and stable isotope study in the Scottish Dalradian. J Geol Soc London 140:577–599

Graham CM, Harmon RS, Sheppard SMF (1984) Experimental hydrogen isotope studies: hydrogen isotope exchange between amphibole and water. Am Mineral 69:128–138

Graham DW, Corliss BH, Bender ML, Keigwin LD (1981) Carbon and oxygen isotopic disequilibrium of recent deep-sea benthic foraminifera. Mar Micropaleontol 6:483–497

Green GR, Ohmoto D, Date J, Takahashi T (1983) Whole-rock oxygen isotope distribution in the Fukazawa-Kosaka Area, Hokuroko District, Japan and its potential application to mineral exploration. Econ Geol Monogr 5:395–411

Gregory RT, Taylor HP (1981) An oxygen isotope profile in a section of Cretaceous oceanic crust, Samail Ophiolite, Oman: evidence for $\delta^{18}O$ buffering of the oceans by deep (> 5 km) seawater-hydrothermal circulation at Mid-Ocean Ridges. J Geophys Res 86:2737–2755

Gregory RT, Taylor HP (1986) Non-equilibrium, metasomatic $^{18}O/^{16}O$ effects in upper mantle mineral assemblages. Contr Mineral Petrol 93:124–135

Grey DC, Jensen ML (1972) Bacteriogenic sulfur in air pollution. Science 177:1099–1100

Grinenko LN, Ukhanov AV (1977) Sulfur levels and isotopic compositions in upper mantle xenoliths from the Obnazhennaya Kimberlite pipe. Geochemistry 14, 6:169–171

Grinenko VA, Dimitriev LV, Migdisov AA, Sharas'Kin AY (1975) Sulfur contents and isotope composition for igneous and metamorphic rocks from mid-ocean ridges. Geochem Int 12, 1:132

Grootenboer J, Schwarcz HP (1969) Experimentally determined sulfur isotope fractionations between sulfide minerals. Earth Planet Sci Lett 7:162–166

Grossman EL (1984a) Carbon isotopic fractionation in live benthic foraminifera – comparison with inorganic precipitate studies. Geochim Cosmochim Acta 48:1505–1512

Grossman EL (1984b) Stable isotope fractionation in live benthic foraminifera from the Southern California borderland. Paleogeogr Paleoclimatol Paleoecol 47:301–327

Haack U, Hoefs J, Gohn E (1982) Constraints on the origin of Damaran granites by Rb/Sr and $\delta^{18}O$ data. Contrib Mineral Petrol 79:279–289

Hackley KC, Anderson TF (1986) Sulfur isotopic variations in low-sulfur coals from the Rocky Mountain region. Geochim Cosmochim Acta 50:1703–1713

Haendel D, Mühle K, Nitzsche HM, Stiehl G, Wand U (1986) Isotopic variations of the fixed nitrogen in metamorphic rocks. Geochim Cosmochim Acta 50: 749–758

Hagemann R, Nief G, Roth E (1970) Absolute isotopic scale for deuterium analysis of natural waters. Absolute D/H ratio for SMOW. Tellus 22:712–715

Haimson M, Knauth LP (1983) Stepwise fluorination — a useful approach for the isotopic analysis of hydrous minerals. Geochim Cosmochim Acta 47:1589–1595

Haines EB (1976) Relation between the stable carbon isotope composition of fiddler crabs, plants and soils in a salt marsh. Limnol Oceanogr 21:880–883

Halliday AN, Stephens WE, Harmon RS (1980) Rb–Sr and O isotopic relationships in 3 zoned Caledonian granitic plutons, Southern Uplands, Scotland: evidence for varied sources and hybridization of magmas. J Geol Soc London 137:329–348

Hamza MS, Epstein S (1980) Oxygen isotopic fractionation between oxygen of different sites in hydroxylbearing silicate minerals. Geochim Cosmochim Acta 44:173–182

Harmon RS, Hoefs J (1984) O-isotope relationships in Cenozoic volcanic rocks: Evidence for a heterogenous mantle source and open-system magmagenesis. In: Proc ISEM Field Conf Open Magmatic System, pp 69–71

Harmon RS, Hoefs J (1986) S-isotope relationships in Late Cenozoic destructive plate margin and continental intraplate volcanic rocks. Terra Cognita 6:182

Harmon RS, Schwarcz HP (1981) Changes of ^2H and ^{18}O enrichment of meteoric water and Pleistocene glaciation. Nature (London) 290:125–127

Harmon RS, Barreiro BA, Moorbath S, Hoefs J, Francis PW, Thorpe RS, Déruelle B, McHugh J, Viglino JA (1984) Regional O-, Sr- and Pb-isotope relationships in late Cenozoic calc-alkaline lavas of the Andean Cordillera. J Geol Soc 141: 803–822

Harmon RS, Hoefs J, Wedepohl KH (1987) Stable isotope (O, H, S) relationships in Tertiary basalts and their mantle xenoliths from the northern Hessian Depression. Contr Mineral Petrol (in press)

Harrison AG, Thode HG (1957a) Kinetic isotope effect in chemical reduction of sulphate. Faraday Soc Trans 53:1648–1651

Harrison AG, Thode HG (1957b) Mechanism of the bacterial reduction of sulphate from isotope fractionation studies. Faraday Soc Trans 54:84–92

Harrison AG, Thode HG (1958) Sulphur isotope abundances in hydrocarbons and source rocks of Uinta Basin, Utah. Bull Am Assoc Petrol Geol 42:2642–2649

Hartmann M, Nielsen H (1969) $\delta^{34}S$-Werte in rezenten Meeressedimenten und ihre Deutung am Beispiel einiger Sedimentprofile aus der westlichen Ostsee. Geol Rundsch 58:621–655

Hattori K, Muehlenbachs K (1982) Oxygen isotope ratios in the Icelandic crust. J Geophys Res 87:6559–6565

Hattori K, Krouse HR, Campbell FA (1983) The start of sulfur oxidation in continental environments: about 2.2×10^9 years ago. Science 221:549–551

Hayes JM (1983) Practice and principles of isotopic measurements in organic geochemistry. In: Organic geochemistry of contemporaneous and ancient sediments, Great Lakes Section, SEPM, Bloomington, Ind, pp 5-1–5-31

Heidenreich JE, Thiemens MH (1983) A non-mass-dependent isotope effect in the production of ozone from molecular oxygen. J Chem Phys 78:892–895

Heidenreich JE, Thiemens MH (1985) The non-mass-dependent oxygen isotope effect in the electro dissociation of carbon dioxide: a step toward understanding NoMaD chemistry. Geochim Cosmochim Acta 49:1303–1306

Heinzinger K (1969) Ein Wasserstoffisotopieeffekt in $CuSO_4 \cdot 5 H_2O$. Z Naturforsch 24:1502

Heilmann H, Lensch G (1977) Sulfur isotope investigations of sulfides and rocks from the Main Basic Series of the Ivrea Zone. Schweiz Mineral Petrogr Mitt 57: 349–360

Heyl AV, Landis GR, Zartman RE (1974) Isotopic evidence for the origin of Mississippi-Valley-type mineral deposits: A review. Econ Geol 69:992–1006

Hilbrecht H, Hoefs J (1986) Geochemical and palaeontological studies of the $\delta^{13}C$-anomaly in Boreal and North Tethyan Cenomanian-Turonian sediments in Germany and adjacent areas. Palaeogeogr-climatol-ecol 53:169–189

Hildreth W, Christiansen RL, O'Neil JR (1984) Catastrophic isotopic modification of rhyolitic magma at times of caldera subsidence, Yellowstone Plateau Volcanic Field. J Geophys Res 89:8339–8369

Hitchon B, Friedman I (1969) Geochemistry and origin of formation waters in the western Canada sedimentary basin. I. Stable isotopes of hydrogen and oxygen. Geochim Cosmochim Acta 33:1321–1349

Hitchon B, Krouse HR (1972) Hydrogeochemistry of the surface waters of the Mackenzie River drainage basin, Canada. III. Stable isotopes of oxygen, carbon and sulfur. Geochim Cosmochim Acta 36:1337–1357

Hoefs J (1970) Kohlenstoff- und Sauerstoff-Isotopenuntersuchungen an Karbonatkonkretionen und umgebendem Gestein. Contrib Mineral Petrol 27:66–79

Hoefs J (1973) Ein Beitrag zur Isotopengeochemie des Kohlenstoffs in magmatischen Gesteinen. Contrib Mineral Petrol 41:277–300

Hoefs J (1981) Isotopic composition of the ocean-atmospheric system in the geologic past. In: O'Connell RJ, Fyfe WS (eds) Evolution of the earth. Geodynamic Series, vol 5. American Geophys. Union, Washington, DC, pp 110–119

Hoefs J, Emmermann R (1983) The oxygen isotope composition of Hercynian granites and pre-Hercynian gneisses from the Schwarzwald, SW Germany. Contrib Mineral Petrol 83:320–329

Hoefs J, Frey M (1976) The isotopic composition of carbonaceous matter in a metamorphic profile from the Swiss Alps. Geochim Cosmochim Acta 40:945–951

Hoefs J, Touret J (1975) Fluid inclusion and carbon isotope study from Bamble granulites (South Norway). Contrib Mineral Petrol 52:165–174

Hoefs J, Faure G, Elliot DH (1980) Correlation of $\delta^{18}O$ and initial $^{87}Sr/^{86}Sr$ ratios in Kirkpatrick basalt on Mt Falla, Transantarctic Mountains. Contrib Mineral Petrol 75:199–203

Hoefs J, Coolen JJM, Touret J (1981) The sulfur and carbon isotope composition of scapolite-rich granulites. Contrib Mineral Petrol 78:332–336

Hoefs J, Müller G, Schuster AK (1982) Polymetamorphic relations in iron ores from the Iron Quadrangle, Brazil: The correlation of oxygen isotope variations with deformation history. Contrib Mineral Petrol 79:241–251

Hoering T (1956) Variations in the nitrogen isotope abundance. Proc 2nd Conf Nucl Process Geol Settings, p 85

Hoering T (1975) The biochemistry of the stable hydrogen isotopes. Carnegie Inst Washington Yearb 74:598

Hoering T, Ford HT (1960) The isotope effect in the fixation of nitrogen by azotobacter. Am Chem J 82:376

Hoering T, Parker PL (1961) The geochemistry of the stable isotopes of chlorine. Geochim Cosmochim Acta 23:186−199

Hoernes S, Friedrichsen HT (1978) Oxygen and hydrogen isotope study of the polymetamorphic area of the Northern Ötztal-Stubai Alps (Tyrol). Contrib Mineral Petrol 67:305−315

Hoernes S, Friedrichsen HT (1980) Oxygen and hydrogen isotopic composition of Alpine and Pre-Alpine minerals of the Swiss Central Alps. Contrib Mineral Petrol 72:19−32

Hoernes S, Hoffer E (1979) Equilibrium relations of prograde metamorphic mineral assemblages. A stable isotope study of rocks of the Damara Orogen, Namibia. Contrib Mineral Petrol 68:377−389

Hoernes S, Hoffer E (1985) Stable isotope evidence for fluid-present and fluid-absent metamorphism in metapelites from the Damara Orogen, Namibia. Contrib Mineral Petrol 90:322−330

Hoffman JH, Hodges RR, McElroy MB, Donahue TM, Kolpin M (1979) Composition and structure of the Venus atmosphere: results from Pioneer Venus. Science 205:49−52

Holm PM, Munksgaard NC (1982) Evidence for mantle metasomatism: an oxygen and strontium isotope study of the Vulsinian district, Central Italy. Earth Planet Sci Lett 60:376−389

Holser WT (1977) Catastrophic chemical events in the history of the ocean. Nature (London) 267:403−408

Holser WT, Kaplan IR (1966) Isotope geochemistry of sedimentary sulfates. Chem Geol 1:93−135

Holser WT, Kaplan IR, Sakai H, Zak I (1979) Isotope geochemistry of oxygen in the sedimentary sulfate cycle. Chem Geol 25:1−17

Horibe Y, Shigehara K, Takakuwa Y (1973) Isotope separation factor of carbon dioxide-water system and isotopic composition of atmospheric oxygen. J Geophys Res 78:2625−2629

Horibe Y, Shigehara K, Langway CJ (1985) Chemical and isotopic composition of air inclusions in a Greenland ice core. Earth Planet Sci Lett 73:207−210

Hubberten HW (1980) Sulfur isotope fractionation in the systems Pb−S, Cu−S and Ag−S. Geochem J 14:177−184

Hubberten HW (1984) Die Fraktionierung der Schwefelisotope bei der Entstehung und Veränderung der ozeanischen Kruste. Habil Univ Karlsruhe

Hubberten HW, Puchelt H (1980) Zur Schwefelisotopengeochemie basaltischer Gesteine. ZFI Mitt Leipzig 30:80−90

Hubberten HW, Nielsen H, Puchelt H (1975) The enrichment of ^{34}S in the solfataras of the Nea Kameni volcano, Santorini archipelago, Greece. Chem Geol 16:197−205

Hudson JD (1977) Stable isotopes and limestone lithification. J Geol Soc London 133:637−660

Hulston JR (1977) Isotope work applied to geothermal systems at the Institute of Nuclear Sciences, New Zealand. Geothermics 5:89−96

Hulston JR (1978) Methods of calculating isotopic fractionation in minerals. In: Stable isotopes in the earth sciences. DSIR Bull 220:211−219

Hulston JR, McCabe WJ (1962) Mass spectrometer measurements in the thermal areas of New Zealand, Part II. Carbon isotopic ratios. Geochim Cosmochim Acta 26:398−410

Hulston JR, Thode HG (1965) Variations in the S^{33}, S^{34} and S^{36} contents of meteorites and their relations to chemical and nuclear effects. J Geophys Res 70: 3475–3484

Ikin NP, Harmon RS (1983) A stable isotope study of serpentinization and metamorphism in the Highland Border Suite Scotland, U.K. Geochim Cosmochim Acta 47:153–167

Irwin H, Coleman M, Curtis C (1977) Isotopic evidence for the source of diagenetic carbonate during burial of organic-rich sediments. Nature (London) 269:209–213

Ivanov MV, Gogotova GI, Matrosov AG, Zyakun AM (1976) Fractionation of sulfur isotopes by phototrophic sulfur bacteria, Ectothiorhodospira slaposhnikovii Microbiology (Englisch translation) 45:655–659

James AT, Baker DR (1976) Oxygen isotope exchange between illite and water at 22 °C. Geochim Cosmochim Acta 40:235–239

James DE (1981) The combined use of oxygen and radiogenic isotopes as indicators of crustal contamination. Annu Rev Earth Planet Sci 9:311–344

Javoy M (1977) Stable isotopes and geothermometry. J Geol Soc 133:609–636

Javoy M (1980) $^{18}O/^{16}O$ and D/H ratios in high temperature peridotites. Coll Int CNRS 272:279–287

Javoy M, Pineau F (1986) Nitrogen isotopes in mantle materials. Terra Cognita 6: 103

Javoy M, Pineau F, Iiyama I (1978) Experimental determination of the isotopic fractionation between gaseous CO_2 and carbon dissolved in the tholeiitic magma. Contrib Mineral Petrol 67:35–39

Javoy M, Pineau F, Demaiffe D (1984) Nitrogen and carbon isotopic composition in the diamonds of Mbuji Mayi (Zaire). Earth Planet Sci Lett 68:399–412

Jeffrey AWA, Pflaum RC, Brooks JM, Sackett WM (1983) Vertical trends in particulate organic carbon $^{13}C/^{12}C$ ratios in the upper water column. Deep Sea Res 30:971–983

Jenkyns HC, Clayton CJ (1986) Black shales and carbon isotopes in pelagic sediments from the Tethyan Lower Jurassic. Sedimentology 33:87–106

Jensen ML, Nakai N (1962) Sulfur isotope meteorite standards, results and recommendations. In: Jensen ML (ed) Biogeochemistry of sulfur isotopes. NSF Symp Vol, p 31

Johns WD, Hoefs J (1985) Maturation of organic matter in Neogene sediments from the Aderklaa Oilfield, Vienna Basin, Austria. TMPM Tschermaks Mineralog Petrogr Mitt 34:143–158

Johnson SJ, Dansgaard W, Clausen HB, Langway CL (1972) Oxygen isotope profiles through the Antarctica and Greenland ice sheets. Nature (London) 235: 429–434

Jouzel J, Merlivat L, Roth E (1975) Isotopic study of hail. J Geophys Res 80: 5015–5030

Jouzel J, Merlivat L, Lorius C (1982) Deuterium excess in an East Antarctic ice core suggests higher relative humidity at the oceanic surface during the last glacial maximum. Nature (London) 299:688–691

Judy C, Meiman JR, Friedman I (1970) Deuterium variations in an annual snowpack. Water Resource Res 6:125

Junk G, Svec H (1958) The absolute abundance of the nitrogen isotopes in the atmosphere and compressed gas from various sources. Geochim Cosmochim Acta 14:234–243

Kajiwara Y, Krouse HR (1971) Sulfur isotope partitioning in metallic sulfide systems. Can J Earth Sci 8:1397–1408

Kanehira K, Yui S, Sakai H, Sasaki A (1973) Sulphide globules and sulphur isotopic ratios in the abyssal tholeiite from the Mid-Atlantic Ridge near 30°N latitude. Geochem J 7:89–96

Kanzaki T, Yoshida M, Momura M, Kakihana H, Ozawa T (1979) Boron isotopic
composition of fumarolic condensates and sassolites from Satsuma Iwo-jima,
Japan. Geochim Cosmochim Acta 43:1859–1863

Kaplan IR (1975) Stable isotopes as a guide to biogeochemical processes. Proc R
Soc London Ser B 189:183–211

Kaplan IR (1983) Stable isotopes of sulfur, nitrogen and deuterium in recent marine
environments. In: Stable isotopes in sedimentary geology. SEPM Short Course
10, Dallas

Kaplan IR, Hulston JR (1966) The isotopic abundance and content of sulfur in
meteorites. Geochim Cosmochim Acta 30:479–496

Kaplan IR, Rafter TA, Hulston JR (1960) Sulphur isotope variations in nature:
Application to some biogeochemical problems. N Z J Sci 3:338

Kaplan IR, Emery KO, Rittenberg SC (1963) The distribution and isotopic abundance of sulphur in recent marine sediments off Southern California. Geochim
Cosmochim Acta 27:297–332

Kaufmann RS, Long A, Bentley H, Davis S (1984) Natural chlorine isotope variations. Nature (London) 309:338–340

Kaufmann RS, Long A, Bentley H, Campbell DJ (1986) Chlorine isotope distribution of formation water in Texas and Louisiana. Bull Am Assoc Petrol Geol (in
press)

Kaye JA, Strobel DF (1983) Enhancement of heavy ozone in the earth's atmosphere? J Geophys Res 88:8447–8452

Keeling CD (1958) The concentration and isotopic abundance of atmospheric carbon dioxide in rural areas. Geochim Cosmochim Acta 13:322–334

Keeling CD (1960) The concentration and isotopic abundances of CO_2 in the atmosphere. Tellus 2:200

Keeling CD (1961) The concentration and isotopic abundances of carbon dioxide
in rural and marine air. Geochim Cosmochim Acta 24:277–298

Keith ML, Weber JN (1964) Carbon and oxygen isotopic composition of selected
limestones and fossils. Geochim Cosmochim Acta 28:1787–1816

Kneith ML, Anderson GM, Eichler R (1964) Carbon and oxygen isotopic composition of mollusk shells from marine and fresh-water environment. Geochim Cosmochim Acta 28:1757–1786

Kelly WC, Rye RO, Livnat A (1986) Saline minewaters of the Keweenaw Peninsula,
Northern Michigan: their nature, origin and relation to similar deep waters in
Precambrian crystalline rocks of the Canadian Shield. Am J Sci 286:281–308

Kelts K, McKenzie JA (1982) Diagenetic dolomite formation in Quaternary anoxic
diatomaceous muds of DSDP Leg 64, Gulf of California. Initial Rep DSDP 64:
553–569

Kemp ALW, Thode HG (1968) The mechanism of the bacterial reduction of sulphate and of sulphite from isotopic fractionation studies. Geochim Cosmochim
Acta 32:71–91

Kerrick R (1980) Archaen gold-bearing chemical sedimentary rocks and veins: A
synthesis of stable isotope and geochemical relations. Ontario Geol Survey Misc
Pap 97:144–175

Kerrick R, Latour TE, Willmore L (1984) Fluid participation in deep fault zones:
evidence from geological, geochemical and $^{18}O/^{16}O$ relations. J Geophys Res
89:4331–4343

Kerridge JF (1983) Isotopic composition of carbonaceous-chondrite kerogen: evidence for an interstellar origin of organic matter in meteorites. Earth Planet Sci Lett 64:186–200

Kerridge JF (1985) Carbon, hydrogen and nitrogen in carbonaceous chondrites: abundances and isotopic compositions in bulk samples. Geochim Cosmochim Acta 49:1707–1714

Kerridge JF, Haymon RM, Kastner M (1983) Sulfur isotope systematics at the 21°N site, East Pacific Rise. Earth Planet Sci Lett 66:91–100

Kharaka YK, Berry FAF, Friedman I (1974) Isotopic composition of oil-field brines from Kettleman North Dome, California and their geologic implications. Geochim Cosmochim Acta 37:1899–1908

Kieffer SW (1982) Thermodynamic and lattice vibrations of minerals: 5. Application to phase equilibria, isotopic fractionation and high-pressure thermodynamic properties. Rev Geophys Space Phys 20:827–849

Killingley JS (1983) Effects of diagenetic recrystallization on $^{18}O/^{16}O$ values of deep-sea sediments. Nature (London) 301:594–597

Kirschenbaum I, Smith JS, Crowell T, Graff J, McKee R (1947) Separation of the nitrogen isotopes by the exchange reaction between ammonia and solutions of ammonium nitrate. J Chem Phys 15:440–446

Kiyosu Y (1973) Sulfur isotopic fractionation among sphalerite, galena and sulfide ions. Geochem J 7:191–199

Kiyosu Y (1983) Hydrogen isotopic compositions of hydrogen and methane from some volcanic areas in northeastern Japan. Earth Planet Sci Lett 62:41–52

Klots CE, Benson BB (1963) Isotope effect in solution of oxygen and nitrogen in distilled water. J Chem Phys 38:890–892

Knauth LP, Epstein S (1975) Hydrogen and oxygen isotope ratios in silica from the JOIDES Deep Sea Drilling Project. Earth Planet Sci Lett 25:1–10

Knauth LP, Epstein S (1976) Hydrogen and oxygen isotope ratios in nodular and bedded cherts. Geochim Cosmochim Acta 40:1095–1108

Knauth LP, Lowe DR (1978) Oxygen isotope geochemistry of cherts from the Onverwacht group (3.4 billion years), Transvaal, South Africa, with implications for secular variations in the isotopic composition of chert. Earth Planet Sci Lett 41:209–222

Knoll AH, Hayes JM, Kaufman AJ, Swett K, Lambert IB (1986) Secular variation in carbon isotope ratios from Upper Proterozoic successions of Svalbard and East Greenland. Nature (London) 321:832–838

Kokubu N, Mayeda TK, Urey HC (1961) Deuterium content of minerals, rocks and liquid inclusions from rocks. Geochim Cosmochim Acta 21:247–256

Kolodny Y, Epstein S (1976) Stable isotope geochemistry of deep sea cherts. Geochim Cosmochim Acta 40:1195–1209

Kolodny Y, Gross S (1974) Thermal metamorphism by combustion of organic matter: Isotopic and petrological evidence. J Geol 82:489–506

Kolodny Y, Kerridge JF, Kaplan IR (1980) Deuterium in carbonaceous chondrites. Earth Planet Sci Lett 46:149–153

Kolodny Y, Luz B, Navon O (1983) Oxygen isotope variations in phosphate of biogenic apatites, I Fish bone apatite – rechecking the rules of the game. Earth Planet Sci Lett 64:393–404

Krichevsky MI, Sesler FD, Friedman I, Newell M (1961) Deuterium fractionation during molecular H_2 formation in a marine pseudomonad. J Biol Chem 236:2520

Kroopnick P (1974a) Correlations between ^{13}C and ΣCO_2 in surface waters and atmospheric CO_2. Earth Planet Sci Lett 22:397–403

Kroopnick P (1974b) The dissolved $O_2 - CO_2 - {}^{13}C$ system in the eastern equatorial Pacific. Deep Sea Res 21:211−227

Kroopnick P (1975) Respiration, photosynthesis, and oxygen isotope fractionation in oceanic surface water. Limnol Oceanogr 20:988−992

Kroopnick P (1985) The distribution of ${}^{13}C$ of ΣCO_2 in the world oceans. Deep Sea Res 32:57−84

Kroopnick P, Craig H (1972) Atmospheric oxygen: isotopic composition and solubility fractionation. Science 175:54−55

Kroopnick P, Craig H (1976) Oxygen isotope fractionation in dissolved oxygen in the deep sea. Earth Planet Sci Lett 32:375−388

Kroopnick P, Weiss RF, Craig H (1972) Total CO_2, ${}^{13}C$ and dissolved oxygen-${}^{18}O$ at Geosecs II in the North Atlantic. Earth Planet Sci Lett 16:103−110

Krouse HR (1977) Sulfur isotope studies and their role in petroleum exploration. J Geochem Explor 7:189−211

Krouse HR (1980) Sulphur isotopes in our environment. In: Fritz P, Fontes JCh (eds) Handbook of environmental isotope geochemistry, vol 1. Elsevier Sci Publ Co, Amsterdam, pp 435−471

Krouse HR, Thode HG (1962) Thermodynamic properties and geochemistry of isotopic compounds of selenium. Can J Chem 40:367

Kung CC, Clayton RN (1978) Nitrogen abundances and isotopic compositions in stony meteorites. Earth Planet Sci Lett 38:421−435

Kuroda Y, Suzuoki T, Matuo S, Kanisawa S (1974) D/H fractionation of coexisting biotite and hornblende in some granitic rock masses. J Jpn Assoc Mineral Petrol Econ Geol 69:95

Kuroda Y, Suzuoki T, Matsuo S, Aoki K (1975) D/H ratios of the coexisting phlogopite and richterite from mica nodules and a peridotite in South African kimberlites. Contrib Mineral Petrol 52:315−318

Kuroda Y, Suzuoki T, Matsuo S (1977) Hydrogen isotope composition of deep-seated water. Contrib Mineral Petrol 60:311−315

Kyser TK, O'Neil JR (1984) Hydrogen isotope systematics of submarine basalts. Geochim Cosmochim Acta 48:2123−2134

Kyser TK, O'Neil JR, Carmichael ISE (1981) Oxygen isotope thermometry of basic lavas and mantle nodules. Contrib Mineral Petrol 77:11−23

Kyser TK, O'Neil JR, Carmichael ISE (1982) Genetic relations among basic lavas and mantle nodules. Contrib Mineral Petrol 81:88−102

Lambert SJ, Epstein S (1980) Stable isotope investigations of an active geothermal system in Valles Caldera, Jemez Mountains, New Mexico. J Volcanol Geotherm Res 8:111−129

Lancet MS, Anders E (1970) Carbon isotope fractionation in the Fischer-Tropsch synthesis and in meteorites. Science 170:980−982

Land LS (1980) The isotopic and trace element geochemistry of dolomite: the state of the art. In: Concepts and models of dolomitization. Soc Econ Paleontol Min Spec Publ 28:87−110

Land LS, Dutton SP (1978) Cementation of a Pennsylvanian deltaic sandstone: isotope data. J Sediment Petrol 48:1167−1176

Lane GA, Dole M (1956) Fractionation of oxygen isotopes during respiration. Science 123:574−576

Lattanzi P, Rye OM, Rice JM (1980) Behavior of ${}^{13}C$ and ${}^{18}O$ in carbonates during contact metamorphism at Marysville, Montana. Am J Sci 280:890−906

Lawrence JR, Taylor HP (1971) Deuterium and oxygen-18 correlation: Clay minerals and hydroxides in Quaternary soils compared to meteoric waters. Geochim Cosmochim Acta 35:993−1003

Lawrence JR, Taylor HP (1972) Hydrogen and oxygen isotope systematics in weathering profiles. Geochim Cosmochim Acta 36:1377−1393

Lawrence JR, Gieskes JM, Broecker WS (1975) Oxygen isotope and cation composition of DSDP pore waters and the alteration of layer II basalts. Earth Planet Sci Lett 27:1−10

Leguy C, Rindsberger M, Zangwil A, Issar A, Gat J (1983) The relation between the ^{18}O and deuterium contents of rain water in the Negev desert and air mass trajectories. Isotope Geosci 1:205−218

Letolle R (1980) Nitrogen-15 in the natural environment. In: Fritz P, Fontes JCh (eds) Handbook of environmental isotope geochemistry. Elsevier, Amsterdam, pp 407−433

Lewan MD (1983) Effects of thermal maturation on stable carbon isotopes as determined by hydrous pyrolysis of Woodford shale. Geochim Cosmochim Acta 47:1471−1480

Lewis RS, Anders E, Wright IP, Norris SJ, Pillinger CT (1983) Isotopically anomalous nitrogen in primitive meteorites. Nature (London) 305:767−771

Lipman PW, Friedman I (1975) Interaction of meteoric water with magma: An oxygen isotope study of ash-flow sheets from Southern Nevada. Geol Soc Am Bull 86:695−702

Liu KK, Epstein S (1984) The hydrogen isotope fractionation between kaolinite and water. Isotope Geosci 2:335−350

Lloyd MR (1967) Oxygen-18 composition of oceanic sulfate. Science 156:1228−1231

Lloyd MR (1968) Oxygen isotope behavior in the sulfate-water system. J Geophys Res 73:6099−6110

Longinelli A (1965) Oxygen isotopic composition of orthophosphate from shells of living marine organisms. Nature (London) 207:716−719

Longinelli A (1966) Ratios of O-18/O-16 in phosphate and carbonate from living and fossil marine organisms. Nature (London) 211:923−927

Longinelli A, Cortecci G (1970) Isotopic abundance of oxygen and sulfur in sulfate ions from river water. Earth Planet Sci Lett 7:376−380

Longinelli A, Craig H (1967) Oxygen-18 variations in sulfate ions in sea-water and saline lakes. Science 156:56−59

Longinelli A, Edmond JM (1983) Isotope geochemistry of the Amazon basin. A reconnaissance. J Geophys Res 88:3703−3717

Longinelli A, Nuti S (1973) Revised phosphate-water isotopic temperature scale. Earth Planet Sci Lett 19:373−376

Longstaffe FJ (1979) The oxygen-isotope geochemistry of Archaen granitoids. In: Barker F (ed) Trondhjemites, dacites and related rocks. Elsevier, New York Amsterdam

Longstaffe FJ (1982) Stable isotopes in the study of granitic pegmatites and related rocks. Mineral Assoc Can Short Course Handb 8:373−404

Longstaffe FJ (1983) Diagenesis 4. Stable isotope studies of diagenesis in clastic rocks. Geosci Can 10:43−58

Longstaffe FJ, Schwarcz HP (1977) $^{18}O/^{16}O$ of Archean clastic metasedimentary rocks: a petrogenetic indicator for Archean gneisses? Geochim Cosmochim Acta 41:1303−1312

Lorius C, Jouzel J, Ritz C, Merlivat L, Barkov NI, Korotkevich YS, Kotlyakov VM (1985) A 150.000 year climatic record from Antarctic ice. Nature (London) 316:591−596

Lowenstam HA (1961) Mineralogy, O-18/O-16 ratios and strontium and magnesium contents of recent and fossil brachiopods and their bearing on the history of the oceans. J Geol 69:241−260

Lyon GL (1974a) Isotopic analyses of gas from the Carico Trench sediments. In: Kaplan IR (ed) Natural gases in marine sediments. Plenum, New York, pp 91–97

Lyon GL (1974b) Geothermal gases. In: Kaplan IR (ed) Natural gases in marine sediments. Plenum, New York, p 41

Maass I, Wand U, Kaemmel T (1978) Die Kohlenstoff-Isotopenzusammensetzung hochinkohlter organischer Substanzen. Z Angew Geol 24:109–120

Macko SE, Estep MLF, Hare PE, Hoering TC (1983) Stable nitrogen and carbon isotopic composition of individual amino acids isolated from cultured microorganisms. Carnegie Inst Washington Yearb 82:404–410

Macnamara J, Thode HG (1950) Comparison of the isotopic constitution of terrestrial and meteoritic sulphur. Phys Rev 78:307

Magaritz M, Heller J (1980) A desert migration indicator – oxygen isotopic composition of land snail shells. Paleogeogr Paleoclimatol Paleoecol 32:153–162

Magaritz M, Taylor HP (1974) Oxygen and hydrogen isotope studies of serpentinization in the Troodos ophiolite complex, Cyprus. Earth Planet Sci Lett 23:8–14

Magaritz M, Taylor HP (1976) $^{18}O/^{16}O$ and D/H along a 500 km traverse across the Coast Range batholith and its country rocks, Central British Columbia. Can J Earth Sci 13:1514–1536

Magaritz M, Taylor HP (1986) Oxygen 18/Oxygen 16 and D/H studies of plutonic granitic and metamorphic rocks across the Cordilleran batholiths of Southern British Columbia. J Geophys Res 91:2193–2217

Magaritz M, Anderson RY, Holser WT, Saltzman ES, Garber J (1983) Isotope shifts in the Late Permian of the Delaware basin, Texas, precisely timed by warved sediments. Earth Planet Sci Lett 66:111–124

Mariotti A (1983) Atmospheric nitrogen is a reliable standard for natural abundance measurements. Nature (London) 303:685–687

Mariotti A, Germon JC, Hubert P, Kaiser P, Letolle R, Tardieux P (1981) Experimental determination of nitrogen kinetic isotope fractionation: some principles, illustration for the denitrification and nitrification processes. Plant Soil 62:413–430

Mariotti A, Germon JC, Leclerc A, Catroux G, Letolle R (1982) Experimental determination of kinetic isotope fractionation of nitrogen isotopes during denitrification. In: Schmidt HL, Förster H, Heinzinger K (eds) Stable isotopes. Elsevier, New York Amsterdam

Mariotti A, Lancelot C, Billen G (1984) Natural isotopic composition of nitrogen as a tracer of origin for suspended organic matter in the Scheldt estuary. Geochim Cosmochim Acta 48:549–555

Marowsky G (1969) Schwefel-, Kohlenstoff- und Sauerstoffisotopenuntersuchungen am Kupferschiefer als Beitrag zur genetischen Deutung. Contrib Mineral Petrol 22:290–334

Marshall B, Taylor BE (1981) Origin of hydrothermal fluids responsible for gold deposition, Alleghany district, Sierra County, California. U S Geol Surv Open-File Rep 81-355:280–293

Marumo K, Nagasawa K, Kuroda Y (1980) Mineralogy and hydrogen isotope geochemistry of clay minerals in Ohnuma geothermal area, northeastern Japan. Earth Planet Sci Lett 47:255–262

Matsubaya O, Sakai H (1973) Oxygen and hydrogen isotopic study on the water of crystallization of gypsum from the Kuroko-type mineralization. Geochem J 7:153–165

Matsuhisa Y (1979) Oxygen isotopic compositions of volcanic rocks from the east Japan island arcs and their bearing on petrogenesis. J Volcanic Geotherm Res 5:271–296

Matsuhisa Y, Goldsmith JR, Clayton RN (1979) Oxygen isotope fractionation in the systems quartz-albite-anorthite-water. Geochim Cosmochim Acta 43:1131–1140

Matsuo S, Friedman I, Smith GI (1972) Studies of Quaternary saline lakes. I. Hydrogen isotope fractionation in saline minerals. Geochim Cosmochim Acta 36:427–435

Matter A, Douglas RG, Perch-Nielsen K (1975) Fossil preservation, geochemistry and diagenesis of pelagic carbonates from Shatsky Rise, northeast Pacific. Initial Rep DSDP 32:891–922

Mattey DP, Carr RH, Wright IP, Pillinger CT (1984) Carbon isotopes in submarine basalts. Earth Planet Sci Lett 70:196–206

Matthews A, Beckinsale RD (1979) Oxygen isotope equilibration systematics between quartz and water. Am Mineral 64:232–240

Matthews A, Katz A (1977) Oxygen isotope fractionation during the dolomitization of calcium carbonate. Geochim Cosmochim Acta 41:1431–1438

Matthews A, Kolodny Y (1978) Oxygen isotope fractionation in decarbonation metamorphism: the Mottled Zone event. Earth Planet Sci Lett 39:179–192

Matthews A, Schliestedt M (1984) Evolution of the blueschist and greenschist facies rocks of Sifnos, Cyclades, Greece. A stable isotope study of subduction-related metamorphism. Contrib Mineral Petrol 88:150–163

Matthews A, Goldsmith JR, Clayton RN (1983a) Oxygen isotope fractionation involving pyroxenes: the calibration of mineral-pair geothermometers. Geochim Cosmochim Acta 47:631–644

Matthews A, Goldsmith JR, Clayton RN (1983b) Oxygen isotope fractionation between zoisite and water. Geochim Cosmochim Acta 47:645–654

Matthews A, Goldsmith JR, Clayton RN (1983c) On the mechanics and kinetics of oxygen isotope exchange in quartz and feldspars at elevated temperatures and pressures. Geol Soc Am Bull 94:396–412

Mauersberger K (1981) Measurement of heavy ozone in the stratosphere. Geophys Res Lett 8:935–937

Mayeda TK, Goldsmith JR, Clayton RN (1986) Oxygen isotope fractionation at high temperature. Terra Cognita 6:261

Mazany T, Lerman JC, Long A (1980) Carbon-13 in tree-ring cellulose as an indicator of past climates. Nature (London) 287:432–435

McCorkle DC, Emerson SR, Quay P (1985) Carbon isotopes in marine porewaters. Earth Planet Sci Lett 74:13–26

McCrea JM (1950) The isotopic chemistry of carbonates and a paleotemperature scale. J Chem Phys 18:849–857

McCready RGL (1975) Sulphur isotope fractionation by Desulfovibrio and Desulfotomaculum species. Geochim Cosmochim Acta 39:1395–1401

McCready RGL, Kaplan IR, Din GA (1974) Fractionation of sulfur isotopes by the yeast Saccharomyces cerevisiae. Geochim Cosmochim Acta 38:1239–1253

McEwing CF, Rees CE, Thode HG (1983) Sulphur isotope ratios in the Canyon Diablo metal spheroids. Meteoritics 18:185–198

McKeegan KD, Walker RM, Zinner E (1985) Ion microprobe isotopic measurements of individual interplanetary dust particles. Geochim Cosmochim Acta 49:1971–1987

McKenzie J (1984) Holocene dolomitization of calcium carbonate sediments from the coastal sabkhas of Abu Dhabi, U.A.E.: A stable isotope study. J Geol 89:185–198

McKenzie J, Kelts KR (1979) A study of interpillow limestones from the M-zero anomaly, DSDP Leg 51, Site 417 D. Initial Rep DSDP 51—53, 2:753—769

McKenzie J, Bernoulli D, Garrison RE (1978) Lithification of pelagic-hemipelagic sediments at DSDP Site 372: oxygen isotope alteration with diagenesis. Initial Rep DSDP 41:473—478

McKenzie WF, Truesdell AH (1977) Geothermal reservoir temperatures estimated from the oxygen isotope compositions of dissolved sulfate and water from hot springs and shallow drillholes. Geothermics 5:51

McKinney CR, McCrea JM, Epstein S, Allen HA, Urey HC (1950) Improvements in mass spectrometers for the measurement of small differences in isotope abundance ratios. Rev Sci Instrum 21:724

McNaughton NJ, Borthwick J, Fallick AE, Pillinger CT (1981) Deuterium/hydrogen ratios in unequilibrated ordinary chondrites. Nature (London) 294:639—641

McNaughton NJ, Fallick AE, Pillinger CT (1982) Deuterium enrichments in type 3 ordinary chondrites. Proc 13th Lunar Planet Sci Conf, J Geophys Res 87:A297—A302 (Suppl)

Mekhtiyeva VL, Pankina GR (1968) Isotopic composition of sulfur in aquatic plants and dissolved sulfates. Geochemistry 5:624

Mekhtiyeva VL, Pankina GR, Gavrilov EYa (1976) Distribution and isotopic composition of forms of sulfur in water animals and plants. Geochem Int 13:82

Melander L (1960) Isotope effects on reaction rates. Ronald, New York

Melander L, Saunders WH (1980) Reaction rates of isotopic molecules. Wiley and Sons, New York

Melton CE, Giardini AA (1974) The composition and significance of gas released from natural diamonds from Africa and Brazil. Am Mineral 59:775—782

Mensing TM, Faure G, Jones LM, Bowman JR, Hoefs J (1984) Petrogenesis of the Kirkpatrick basalt, Solo Nunatek, Northern Victoria Land, Antarctica based on isotopic composition of strontium, oxygen and sulfur. Contrib Mineral Petrol 87:101—108

Michaelis J, Usdowski E, Menschel G (1985) Partitioning of ^{13}C and ^{12}C on the degassing of CO_2 and the precipitation of calcite: Rayleigh-type fractionation and a kinetic model. Am J Sci 285:318—327

Michard-Vitrac A, Albarede F, Dupuis C, Taylor HP (1980) The genesis of Variscan (Hercynian) plutonic rocks. Inferences from Sr, Pb and O studies on the Maladeta igneous complex, Central Pyrenees, Spain. Contrib Mineral Petrol 72:57—72

Milliken KL, Land LS, Loucks RG (1981) History of burial diagenesis determined from isotopic geochemistry, Frio Formation, Brazoria County, Texas. Am Assoc Petrol Geol Bull 65:1397—1413

Mills GA, Urey HC (1940) The kinetics of isotopic exchange between carbon dioxide, bicarbonate ion, carbonate ion and water. J Am Chem Soc 62:1019

Minson DJ, Ludlow MM, Throughton JH (1975) Differences in natural carbon isotope ratios of milk and hair from cattle grazing tropical and temperature pastures. Nature (London) 256:602

Miyake Y, Wada E (1971) The isotope effect on the nitrogen in biochemical oxidation-reduction reactions. Res Oceanogr Works Jpn 11:1

Mizutani Y, Rafter TA (1969) Isotopic composition of sulphate in rain water, Gracefield, New Zealand. N Z J Sci 12:69

Moldovanyi EP, Lohmann KC (1984) Isotopic and petrographic record of phreatic diagenesis: Lower Cretaceous Sligo and Cupido Formations. J Sediment Petrol 54:972—985

Monson KD, Hayes JM (1982) Carbon isotopic fractionation in the biosynthesis of bacterial fatty acids. Ozonolysis of unsaturated fatty acids as a means of determining the intramolecular distribution of carbon isotopes. Geochim Cosmochim Acta 46:139–149

Monster J (1972) Homogeneity of sulfur and carbon isotope ratios, $^{34}S/^{32}S$ and $^{13}C/^{12}C$, in petroleum. Bull Am Assoc Petrol Geol 56:941–949

Monster J, Anders E, Thode HG (1965) $^{34}S/^{32}S$ ratios for the different forms of sulphur in the Orgueil meteorite and their mode of formation. Geochim Cosmochim Acta 29:773–779

Monster J, Appel PWU, Thode HG, Schidlowski M, Carmichael CM, Bridgwater D (1979) Sulfur isotope studies in early Archean sediments from Isua, West Greenland: implications for the antiquity of bacterial sulfate reduction. Geochim Cosmochim Acta 43:405–413

Mook WG, van der Hoek S (1983) The N_2O correction in the carbon and oxygen isotopic analyses of atmospheric CO_2. Isotope Geosci 1:237–242

Mook WG, Bommerson JC, Staverman WH (1974) Carbon isotope fractionation between dissolved bicarbonate and gaseous carbon dioxide. Earth Planet Sci Lett 22:169–176

Mook WG, Koopman M, Carter AF, Keeling CD (1983) Seasonal, latitudinal and secular variations in the abundance and isotopic ratios of atmospheric carbon dioxide. I. Results from land stations. J Geophys Res 88:10915–10933

Moore JG (1970) Water content of basalt erupted on the ocean floor. Contr Mineral Petrol 28:272–279

Moore JG, Bachelder JN, Cunningham CG (1977) CO_2-filled vesicles in mid-ocean basalt. J Volcanol Geotherm Res 2:309–327

Morikiyo T (1984) Carbon isotopic study on coexisting calcite and graphite in the Ryoke metamorphic rocks, northern Kiso district, central Japan. Contrib Mineral Petrol 87:251–259

Muehlenbachs K, Byerly G (1982) ^{18}O enrichment of silicic magmas caused by crystal fractionation at the Galapagos Spreading Center Contrib Mineral Petrol 79:76–79

Muehlenbachs K, Clayton RN (1972a) Oxygen isotope studies of fresh and weathered submarine basalts. Can J Earth Sci 8:1591–1594

Muehlenbachs K, Clayton RN (1972b) Oxygen isotope geochemistry of submarine greenstones. Can J Earth Sci 9:471–478

Muehlenbachs K, Clayton RN (1976) Oxygen isotope composition of the oceanic crust and its bearing on seawater. J Geophys Res 81:4365–4369

Muehlenbachs K, Kushiro I (1974) Oxygen isotope exchange and equilibrium of silicates with CO_2 or O_2. Geophys Lab Yearb 73:232

Murata KJ, Friedman I, Gleason JD (1977) Oxygen isotope relations between diagenetic silica minerals in Monterey Shale, Temblor Range, California. Am J Sci 277:259–272

Murozumi M (1961) Bull Geol Surv Jpn 12:183

Nabelek PI, O'Neil JR, Papike JJ (1983) Vapor phase exsolution as a controlling factor in hydrogen isotope variation in granitic rocks. The Notch Peak granitic stock, Utah. Earth Planet Sci Letters 66:137–150

Nabelek PI, Labotka TC, O'Neil JR, Papike JJ (1984) Constrasting fluid/rock in-interaction between the Notch Peak granitic intrusion and argillites and limestones in western Utah: evidence from stable isotopes and phase assemblages. Contrib Mineral Petrol 86:25–43

Nagy KL, Parmentier EM (1982) Oxygen isotope exchange at an igneous intrusive contact. Earth Planet Sci Lett 59:1–10

Nakai N, Jensen ML (1964) The kinetic isotope effect in the bacterial reduction and oxidation of sulfur. Geochim Cosmochim Acta 28:1893–1912

Navon O, Wasserburg GJ (1985) Self-shielding in O_2 – a possible explanation for oxygen isotope anomalies in meteorites. Earth Planet Sci Lett 73:1–16

Newton RC, Smith JV, Windley BF (1980) Carbonic metamorphism, granulites and crustal growth. Nature (London) 288:45–50

Nielsen H (1965) S-Isotope im marinen Kreislauf und das $\delta^{34}S$ der früheren Meere. Geol Rundsch 55:160–172

Nielsen H (1972) Sulphur isotopes and the formation of evaporite deposits. In: Geology of saline deposits, Proc Hannover Symp 1968, Earth Sci 7:91, UNESCO 1972

Nielsen H (1974) Isotopic composition of the major contributors to atmospheric sulfur. Tellus 26:213

Nielsen H (1978) Sulfur isotopes. In: Wedepohl KH (ed) Handbook of geochemistry. Springer, Berlin Heidelberg New York

Nielsen H (1979) Sulfur isotopes. In: Jäger E, Hunziker J (eds) Lectures in isotope geology. Springer, Berlin Heidelberg New York, pp 283–312

Nielsen H (1985a) Isotope in der Lagerstättenforschung. In: Bender F (ed) Angewandte Geowissenschaften. Enke, Stuttgart

Nielsen H (1985b) Sulfur isotopes in stratabound mineralizations of Central Europe. Geol Jahrb D70:225–262

Nielsen H, Ricke W (1964) S-Isotopenverhältnisse von Evaporiten aus Deutschland. Ein Beitrag zur Kenntnis von $\delta^{34}S$ im Meerwasser Sulfat. Geochim Cosmochim Acta 28:577–591

Nier AO (1950) A redetermination of the relative abundances of the isotopes of carbon, nitrogen, oxygen, argon and potassium. Phys Rev 77:789

Nier AO, Ney EP, Inghram MG (1947) A null method for the comparison of two ion currents in a mass spectrometer. Rev Sci Instrum 18:294

Nier AO, McElroy MB, Yung YL (1976) Isotopic composition of the Martian atmosphere. Science 194:68–70

Nissenbaum A (1974) Deuterium content of humic acids from marine and non-marine environments. Mar Chem 2:59

Nissenbaum A, Kaplan IR (1972) Chemical and isotopic evidence for the in situ origin of marine humic substances. Limnol Oceanogr 17:570–582

Nissenbaum A, Schallinger KM (1974) The distribution of stable carbon isotopes ($^{13}C/^{12}C$) in fractions of soil organic matter. Geoderma 11:137–145

Nissenbaum A, Presley BJ, Kaplan IR (1972) Early diagenesis in a reducing Fjord, Saanich Inlet, British Columbia. I: Chemical and isotopic changes in major components of interstitial water. Geochim Cosmochim Acta 36:1007–1027

Nitzsche HM, Stiehl G (1984) Untersuchungen zur Isotopenfraktionierung des Stickstoffs in den Systemen Ammonium/Ammoniak und Nitrid/Stickstoff. ZFI Mitt 84:283–291

Nomura M, Kanzaki T, Ozawa T, Okamoto M, Kakihana H (1982) Boron isotopic composition of fumarolic condensates from some volcanoes in Japanese island arcs. Geochim Cosmochim Acta 46:2403–2406

Northrop DA, Clayton RN (1966) Oxygen isotope fractionations in systems containing dolomite. J Geol 74:174–196

Nriagu J (1974) Fractionation of sulfur isotopes by sediment adsorption of sulfate. Earth Planet Sci Lett 22:366–370

Ohmoto H (1972) Systematics of sulfur and carbon isotopes in hydrothermal ore deposits. Econ Geol 67:551–578

Ohmoto H, Lasaga AC (1982) Kinetics of reactions between aqueous sulfates and sulfides in hydrothermal systems. Geochim Cosmochim Acta 46:1727–1745

Ohmoto H, Rye RO (1979) Isotopes of sulfur and carbon. In: Geochemistry of hydrothermal ore deposits, 2nd edn. Holt Rinehart and Winston, New York

Ohmoto H, Skinner BJ (1983) The Kuroko and related deposits. Econ Geol Monograph 5

O'Leary MH (1981) Carbon isotope fractionation in plants. Phytochemistry 20: 553–567

O'Neil JR (1968) Hydrogen and oxygen isotopic fractionation between ice and water. J Phys Chem 72:3683

O'Neil JR (1977) Stable isotopes in mineralogy. Phys Chem Mineral 2:105

O'Neil JR, Clayton RN (1964) Oxygen isotope thermometry. In: Craig H, Miller SL, Wasserburg GJ (eds) Isotopic and cosmic chemistry. North Holland Publ Co, Amsterdam, pp 157–168

O'Neil JR, Epstein S (1966) A method for oxygen isotope analysis of milligram quantities of water and some of its applications. J Geophys Res 71:4955–4961

O'Neil JR, Hay RL (1973) $^{18}O/^{16}O$ ratios in cherts associated with the saline lake deposits of East Africa. Earth Planet Sci Lett 19:257–266

O'Neil JR, Kharaka YK (1976) Hydrogen and oxygen isotope exchange reactions between clay minerals and water. Geochim Cosmochim Acta 40:241–246

O'Neil JR, Silberman ML (1974) Stable isotope relations in epithermal Au–Ag deposits. Econ Geol 69:902–909

O'Neil JR, Taylor HP (1967) The oxygen isotope and cation exchange chemistry of feldspars. Am Mineral 52:1414–1437

O'Neil JR, Taylor HP (1969) Oxygen isotope equilibrium between muscovite and water. J Geophys Res 74:6012–6022

O'Neil JR, Clayton RN, Mayeda TK (1969) Oxygen isotope fractionation in divalent metal carbonates. J Chem Phys 51:5547

O'Neil JR, Adami LH, Epstein S (1975) Revised value for the ^{18}O fractionation between CO_2 and H_2O at 25 °C. J Res US Geol Surv 3:623

Onuma N, Clayton RN, Mayeda TK (1970a) Apollo 11 rocks: Oxygen isotope fractionation between minerals and an estimate of the temperature of formation. Proc Apollo 11 Lunar Sci Conf Geochim Cosmochim Acta Suppl 2:1429–1434

Onuma N, Clayton RN, Mayeda TK (1970b) Oxygen isotope fractionation between minerals and an estimate of the temperature of formation. Science 167:536–538

Orr WL (1974) Changes in sulfur content and isotopic ratios of sulfur during petroleum maturation. Study of Big Horn Basin Paleozoic oils. Am Assoc Petrol Geol Bull 58:2295–2318

Osmond CB, Ziegler H (1975) Schwere Pflanzen und leichte Pflanzen: Stabile Isotope im Photosynthesestoffwechsel und in der biochemischen Ökologie. Naturwiss Rundsch 28:323

Owen T, Biemann K, Rushneck DR, Biller JE, Howarth DW, Lafleur AL (1977) The composition of the atmosphere at the surface of Mars. J Geophys Res 82:4635–4639

Panichi C, Gonfiantini R (1978) Environmental isotopes in geothermal studies. Geothermics 6:143–161

Panichi C, Ferrara GC, Gonfiantini R (1977) Isotope geochemistry in the Larderello geothermal fields. Geothermics 5:81–88

Parada CB, Long A, Davis SN (1983) Stable isotopic composition of soil carbon dioxide in the Tuscon basin, Arizona, USA. Isotope Geosci 1:219–236

Pardue JW, Scalan RS, van Baalen C, Parker PL (1976) Maximum carbon isotope fractionation in photosynthesis by blue-green algae and a green alga. Geochim Cosmochim Acta 40:309–312

Park R, Epstein S (1960) Carbon isotope fractionation during photosynthesis. Geochim Cosmochim Acta 21:110–126

Parker PL (1964) The biogeochemistry of the stable isotopes of carbon in a marine bay. Geochim Cosmochim Acta 28:1155−1164

Paterson WSB, Koerner RM, Fisher D, Johnsen SJ, Clausen HB, Dansgaard W, Bucher P, Oeschger H (1977) An oxygen isotope climatic record from the Devon Island Ice Cap, Arctic Canada. Nature (London) 266:508−511

Penzias AA (1980) Nuclear processing and isotopes in the galaxy. Science 208: 663−669

Perry EA, Gieskes JM, Lawrence JR (1976) Mg, Ca and $^{18}O/^{16}O$ exchange in the sediment-pore water system, Hole 149, DSDP. Geochim Cosmochim Acta 40: 413−423

Perry EC (1967) The oxygen isotope chemistry of ancient cherts. Earth Planet Sci Lett 3:62−66

Perry EC, Tan FC (1972) Significance of oxygen and carbon isotope variations in early Precambrian cherts and carbonate rocks of Southern Africa. Bull Geol Soc Am 83:647−664

Peters KE, Rohrback BG, Kaplan IR (1981) Carbon and hydrogen stable isotope variations in kerogen during laboratory-simulated thermal maturation. Am Assoc Petrol Geol Bull 65:501−508

Pillinger CT (1984) Light element stable isotopes in meteorites − from grams to picograms. Geochim Cosmochim Acta 48:2739−2768

Pinckney DM, Rye RO (1972) Variation of $^{18}O/^{16}O$, $^{13}C/^{12}C$, texture and mineralogy in altered limestone in the Hill Mine, Cave-in-District, Illinois. Econ Geol 67:1−18

Pineau F, Javoy M (1983) Carbon isotopes and concentrations in mid-ocean ridge basalts. Earth Planet Sci Lett 62:239−257

Pineau F, Javoy M, Bottinga Y (1976a) $^{13}C/^{12}C$ ratios of rocks and inclusions in popping rocks of the Mid-Atlantic Ridge and their bearing on the problem of isotopic composition of deep-seated carbon. Earth Planet Sci Lett 29:413−421

Pineau F, Javoy M, Hawkins JW, Craig H (1976b) Oxygen isotope variations in marginal basin and ocean-ridge basalts. Earth Planet Sci Lett 28:299−307

Pisciotti KA, Mahoney JJ (1981) Isotopic survey of diagenetic carbonates, DSDP Leg 63. Initial Rep DSDP 63:595−609

Presley BJ, Kaplan IR (1968) Changes in dissolved sulfate, calcium and carbonate from interstitial water of near-shore sediments. Geochim Cosmochim Acta 32: 1037−1048

Price FT, Shieh YN (1979) The distribution and isotopic composition of sulfur in coals from the Illinois Basin. Econ Geol 74:1445−1461

Prombo CA, Clayton RN (1985) A striking nitrogen isotope anomaly in the Bencubbin and Weatherford meteorites. Science 230:935−937

Puchelt H, Hubberten HW (1980) Preliminary results of sulfur isotope investigations on deep sea drilling project cores from legs 52 and 53. Initial Rep DSDP 51, 52, 53, Part 2:1145−1148

Puchelt H, Sabels BR, Hoering TC (1971) Preparation of sulfur hexafluoride for isotope geochemical analysis. Geochim Cosmochim Acta 35:625−628

Rabinowitch EI (1945) Photosynthesis and related processes, vol I. Interscience, New York, p 10

Rabinovitch AL, Grinenko VA (1979) Sulfate sulfur isotope ratios for USSR river water. Geochemistry 16, 2:68−79

Rafter TA (1957) Sulphur isotopic variations in nature, P 1: The preparation of sulphur dioxide for mass spectrometer examination. N Z J Sci Tech B38:849

Raiswell R (1982) Pyrite tecture, isotopic composition and the availability of iron. Am J Sci 282:1244−1236

Rakestraw NM, Rudd DP, Dole M (1951) Isotopic composition of oxygen in air dissolved in Pacific Ocean water as a function of depth. J Am Chem Soc 73: 2976

Rashid K, Krouse HR, McCready RGL (1978) Selenium isotope fractionation during bacterial selenite reduction. In: Short Pap 4th Int Conf Geochronol Cosmochronol Isotope Geol, p 347

Rau GH, Sweeney RE, Kaplan IR (1982) Plankton $^{13}C/^{12}C$ ratio changes with latitude: differences between northern and southern oceans. Deep Sea Res 29: 1035–1039

Rayleigh JWS (1896) Theoretical considerations respecting the separation of gases by diffusion and similar processes. Philos Mag 42:493

Redding CE, Schoell M, Monin JC, Durand B (1980) Hydrogen and carbon isotopic composition of coals and kerogen. In: Douglas AG, Maxwell JR (eds) Phys Chem Earth 12:711–723

Redfield AC, Friedman I (1965) Factors affecting the distribution of deuterium in the ocean. In: Symp Rhode Island Occ Publ 3:149

Rees CE (1970) The sulphur isotope balance of the ocean: an improved model. Earth Planet Sci Lett 7:366–370

Rees CE (1978) Sulphur isotope measurements using SO_2 and SF_6. Geochim Cosmochim Acta 42:383–389

Rees CE, Thode HG (1966) Selenium isotope effects in the reduction of sodium selenite and of sodium selenate. Can J Chem 44:419

Rees CE, Jenkins WJ, Monster J (1978) The sulphur isotopic composition of ocean water sulphate. Geochim Cosmochim Acta 42:377–381

Reibach PH, Benedict CR (1977) Fractionation of stable carbon isotopes by PEP carboxylase from C_4 plants. Plant Physiol 59:564–568

Rex RW, Syers JK, Jackson JK, Clayton RN (1969) Eolian origin of quartz in soils of Hawaiian Islands and in Pacific pelagic sediments. Science 163:277–279

Rice CM, Harmon RS, Shepherd TJ (1985) Central City, Colorado: The upper part of an alkaline porphyry molybdenum system. Econ Geol 80:1769–1796

Rice DD (1983) Relation of natural gas composition to thermal maturity and source rock type in San Juan Basin, northwestern New Mexico and southwestern Colorado. Am Assoc Petrol Geol Bull 67:1199–1218

Rice DD, Claypool GE (1981) Generation, accumulation and resource potential of biogenic gas. Am Assoc Petrol Geol Bull 65:5–25

Richet P, Bottinga Y, Javoy M (1977) A review of H, C, N, O, S, and Cl stable isotope fractionation among gaseous molecules. Annu Rev Earth Planet Sci 5:65–110

Ricke W (1964) Präparation von Schwefeldioxid zur massenspektrometrischen Bestimmung des S-Isotopenverhältnisses in natürlichen S-Verbindungen. Z Anal Chemie 199:401

Robert F, Epstein S (1980) Carbon, hydrogen and nitrogen isotopic composition of the Renazzo and Orgeuil organic components. Meteoritics 15:351

Robert F, Epstein S (1982) The concentration and isotopic composition of hydrogen, carbon and nitrogen carbonaceous meteorites. Geochim Cosmochim Acta 46:81–95

Robert F, Merlivat L, Javoy M (1978) Water and deuterium content in ordinary chondrites. Meteoritics 12:349–354

Robert F, Merlivat L, Javoy M (1979a) Water and deuterium content in the Chainpur meteorite. Meteoritics 13:613–615

Robert F, Merlivat L, Javoy M (1979b) Deuterium concentration in the early solar system: a hydrogen and oxygen isotope study. Nature (London) 282:785–789

Robinson BW (1973) Sulphur isotope equilibrium during sulphur hydrolysis at high temperatures. Earth Planet Sci Lett 18:443—450

Robinson BW (1975) Carbon and oxygen isotopic equilibria in hydrothermal calcites. Geochem J 9:43—46

Robinson BW (1978) Sulfate-water and H_2S isotopic thermometry in the New Zealand geothermal systems. In: Short Pap 4th Int Con Geochronol Cosmochronol Isotope Geol, p 354—356

Robinson BW, Kusakabe M (1975) Quantitative preparation of sulphur dioxide for $^{34}S/^{32}S$ analyses from sulphides by combustion with cuprous oxide. Anal Chem 47:1179

Roginsky SS (1962) Theoretische Grundlagen der Isotopenchemie. Deutscher Verlag der Wissenschaften, Berlin

Rosenbaum J, Sheppard SMF (1986) An isotopic study of siderites, dolomites and ankerites at high temperatures. Geochim Cosmochim Acta 50:1147—1150

Ross PJ, Martin EA (1970) Rapid procedure for preparing gas samples for nitrogen-15 determinations. Analyst 95:817—822

Rothe P, Hoefs J (1977) Isotopengeochemische Untersuchungen an Karbonaten der Ries-See-Sedimente der Forschungsbohrung Nördlingen 1973. Geol Bavaria 75:59—66

Rubinson M, Clayton RN (1969) Carbon-13 fractionation between aragonite and calcite. Geochim Cosmochim Acta 33:997—1002

Rumble D III (1978) Mineralogy, petrology and oxygen isotope geochemistry of the Clough Formation, Black Mountain, western New Hampshire, USA. J Petrol 19:317—340

Rumble D III (1982) Stable isotope fractionation during metamorphic devolatilization reactions. In: Characterization of metamorphism through mineral equilibria. Rev Mineral 10:327—353

Rumble D III, Spear FS (1983) Oxygen-isotope equilibration and permeability enhancement during regional metamorphism. J Geol Soc London 140:619—628

Rumble D III, Ferry JM, Hoering TC, Boucot AJ (1982) Fluid flow during metamorphism at the Beaver Brook fossil locality. Am J Sci 282:886—919

Russell WA, Papanastassiou DA, Tombrello TA (1978) Ca isotope fractionation on the Earth and other solar system materials. Geochim Cosmochim Acta 42:1075—1090

Rye RO (1974) A comparison of sphalerite-galena sulfur isotope temperatures with filling-temperatures of fluid inclusions. Econ Geol 69:26—32

Rye RO, Ohmoto H (1974) Sulfur and carbon isotopes and ore genesis. A review. Econ Geol 69:826—842

Rye RO, O'Neil JR (1968) The ^{18}O-content of water in primary fluid inclusions from Providencia, North Central Mexico. Econ Geol 63:232—238

Rye RO, Sawkins FJ (1974) Fluid inclusion and stable isotope studies on the Casapalca Ag—Pb—Zn—Cu deposit, central Andes, Peru. Econ Geol 69:181—205

Rye RO, Hall WE, Ohmoto H (1974) Carbon, hydrogen, oxygen and sulfur isotope study of the Darwin lead-silver-zinc deposit, southern California. Econ Geol 69:468—481

Rye RO, Schuiling RD, Rye DM, Jansen JBH (1976) Carbon, hydrogen and oxygen isotope studies of the regional metamorphic complex at Naxos, Greece. Geochim Cosmochim Acta 40:1031—1049

Sackett WM (1978) Carbon and hydrogen isotope effects during the thermocatalytic production of hydrocarbons in laboratory simulation experiments. Geochim Cosmochim Acta 42:571—580

Sackett WM, Eadie BJ, Exner ME (1973) Stable isotope composition of organic carbon in Recent Antarctic sediments. Adv Org Geochem 1973:661

Saino T, Hattori A (1980) ^{15}N natural abundance in oceanic suspended particulate organic matter. Nature (London) 283:752–754

Sakai H (1957) Fractionation of sulphur isotopes in nature. Geochim Cosmochim Acta 12:150–169

Sakai H (1968) Isotopic properties of sulfur compounds in hydrothermal processes. Geochem J 2:29–49

Sakai H (1977) Sulfate-water isotope thermometry applied to geothermal systems. Geothermics 5:67–74

Sakai H, Krouse HR (1971) Elimination of memory effect in ^{18}O/^{16}O determinations in sulfates. Earth Planet Sci Lett 11:369–373

Sakai H, Matsubaya O (1974) Isotopic geochemistry of the thermal waters of Japan and its bearing on the Kuroko ore solutions. Econ Geol 69:974–991

Sakai H, Tsutsumi M (1978) D/H fractionation factors between serpentine and water at 100 to 500 °C and 2000 bar water pressure and the D/H ratios of natural serpentines. Earth Planet Sci Lett 40:231–242

Sakai H, Ueda A, Field CW (1978) δ^{34}S and concentration of sulfide and sulfate sulfurs in some ocean-floor basalts and serpentinites. Short Pap 4th Int Conf Geochronol Cosmochromol Isotope Geol, Geol Surv Open-File Rep 78-701, p 371

Sakai H, Gunnlaugson E, Tomasson J, Rouse JE (1980) Sulfur isotope systematics in Icelandic geothermal systems and influence of seawater circulation at Reykjanes. Geochim Cosmochim Acta 44:1223–1231

Sakai H, Casadevall TJ, Moore JG (1982) Chemistry and isotope ratios of sulfur in basalts and volcanic gases at Kilauea volcano, Hawaii. Geochim Cosmochim Acta 46:729–738

Sakai H, DesMarais DJ, Ueda A, Moore JG (1984) Concentrations and isotope ratios of carbon, nitrogen and sulfur in ocean-floor basalts. Geochim Cosmochim Acta 48:2433–2441

Sangster DF (1968) Relative sulphur isotope abundances of ancient seas and stratabound sulphide deposits. Geol Assoc Can Proc 19:79

Sangster DF (1976) Sulphur and lead isotopes in stratabound deposits. In: Wolf KH (ed) Handbook of stratabound and stratiform ore deposits, vol 2, pp 219–266

Sasaki A, Arikawa Y, Folinsbee RE (1979) Kiba reagent method of sulfur extraction applied to isotopic work. Bull Geol Surv Jpn 30:241

Sass E, Kolodny Y (1972) Stable isotopes, chemistry and petrology of carbonate concretions (Mishash formation, Israel). Chem Geol 10:261–286

Satake H, Matsuo S (1984) Hydrogen isotopic fractionation factor between brucite and water in the temperature range from 100 to 510 °C. Contrib Mineral Petrol 86:19–24

Savin SM (1977) The history of the earth's surface temperature during the past 100 million years. Annu Rev Earth Planet Sci 5:319–355

Savin SM, Epstein S (1970a) The oxygen and hydrogen isotope geochemistry of clay minerals. Geochim Cosmochim Acta 34:25–42

Savin SM, Epstein S (1970b) The oxygen and hydrogen isotope geochemistry of ocean sediments and shales. Geochim Cosmochim Acta 34:43–63

Savin SM, Epstein S (1970c) The oxygen isotopic composition of coarse grained sedimentary rocks and minerals. Geochim Cosmochim Acta 34:323–329

Savin SM, Yeh HW (1981) Stable isotopes in ocean sediments. In: Emiliani C (ed) The sea, vol 7. Wiley-Interscience, New York, pp 1521–1554

Schidlowski M, Hayes JM, Kaplan IR (1983) Isotopic inferences of ancient biochemistries: carbon, sulfur, hydrogen and nitrogen. In: Schopf JW (ed) Earth's earliest biosphere: Its origin and evolution. Princeton Univ Press, pp 149–186

Schiegl WE, Vogel JV (1970) Deuterium content of organic matter. Earth Planet Sci Lett 7:307—313

Schneider A (1970) The sulfur isotope composition of basaltic rocks. Contrib Mineral Petrol 25:95—124

Schoell M (1980) The hydrogen and carbon isotopic composition of methane from natural gases of various origins. Geochim Cosmochim Acta 44:649—661

Schoell M (1984a) Recent advances in petroleum isotope geochemistry. Organ Geochem 6:645—663

Schoell M (1984b) Wasserstoff- und Kohlenstoffisotope in organischen Substanzen, Erdölen und Erdgasen. Geol Jahrb R D, H 67

Schoell M, Faber E, Coleman ML (1983) Carbon and hydrogen isotopic compositions of the NBS 22 and NBS 21 stable isotope reference materials: An interlaboratory comparison. Organ Geochem 5:3—6

Schoeller DA, Peterson DW, Hayes JM (1983) Double-comparison method for mass spectrometric determination of hydrogen isotopic abundances. Anal Chem 55:827—832

Schoeninger MJ, DeNiro MJ (1984) Nitrogen and carbon isotopic composition of bone collagen from marine and terrestrial animals. Geochim Cosmochim Acta 48:625—639

Scholle PA, Arthur MA (1980) Carbon isotope fluctuations in Cretaceous pelagic limestones: potential stratigraphic and petroleum exploration tool. Am Assoc Petrol Geol Bull 64:67—87

Schwarcz HP, Agyei EK, McCullen CC (1969) Boron isotopic fractionation during clay adsorption from seawater. Earth Planet Sci Lett 6:1—5

Seckbach J, Kaplan IR (1973) Growth pattern and $^{13}C/^{12}C$ isotope fractionation of Cyanidium calcarium and hot spring algal mats. Chem Geol 12:161—169

Shackleton N (1968) Depth of pelagic foraminifera and isotopic changes in Pleistocene Oceans. Nature (London) 218:79—80

Shackleton NJ (1977a) The oxygen isotope stratigraphic record of the late Pleistocene. Philos Trans R Soc London Ser B 280:169—182

Shackleton NJ (1977b) Carbon-13 in Uvigerina: tropical rainforest history and the equatorial Pacific carbonate dissolution cycles. In: The fate of fossil fuel CO_2 in the oceans. Plenum, New York, pp 401—428

Shackleton NJ, Kennett JP (1975) Paleotemperature history of the Cenozoic and initiation of Antarctic glaciation: oxygen and carbon isotope analyses in DSDP sites 277, 279 and 281. Initial Rep DSDP 29:743—755

Shackleton NJ, Opdyke ND (1973) Oxygen isotope and paleomagnetic stratigraphy of equatorial Pacific core V 28—V39: Oxygen isotope temperatures and ice volumes on a 10^5 and 10^6 year scale. Q Res 3:39

Shackleton NJ, Hall MA, Line J, Shuxi C (1983) Carbon isotope data in core V19-30 confirm reduced carbon dioxide concentrations in the ice age atmosphere. Nature (London) 306:319—322

Sharma T, Clayton RN (1965) Measurement of O^{18}/O^{16} ratios of total oxygen of carbonates. Geochim Cosmochim Acta 29:1347—1353

Shelton KL, Rye DM (1982) Sulfur isotopic compositions of ores from Mines Gaspé, Quebec: An example of sulfate-sulfide isotopic disequilibria in ore-forming fluids with applications to other porphyry-type deposits. Econ Geol 77:1688—1709

Shemesh A, Kolodny Y, Luz B (1983) Oxygen isotope variations in phosphate of biogenic apatites, II. Phosphorite rocks. Earth Planet Sci Lett 64:405—416

Sheppard SMF (1984) Isotopic geothermometry. In: Lagache M (ed) Thermométrie et barométrie géologiques. Soc Fr Mineral Cristallogr, pp 349—412

Sheppard SMF, Epstein S (1970) D/H and O^{18}/O^{16} ratios of minerals of possible mantle or lower crustal origin. Earth Planet Sci Lett 9:232–239

Sheppard SMF, Harris C (1985) Hydrogen and oxygen isotope geochemistry of Ascension Island lavas and granites: variation with crystal fractionation and interaction with sea water. Contrib Mineral Petrol 91:74–81

Sheppard SMF, Nielsen RL, Taylor HP (1969) Oxygen and hydrogen isotope ratios of clay minerals from Porphyry Copper Deposits. Econ Geol 64:755–777

Sheppard SMF, Nielsen RL, Taylor HP (1971) Hydrogen and oxygen isotope ratios in minerals from Porphyry Copper Deposits. Econ Geol 66:515–542

Shieh YN, Schwarcz HP (1974) Oxygen isotope studies of granite and migmatite, Grenville province of Ontario, Canada. Geochim Cosmochim Acta 38:21–45

Shieh YN, Taylor HP (1969a) Oxygen and hydrogen isotope studies of contact metamorphism in the Santa Rosa Range, Nevada and other areas. Contrib Mineral Petrol 20:306–356

Shieh YN, Taylor HP (1969b) Oxygen and carbon isotope studies of contact metamorphism of carbonate rocks. J Petrol 10:307–331

Shima M (1986) A summary of extremes of isotopic variations in extra-terrestrial materials. Geochim Cosmochim Acta 50:577–584

Siegenthaler U (1979) Stable hydrogen and oxygen isotopes in the water cycle. In: Jäger E, Hunziker JC (eds) Lectures in isotope geology. Springer, Berlin Heidelberg New York, pp 264–273

Silverman SR (1964) Investigations of petroleum origin and evolution mechanisms by carbon isotope studies. In: Isotopic and cosmic chemistry. Elsevier/North Holland Biomedical Press, Amsterdam, p 92

Silverman SR (1965) Migration and segregation of oil and gas. In: Fluids in subsurface environments. AAPG Mem 4:53

Silverman SR (1967) Carbon isotopic evidence for the role of lipids in petroleum. J Am Oil Chem Soc 44:691

Silverman SR, Epstein S (1958) Carbon isotopic compositions of petroleum and other sedimentary organic materials. Bull Am Assoc Petrol Geol 42:998

Skirrow R, Coleman ML (1982) Origin of sulfur and geothermometry of hydrothermal sulfides from the Galapagos Rift, 86°W. Nature (London) 249:142–144

Smith BN, Epstein S (1970) Biochemistry of the stable isotopes of hydrogen and carbon in salt marsh biota. Plant Physiol 46:738

Smith BN, Epstein S (1971) Two categories of $^{13}C/^{12}C$ ratios for higher plants. Plant Physiol 47:380

Smith JW, Batts BD (1974) The distribution and isotopic composition of sulfur in coal. Geochim Cosmochim Acta 38:121–123

Smith JW, Gould KW, Rigby D (1982) The stable isotope geochemistry of Australian coals. Org Geochem 3:111–131

Sofer Z (1978) Isotopic composition of hydration water in gypsum. Geochim Cosmochim Acta 42:1141–1149

Sofer Z (1984) Stable carbon isotope compositions of crude oils: Application to source depositional environments and petroleum alteration. Am Assoc Petrol Geol Bull 68:31–49

Sofer Z, Gat JR (1972) Activities and concentrations of oxygen-18 in concentrated aqueous salt solutions: Analytical and geophysical implications. Earth Planet Sci Lett 15:232–238

Spivack AJ (1985) Boron isotope marine geochemistry (Abstr.). Conf Int Les isotopes dans le cycle sedimentaire, Obernai, Fr 1–5 July 1985

Spivack AJ, Edmond JM (1986) Determination of boron isotope ratios by thermal ionization mass spectrometry of the dicesium metaborate cation. Anal Chem 58:31–35

Spooner ETC, Beckinsale RD, Fyfe WS, Snewing JD (1974) O^{18}-enriched ophiolitic metabasic rocks from E. Liguria (Italy), Pindos (Greece) and Troodos (Cyprus). Contrib Mineral Petrol 47:41–62

Stahl W (1977) Carbon and nitrogen isotopes in hydrocarbon research and exploration. Chem Geol 20:121–149

Stahl W (1979) Carbon isotopes in petroleum geochemistry. In: Jäger E, Hunziker JC (eds) Lectures in isotope geology. Springer, Berlin Heidelberg New York

Stahl W (1980) Compositional changes and $^{13}C/^{12}C$ fractionations during the degradation of hydrocarbons by bacteria. Geochem Cosmochim Acta 44:1903–1907

Stahl W, Aust H, Dounas A (1974) Origin of artesian and thermal waters determined by oxygen, hydrogen and carbon isotope analyses of water samples from the Sperklios Valley, Greece. In: Isotope techniques in groundwater hydrology, IAEA Vienna, 1:317

Stahl W, Wollanke G, Boigk H (1977) Carbon and nitrogen isotope data of Upper Carboniferous and Rotliegend natural gases from North Germany and their relationship to the maturity of the organic source material. In: Campos R, Goni J (eds) Advances in organic geochemistry. Madrid, p 539

Stern MJ, Spindel W, Monse EU (1968) Temperature dependence of isotope effects. J Chem Phys 48:2908

Stevens LM, Krout L, Walling D, Venters A, Engelkemeir A, Ross LE (1972) The isotopic composition of atmospheric carbon monoxide. Earth Planet Sci Lett 16:147–165

Stewart MK (1974) Hydrogen and oxygen isotope fractionation during crystallization of mirabilite and ice. Geochim Cosmochim Acta 38:167–172

Strauss H (1986) Carbon and sulfur isotopes in Precambrian sediments from the Canadian Shield. Geochim Cosmochim Acta 50:2653–2662

Styrt MM, Brackmann AJ, Holland HD, Clark BC, Pisutha-Arnold U, Eldridge CS, Ohmoto H (1981) The mineralogy and the isotopic composition of sulfur in hydrothermal sulfide/sulfate deposits on the East Pacific Rise, 21°N latitude. Earth Planet Sci Lett 53:382–390

Suchecki RK, Land LS (1983) Isotopic geochemistry of burial metamorphosed volcanogenic sediments, Great Valley sequence, northern California. Geochim Cosmochim Acta 47:1487–1500

Suzuoki T, Epstein S (1976) Hydrogen isotope fractionation between OH-bearing minerals and water. Geochim Cosmochim Acta 40:1229–1240

Swart PK, Grady MM, Pillinger CT (1982) Isotopically distinguishable carbon phases in the Allende meteorite. Nature (London) 297:381–383

Swart PK, Grady MM, Pillinger CT, Lewis RS, Anders E (1983) Interstellar carbon in meteorites. Science 220:406–410

Sweeney RE, Liu KK, Kaplan IR (1978) Oceanic nitrogen isotopes and their use in determining the source of sedimentary nitrogen. In: Robinson BW (ed) DSIR Bull 220:9–26

Swihart GH, Moore PB, Callis EL (1986) Boron isotopic composition of marine and non-marine evaporite borates. Geochim Cosmochim Acta 50:1297–1301

Tarutani T, Clayton RN, Mayeda TK (1969) The effect of polymorphism and magnesium substitution on oxygen isotope fractionation between calcium carbonate and water. Geochim Cosmochim Acta 33:987–996

Taube H (1954) Use of oxygen isotope effects in the study of hydration ions. J Phys Chem 58:523

Taylor BE, Friedrichsen H (1983) Light stable isotope systematics of granitic pegmatites from North America and Norway. Isotope Geosci 1:127−167

Taylor BE, O'Neil JR (1977) Stable isotope studies of metasomatic Ca−Fe−Al−Si skarns and associated metamorphic and igneous rocks, Osgood Mountains, Nevada. Contrib Mineral Petrol 63:1−49

Taylor BE, Foord EE, Friedrichsen H (1979) Stable isotope and fluid inclusion studies of gem-bearing granitic pegmatite aplite dikes, San Diego Co., California. Contrib Mineral Petrol 68:187−205

Taylor BE, Eichelberger JC, Westrich HR (1983) Hydrogen isotopic evidence of rhyolitic magma degassing during shallow intrusion and eruption. Nature (London) 306:541−545

Taylor HP (1967) Oxygen isotope studies of hydrothermal mineral deposits. In: Barnes HL (ed) Geochemistry of hydrothermal ore deposits. Holt Rinehart and Winston, New York

Taylor HP (1968) The oxygen isotope geochemistry of igneous rocks. Contrib Mineral Petrol 19:1−71

Taylor HP (1974a) The application of oxygen and hydrogen isotope studies to problems of hydrothermal alteration and ore deposition. Econ Geol 69:843−883

Taylor HP (1974b) Oxygen and hydrogen isotope evidence for large-scale circulation and interaction between groundwaters and igneous intrusions with particular reference to the San Juan volcanic field, Colorado. In: Geochemical transport and kinetics. Carnegie Inst Washington 634:299−323

Taylor HP (1977) Water/rock interactions and the origin of H_2O in granitic batholiths. J Geol Soc 133:509−558

Taylor HP (1978) Oxygen and hydrogen isotope studies of plutonic granitic rocks. Earth Planet Sci Lett 38:177−210

Taylor HP (1980) The effects of assimilation of country rocks by magmas on $^{18}O/^{16}O$ and $^{87}Sr/^{86}Sr$ systematics in igneous rocks. Earth Planet Sci Lett 47:243−254

Taylor HP, Epstein S (1961) $^{18}O/^{16}O$ ratios of feldspars and quartz in zoned granitic pegmatites. Geol Soc Am Spec Pap 68:183

Taylor HP, Epstein S (1962) Relationship between $^{18}O/^{16}O$ ratios in coexisting minerals of igneous and metamorphic rocks, Part I: Principles and experimental results. Bull Geol Soc Am 73:461−480

Taylor HP, Epstein S (1964) Comparison of oxygen isotope analyses of tektites, soils and impactite glasses. In: Cosmic and isotopic chemistry. Elsevier/North Holland Biomedical Press, Amsterdam

Taylor HP, Epstein S (1966) Oxygen isotope studies of Ivory Coast tektites and impactite glass from the Bosumtwi crater, Ghana. Science 153:173−175

Taylor HP, Epstein S (1969) Correlations between O^{18}/O^{16} ratios and chemical composition of tektites. J Geophys Res 74:6834−6844

Taylor HP, Forester RW (1971) Low-^{18}O igneous rocks from the intrusive complexes of Skye, Mull and Ardnamurchan, Western Scotland. J Petrol 12:465−497

Taylor HP, Forester RW (1979) An oxygen and hydrogen isotope study of the Skaergaard Intrusion and its country rocks: a description of a 55 M.Y. old fossil hydrothermal system. J Petrol 20:355−419

Taylor HP, Silver LT (1978) Oxygen isotope relationships in plutonic igneous rocks of the Peninsular Ranges Batholith, Southern and Baja California. Short papers of the 4th Intern Conf Geochronology, Cosmochronology, Isotope Geology. U.S. Geological Survey Open-File Report 78-701, pp 423−426

Taylor HP, Albee AL, Epstein S (1963) O^{18}/O^{16} ratios of coexisting minerals in three assemblages of kyanite zone pelitic schists. J Geol 71:513–522

Taylor HP, Gianetti B, Turi B (1979) Oxygen isotope geochemistry of the potassic igneous rocks from Roccamonfina volcano, Roman comagmatic region. Earth Planet Sci Lett 46:81–106

Taylor HP, Turi B, Cundari A (1984) $^{18}O/^{16}O$ and chemical relationships in K-rich volcanic rocks from Australia, East Africa, Antarctica and San Venanzo-Cupaello, Italy. Earth Planet Sci Lett 69:263–276

Thiemens MH, Heidenreich JE (1983) The mass independent fractionation of oxygen – A novel isotope effect and its cosmochemical implications. Science 219: 1073–1075

Thierstein HR, Geizzenauer KR, Molfino B, Shackleton NJ (1977) Global synchroneity of late Quaternary coccolith datum levels: validation by oxygen isotopes. Geology 5:400–404

Thode HG (1970) Sulphur isotope geochemistry and fractionation between coexisting sulphide minerals. Mineral Soc Am Spec Pap 3:133

Thode HG (1981) Sulfur isotope ratios in petroleum research and exploration: Williston Basin. Am Assoc Petrol Geol Bull 65:1527–1535

Thode HG, Goodwin AM (1983) Further sulfur and carbon isotope studies of late Archean iron-formations of the Canadian Shield and the rise of sulfate reducing bacteria. Precambrian Res 20:337–356

Thode HG, Monster J (1964) The sulfur isotope abundances in evaporites and in ancient oceans. In: Vinogradov AP (ed) Proc Geochem Conf Commemorating the Centenary of V I Vernadskii's Birth, vol 2, 630 p

Thode HG, Macnamara J, Collins CB (1949) Natural variations in the isotopic content of sulphur and their significance. Can J Res 27B:361

Thode HG, Monster J, Dunford HB (1958) Sulphur isotope abundances in petroleum and associated materials. Am Assoc Petrol Geol Bull 42:2619–2641

Thode HG, Harrsion AG, Monster J (1960) Sulphur isotope fractionation in early diagenesis of recent sediments of Northeast Venezuela. Am Assoc Petrol Geol Bull 44:1809–1817

Thode HG, Monster J, Dunford HB (1961) Sulphur isotope geochemistry. Geochim Cosmochim Acta 25:159–174

Thode HG, Cragg CB, Hulston JR, Rees CE (1971) Sulphur isotope exchange between sulphur dioxide and hydrogen sulphide. Geochim Cosmochim Acta 35:35–45

Thompson AB (1983) Fluid absent metamorphism. J Geol Soc London 140: 533–547

Touret J (1971) Le faciès granulite en Norvège méridionale. Les inclusions fluids. Lithos 4:423–436

Tracy RJ, Rye DM, Hewitt DA, Schiffries CM (1983) Petrologic and stable isotopic studies of fluid-rock interactions, south central Connecticut. I. The role of infiltration in producing reaction assemblages in impure marbles. Am J Sci 238A:589–616

Trofimov A (1949) Isotopic constitution of sulfur in meteorites and in terrestrial objects. Dokl Akad Nauk SSSR 66:181 (in Russian)

Trudinger PA, Chambers LA (1973) Reversibility of bacterial sulfate reduction and its relevance to isotope fractionation. Geochim Cosmochim Acta 37:1775–1778

Truesdell AH (1974) Oxygen isotope activities and concentrations in aqueous salt solution at elevated temperatures: Consequences for isotope geochemistry. Earth Planet Sci Lett 23:387–396

Truesdell AH, Hulston JR (1980) Isotopic evidence on environments of geothermal systems. In: Fritz P, Fontes J (eds) Handbook of environmental isotope geochemistry, vol 1. Elsevier, New York Amsterdam, pp 179–226

Truesdell AH, Nathenson M, Rye RO (1977) The effects of subsurface boiling and dilution on the isotopic compositions of Yellowstone thermal waters. J Geophys Res 82:3694–3704

Tsai HM, Shieh Y, Meyer HOA (1979) Mineralogy and $^{34}S/^{32}S$ ratios of sulfides associated with kimberlite xenoliths and diamonds. In: Boyd FR, Meyer HOA (eds) The mantle samples: inclusions in kimberlites and other volcanics. AGU, Washington, pp 87–103

Tucker ME (1983) Diagenesis, geochemistry and origin of a Precambrian dolomite: The Beck spring dolomite of eastern California. J Sediment Petrol 53:1097–1119

Tudge AP (1960) A method of analysis of oxygen isotopes in orthophosphate – its use in the measurement of paleotemperatures. Geochim Cosmochim Acta 18:81–93

Tudge AP, Thode HG (1950) Thermodynamic properties of isotope compounds of sulphur. Can J Res 28:567

Turner JV (1982) Kinetic fractionation of carbon-13 during calcium carbonate precipitation. Geochim Cosmochim Acta 46:1183–1192

Ueda A, Sakai H (1983) Simultaneous determinations of the concentration and isotope ratio of sulfate- and sulfide-sulfur and carbonate-carbon in geological samples. Geochemical J 17:185–196

Ueda A, Sakai S (1984) Sulfur isotope study of Quaternary volcanic rocks from the Japanese Island Arc. Geochim Cosmochim Acta 48:1837–1848

Urey HC (1947) The thermodynamic properties of isotopic substances. J Chem Soc 1947:562

Urey HC, Brickwedde FG, Murphy GM (1932a) An isotope of hydrogen of mass 2 and its concentration (Abstr). Phys Rev 39:864

Urey HC, Brickwedde FG, Murphy GM (1932b) A hydrogen isotope of mass 2 and its concentration. Phys Rev 40:1

Urey HC, Lowenstam HA, Epstein S, McKinney CR (1951) Measurement of paleotemperatures and temperatures of the Upper Cretaceous of England, Denmark and the Southeastern United States. Bull Geol Soc Am 62:399–416

Usdowski HE (1982) Reactions and equilibria in the systems $CO_2–H_2O$ and $CaCO_3–CO_2–H_2O(0°-50°)$. A review. N Jahrb Miner Abh 144:148–171

Usdowski HE, Hoefs J, Menschel G (1979) Relationship between ^{13}C and ^{18}O fractionation and changes in major element composition in a recent calcite-depositing spring – a model of chemical variations with inorganic calcite precipitation. Earth Planet Sci Lett 42:267–276

Valley JW (1986) Stable isotope geochemistry of metamorphic rocks. In: Valley JW, Taylor HP, O'Neil JR (eds) Stable isotopes in high temperature geological processes. MSA Rev Mineral 16 (in press)

Valley JW, O'Neil JR (1981) $^{13}C/^{12}C$ exchange between calcite and graphite: a possible thermometer in Grenville marbles. Geochim Cosmochim Acta 45:411–419

Valley JW, O'Neil JR (1984) Fluid heterogeneity during granulite facies metamorphism in the Adirondacks: stable isotope evidence. Contrib Mineral Petrol 85:158–173

Veizer J, Hoefs J (1976) The nature of O^{18}/O^{16} and C^{13}/C^{12} secular trends in sedimentary carbonate rocks. Geochim Cosmochim Acta 40:1387–1395

Veizer J, Holser WT, Wilgus CK (1980) Correlation of $^{13}C/^{12}C$ and $^{34}S/^{32}S$ secular variations. Geochim Cosmochim Acta 44:579–587

Viglino JA, Harmon RS, Borthwick J, Nehring NL, Motyka RJ, White LD, Johnston DA (1985) Stable-isotope evidence for a magmatic component in fumarole condensates from Augustine volcano, Cook Inlet, Alaska, USA. Chem Geol 49: 141–157

Vincent E, Killingley JS, Berger WS (1981) Stable isotope composition of benthic foraminifera from the equatorial Pacific. Nature (London) 289:639–643

Vinogradov AP, Grinenko VA, Ustinov VI (1962) Isotopic composition of sulfur compounds in the Black Sea. Geochemistry 1962:973

Vogel DE, Garlick GD (1970) Oxygen isotope ratios in metamorphic eclogites. Contr Mineral Petrol 28:183–191

Vogel JC, Urk H van (1975) Isotopic composition of groundwater in semiarid regions of South Africa. J Hydrol 25:23

Wachter EA, Hayes JM (1985) Exchange of oxygen isotopes in carbon dioxide-phosphoric acid systems. Chem Geol Isotope Geosci Sect 52:365–374

Wada H, Suzuki K (1983) Carbon isotopic thermometry calibrated by dolomite-calcite solvus temperatures. Geochim Cosmochim Acta 47:697–706

Warren CG (1972) Sulfur isotopes as a clue to the genetic geochemistry of a roll-type uranium deposit. Econ Geol 67:759–767

Way K, Fano L, Scott MR, Thew K (1950) Nuclear data. A collection of experimental values of halflifes, radiation energies, relative isotopic abundances, nuclear moments and cross-sections. Natl Bur Stand U S Circ 499

Weber JN (1968) Fractionation of the stable isotopes of carbon and oxygen in calcareous marine invertebrates – the Asteroidea, Ophiuroidea and Crinoidea. Geochim Cosmochim Acta 32:33–70

Weber JN, Raup DM (1966a) Fractionation of the stable isotopes of carbon and oxygen in marine calcareous organisms – the Echinoidea. I. Variation of ^{13}C and ^{18}O content within individuals. Geochim Cosmochim Acta 30:681–703

Weber JN, Raup DM (1966b) Fractionation of the stable isotopes of carbon and oxygen in marine calcareous organisms – the Echinoidea. II. Environmental and genetic factors. Geochim Cosmochim Acta 30:705–736

Wedeking KW, Hayes JM, Matzigkeit U (1983) Procedures of organic geochemical analysis. In: Schopf JW (ed) Earth's earliest biosphere: Its origin and evolution. Princeton Univ Press

Wellmann RP, Cook FD, Krouse HR (1968) Nitrogen-15: Microbiological alteration of abundance. Science 161:269–270

Welte DH, Kalkreuth W, Hoefs J (1975) Age-trend in carbon isotopic composition in Paleozoic sediments. Naturwissenschaften 62:482–483

Wenner DB, Taylor HP (1973) Oxygen and hydrogen isotope studies of the serpentinization of ultramafic rocks in oceanic environments and continental ophiolite complexes. Am J Sci 273:207–239

Wenner DB, Taylor HP (1974) D/H and $^{18}O/^{16}O$ ratios of serpentinization of ultramafic rocks. Geochim Cosmochim Acta 38:1255–1286

Whalen M, Yoshinara T (1985) Oxygen isotope ratios of N_2O from different environments. Nature (London) 313:697–782

White DE (1974) Diverse origins of hydrothermal ore fluids. Econ Geol 69:954–973

Whiticar MJ, Faber E, Schoell M (1986) Biogenic methane formation in marine and freshwater environments: CO_2 reduction vs. acetat fermentation – Isotopic evidence. Geochim Cosmochim Acta 50:693–709

Wickham SM, Taylor HR (1985) Stable isotope evidence for large-scale seawater infiltration in a regional metamorphic terrane; the Trois Seigneurs Massif, Pyrenees, France. Contrib Mineral Petrol 91:122–137

Wickman FE (1952) Variation in the relative abundance of carbon isotopes in plants. Geochim Cosmochim Acta 2:243–254

Willan RCR, Coleman ML (1983) Sulfur isotope study of the Aberfeldy barite, zinc, lead deposit and minor sulfide mineralizations in the Dalradian metamorphic terrain, Scotland. Econ Geol 78:1619–1656

Williams DF, Sommer MA, Bender ML (1977) Carbon isotopic compositions of recent planktonic foraminifera of the Indian Ocean. Earth Planet Sci Lett 36: 391–403

Williams DF, Röttger R, Schmaljohann R, Keigwin L (1981) Oxygen and carbon isotopic fractionation and algal symbiosis in the benthic foraminifera "Hetero stegina depressa". Palaeogeogr Palaeoclim Palaeoecol 33:231–251

Wilson AF, Baski AK (1983) Widespread ^{18}O-depletion in some Precambrian granulites of Australia. Precambrian Res 23:33–56

Wilson AF, Green DC, Davidson LR (1970) The use of oxygen isotope geothermometry on the granulites and related intrusives, Musgrave Ranges, Central Australia. Contrib Mineral Petrol 27:166–178

Wong WW, Sackett WM (1978) Fractionation of stable carbon isotopes by marine phytoplankton. Geochim Cosmochim Acta 42:1809–1815

Wong WW, Benedict CR, Kohel JR (1979) Enzymatic fractionation of the stable isotope of carbon dioxide by RudP-carboxylase. Plant Physiol 63:852–856

Wright IP, McNaughton NJ, Fallick AE, Gardiner LR, Pillinger CT (1983) A high-sensitivity-high precision stable isotope mass-spectrometer. J Phys (E) 16:497–504

Wyllie PJ (1979) Mantle fluid compositions buffered in peridotite–CO_2–H_2O by carbonates, amphibole and phlogopite. J Geol 86:687–713

Yang J, Epstein S (1982) On the origin and composition of hydrogen and carbon in meteorites. Meteoritics 17:301

Yang J, Epstein S (1983) Interstellar organic matter in meteorites. Geochim Cosmochim Acta 47:2199–2216

Yang J, Epstein S (1984) Relic interstellar grains in Murchison meteorite. Nature (London) 311:544–547

Yapp CJ (1979) Oxygen and carbon isotope measurements of land snail shell carbonate. Geochim Cosmochim Acta 43:629–635

Yeh HW (1980) D/H ratios and late stage hydration of shales during burial. Geochim Cosmochim Acta 45:341–352

Yeh HW, Epstein S (1978) Hydrogen isotope exchange between clay minerals and seawater. Geochim Cosmochim Acta 42:140–143

Yeh HW, Epstein S (1981) Hydrogen and carbon isotopes of petroleum and related organic matter. Geochim Cosmochim Acta 45:753–762

Yeh HW, Savin SM (1976) The extent of oxygen isotope exchange between clay minerals and seawater. Geochim Cosmochim Acta 40:743–748

Yeh HW, Savin SM (1977) Mechanism of burial metamorphism of argillaceous sediments. 3. O-isotope evidence. Geol Soc Am Bull 88:1321–1330

Yoshida N, Matsuo S (1983) Nitrogen isotope ratio of atmospheric N_2O as a key to the global cycle of N_2O. Geochem J 17:231–239

Yoshida N, Hattori A, Saino T, Matsuo S, Wada E (1984) $^{15}N/^{14}N$ ratio of dissolved N_2O in the eastern tropical Pacific Ocean. Nature (London) 307:442–444

Yuen G, Blair N, DesMarais DJ, Chang S (1984) Carbon isotopic composition of individual, low molecular weight hydrocarbons and monocarboxylic acids from Murchison meteorite. Nature (London) 308:252–254

Yund RA, Anderson TF (1974) Oxygen isotope exchange between potassium feld-spar and KCl solution. In: Geochemical transport and kinetics. Carnegie Inst Washington, Publ 634, pp 99−105

Yurtsever Y (1975) Worldwide survey of stable isotopes in precipitation. Rep Sect Isotope Hydrol IAEA, November 1975, 40 pp

Zierenberg RA, Shanks WC, Bischoff JL (1984) Massive sulfide deposit at 21°N, East Pacific Rise: chemical composition, stable isotopes, and phase equilibria. Bull Geol Soc Am 95:922−929

ZoBell CE (1958) Ecology of sulfate-reducing bacterial. Prod Mongr 22:12

Subject Index

ANDREW THOMPSON
UNIVERSITY OF GLAMORGAN
MARCH 2004

Also by Umut Özkırımlı

CONTEMPORARY NATIONALISM

CONTESTED TERRAINS: A Comparative Study of Greek and Turkish Nationalist Imaginaries (*with Spyros A. Sofos*)

THEORIES OF NATIONALISM: A Critical Introduction

Nationalism and its Futures

Nationalism and its Futures

Edited by

Umut Özkırımlı
Department of International Relations
Istanbul Bilgi University
Turkey

First published 2003 by
PALGRAVE MACMILLAN
Houndmills, Basingstoke, Hampshire RG21 6XS and
175 Fifth Avenue, New York, N.Y. 10010
Companies and representatives throughout the world

PALGRAVE MACMILLAN is the global academic imprint of
the Palgrave Macmillan division of St. Martin's Press, LLC and of
Palgrave Macmillan Ltd. Macmillan® is a registered trademark in the
United States, United Kingdom and other countries. Palgrave is a
registered trademark in the European Union and other countries.

ISBN 1–4039–1713–2

This book is printed on paper suitable for recycling and made from fully
managed and sustained forest sources.

A catalogue record for this book is available from the British Library.

Library of Congress Cataloging-in-Publication Data
Nationalism and its futures/edited by Umut Özkırımlı
        p. cm.
    Includes bibliographical references and index.
    ISBN 1–4039–1713–2 (cloth)
    1. Nationalism. I. Özkırımlı, Umut.

JC311.N2953 2003
320.54—dc21   2003053268

10 9 8 7 6 5 4 3 2 1
12 11 10 09 08 07 06 05 04 03

Printed and bound in Great Britain by
Antony Rowe Ltd, Chippenham and Eastbourne

Contents

Acknowledgements

This book grew out of a conference I organized on 10 May 2001, at Istanbul Bilgi University. My greatest debt is to Bilgi University, which sponsored this conference, and, later on, granted me a generous research fellowship for the 2001–2002 academic year during which I was able to complete the editing of this book. I would also like to express my gratitude to all the participants of this conference, Craig Calhoun, Partha Chatterjee, John A. Hall, John Hutchinson and Nira Yuval-Davis, for kindly allowing me to assemble their papers in an edited volume, and to Fred Halliday and Will Kymlicka for contributing to this collective endeavour. I would also like to thank the anonymous reviewers of Palgrave Macmillan for their insightful comments and criticisms on an earlier draft of this book.

The extract from 'The Times They are a-Changin'' by Bob Dylan is quoted by kind permission of Special Rider Music, New York.

Notes on the Contributors

Craig Calhoun is President of the Social Science Research Council, and Professor of Sociology and History at New York University. He is the author of *Nationalism*, and a number of books on social and cultural theory. Among his recent publications are *Understanding September 11* (with P. Price and A. Timmer), *The Contemporary Social Theory Reader* (with J. Gerteis, J. Moody, S. Pfaff and I. Virk) and *Dictionary of the Social Sciences*.

Partha Chatterjee is Director and Professor of Political Science at the Center for Studies in Social Sciences, Calcutta, and Visiting Professor of Anthropology at Columbia University. He is the author of *Nationalist Thought and the Colonial World* and *The Nation and Its Fragments*. His most recent book is *A Princely Impostor? The Strange and Universal History of the Kumar of Bhawal*.

John A. Hall is Professor of Sociology at McGill University, and a Research Professor of Sociology at The Queen's University of Belfast. He is the editor of *The State of the Nation: Ernest Gellner and the Theory of Nationalism*, and the author of a number of books on the state, civil society and social theory. His forthcoming publications include *Ernest Gellner* (co-author Brendan O'Leary) and *The Nation-State in Question* (co-edited with T.V. Paul and G.J. Ikenberry).

Fred Halliday is Professor of International Relations at the London School of Economics. His recent publications include *Revolutions and World Politics*, *Nation and Religion in the Middle East* and *Two Hours that Shook the World*.

John Hutchinson is Senior Lecturer in Nationalism at the London School of Economics. He is the author and editor of a number of books on nationalism, including *Modern Nationalism*, *Understanding Nationalism* (with Monserrat Guibernau), and *Nationalism*, (with Anthony D. Smith). He is currently working on a book entitled *Nations as Zones of Conflict*.

Will Kymlicka is Professor of Philosophy at Queen's University, Kingston, Ontario. His publications include *Multicultural Citizenship, Contemporary Political Philosophy* and *Politics in the Vernacular: Nationalism, Multiculturalism and Citizenship.* He is currently editing a volume entitled *Ethnicity and Democracy in Africa* (with B. Berman and D. Eyoh), and a book called *Language Rights and Political Theory* (with A. Patten).

Umut Özkırımlı is Assistant Professor of International Relations at Istanbul Bilgi University. He is the author of *Theories of Nationalism: A Critical Introduction.* His forthcoming publications include *Contemporary Nationalism* and *Contested Terrains: A Comparative Study of Greek and Turkish Nationalist Imaginaries* (with Spyros A. Sofos).

Nira Yuval-Davis has been, until recently, a Professor at the Department of Sociology at the University of Greenwich, London, and is now a Visiting Professor at the Department of Cultural Studies at the University of East London where she is resurrecting her graduate course in Gender, Sexualities and Ethnic Studies. She is the author of *Gender and Nation,* and the editor of many books including *Woman-Nation-State* and *Racialized Boundaries* (both with Floya Anthias), and more recently *Women, Citizenship & Difference* (with Pnina Werbner).

1
Introduction
Umut Özkırımlı

> Come writers and critics/Who prophesize with your pen
> And keep your eyes wide/The chance won't come again
> And don't speak too soon/For the wheel's still in spin
> And there's no tellin' who/That it's namin'
> For the loser now/Will be later to win
> For the times they are a-changin'
> Bob Dylan, 1964

Times were definitely a-changin' in the early 1990s.[1] The general mood, at least in the capital of the world's only remaining super-power, was one of optimism and triumphalism. There was no dearth of writers and critics prophesizing. One of them, a deputy director of the State Department's policy planning staff and former analyst at the RAND Corporation, had even proclaimed 'the end of history'. For him, what we were witnessing was not 'just the end of the Cold War, or the passing of a particular period of post-war history ... but the end point of mankind's ideological evolution and the universalization of Western liberal democracy as the final form of human government' (Fukuyama 1989: 4; see also Fukuyama 1992, 2001; Achenbach 2001).

Yet all this talk of 'the end of history' was no more than florid rhetoric for those caught up in the wave of ethnic and nationalist conflicts that swamped much of the world at roughly the same time. For these people, there was nothing but 'despair, nausea, and horror'. Their ordeal was eloquently articulated by the Croatian actress Mira Furlan, herself the victim of a merciless propaganda campaign waged against her in Croatia for performing in an international theatre festival in Belgrade. 'I played in those performances', wrote Furlan in an open letter to her co-citizens, 'for those anguished people who were not "Serbs", but human beings, human beings like me, human beings who recoil before this monstrous Grand Guignol farce in which dead heads are flying':

> To whom am I addressing this letter? Who will read it?...
> Everyone is so caught up by the great cause that small personal fates are not important any more...I am sorry, my system of values is different. For me there have always existed, and always will exist, only human beings, individual people, and those human beings (God, how few of them there are!) will always be excepted from generalizations of any kind, regardless of events, however catastrophic. I, unfortunately, shall never be able to 'hate all Serbs', nor even understand what that means.[2]

The story of Mira Furlan is a familiar one, at least for those not confined to the secluded offices of Washington think tanks. Nationalism continues to be one of the major actors of the social and political landscape, and a central part of the fabric of our everyday lives. Yet it plays only a limited role in the scenarios concocted by the prophets of a 'post-historical' world. Nationalism, Fukuyama argues, cannot be an ideological competitor to liberal democracy because it does not offer a comprehensive agenda for socio-economic organization. As such, it is compatible with doctrines and ideologies that do offer such agendas. In any case, Fukuyama hastens to add, nationalism is on the rise in regions such as Eastern Europe and the former Soviet Union where peoples have long been denied their national identities. Within the world's oldest and most secure nationalities, nationalism has been domesticated and made compatible with universal recognition, much like religion three or four centuries earlier (Fukuyama 1989, 1992).

Naivety? Complacency? Or blatant Eurocentrism? Finding the appropriate term to describe Fukuyama's account is more difficult than exposing its shallowness. First, only movements with a particular socio-economic agenda qualify as ideologies for Fukuyama. This enables him to claim that there have only been two major challenges to liberalism in the twentieth century, those of fascism and communism. This, however, is a very limited, if not reductionist, interpretation which downplays the political and cultural elements of ideologies. Nationalism does have a political and cultural agenda of its own: it envisages a particular world order, one organized according to the principle of national self-determination, where individual units, that is nation-states, should possess as much political, cultural, and yes, economic, autonomy as possible. Second, although it is true that nationalism usually mixes with other ideologies, including liberalism, it leaves its mark on the resulting amalgam. In other words, liberalism is one thing, nationalist liberalism is quite another. This has several policy implications on a number of vital policy issues, domestic or international, ranging from immigration and citizenship laws to foreign aid. Finally, the claim that nationalism has been domesticated within the world's oldest and most secure nationalities is hard to square with the realities on the ground. Some of the fiercest struggles for recognition take place in the West, including such countries as Britain, Canada, Belgium and Spain – a rather obvious point to be sure, for which ample evidence can be found in newspaper headlines, not to mention the burgeoning literature on multiculturalism and minority rights.

Yet Fukuyama (1989) is not the only one predicting the growing 'Common Marketization' of international relations and 'the ineluctable spread of Western consumerist culture'. 'The crisis of the nation-state' is one of the most worn-out clichés of the last decade, and the future of nationalism appears to many to be more uncertain than ever under the twin pressures of globalization and identity politics.

The aim of the present volume is to explore the challenges posed by and to nationalism at the turn of a new millennium. Each chapter will engage with a different challenge, on a different level of analysis (international or state-level). Some of the questions that will be addressed in the book are:

- What is national homogenization? What were the actual practices designed to create national unity? Have the conditions which gave nationalism such an appalling reputation changed? What are the different options of belonging today? Will national homogenization continue to be the norm of world politics or will other, possibly more benign, models of integration emerge?

- Is there a difference between nationalism and ethnic politics? What are the pitfalls of the dominant view of modernity which is based on ethical universalism? What are the alternatives to the politics of homogeneity?
- What is the record of nationalism with regard to human rights? What are the major obstacles to a compatibility between nationalism and universal principles? Is there a way of overcoming these obstacles?
- What are the implications of the processes of globalization for nationalism? Can we talk of a 'crisis of the nation-state'? Are there any indications that the nation is being superseded by other forms of collective identification? Will interstate conflicts be replaced by a clash of civilizations?
- Is cosmopolitanism an alternative to nationalism? What are the strengths and weaknesses of 'actually existing' cosmopolitanism? What is the nature of the relationship between cosmopolitanism and local democracy? How can cosmopolitan democracy flourish?
- What does 'belonging' mean? What are the alternative narratives of belonging to that of the nation-state? Is it possible to formulate a model of belonging that encompasses both identity and citizenship?

Let me now briefly elaborate on the general themes around which the following chapters revolve and outline the main arguments.[3]

National unity and homogenization

Nationalism is about homogenization, suggests John A. Hall, reminding us of Gellner's famous definition of nationalism as 'a political doctrine which holds that the political and national unit should be congruent' (1983: 1). Not surprisingly, geopolitical realities were a far cry from this ideal and the history of nationalism has been an endless drive to create that perfect fit between nationality and politics. The strategies followed by nationalizing states varied, ranging from voluntary assimilation to the more vicious practices of ethnic cleansing, population transfer and genocide.

According to Hall, the maximal point of national homogenization was reached in Europe, with the adoption of more vicious practices from the end of the nineteenth century onwards. There were various reasons for that. The first was the problem of nationalities. Unification, when it took place over a long period, was more voluntary and benign, as in the case of England which had a centralized polity from 1066. But the great European land empires of the nineteenth century, the Ottomans, the Romanovs and the Habsburgs, were not that lucky. They were the latecomers, and when they tried to rationalize their possessions, they ran into the problem of nationalities who were able to protect themselves as they had already codified their languages and established their own educational systems. The other reason was the need for geopolitical autonomy. This in turn required territorial aggrandizement, hence the link between nationalism and imperialism. Combined, these factors made nationalism aggressive and expansionist.

Now, however, times have changed. Europe's security dilemma is solved and the link between nationalism and imperialism is broken. Today it is possible to distinguish between four options of belonging. The first two, that is ethnic and civic nationalism, are not particularly attractive, claims Hall, as they can be quite intolerant of diversity. The third option, or civil nationalism, which Hall defines as 'the acceptance of diverse positions and cultures', is more promising. But that diversity should be limited by a consensus on shared values. If groups have rights over individuals, this would lead to 'social caging', the fourth option, and become repulsive.

Are there any hopes for the future? Yes, says Hall, and this hope rests 'on the possibility that the developing world...will build on their institutional accomplishments so as not to imitate those who blithely consider themselves advanced'.

Partha Chatterjee picks up where Hall stops and explores homogeneity in the context of the postcolonial world. His main target is the distinction Benedict Anderson developed in his recent book, *The Spectre of Comparisons* (1998). In this book, Anderson identifies two kinds of seriality: the 'unbound seriality' of the everyday universals of modern social thought, such as nations, citizens, intellectuals and so on, and the 'bound seriality' of governmentality, 'the finite totals of enumerable classes of population produced by the modern census and electoral systems'. Anderson uses this distinction, Chatterjee notes, to make an argument about the goodness of nationalism and the nastiness of ethnic politics, believing that the two arise on different sites and mobilize on different sentiments. According to Chatterjee, what underlies this attempt is a particular – standardized – conception of politics, politics inhabiting the empty homogeneous time of modernity.

Yet this view of politics and modernity is mistaken, because it is one-sided, says Chatterjee. 'Empty homogeneous time is the utopian time of capital', it is not located anywhere in real space. Modern life, by contrast, is heterogeneous, unevenly dense. Chatterjee cites examples from the postcolonial world to support his argument. There, one could find 'industrial capitalists delaying the closing of a business deal because they had not yet had word from their respective astrologers, or industrial workers who would not touch a new machine until it had been consecrated with appropriate religious rites'. These examples do not attest to the co-presence of several times, the time of the modern and the premodern, but to the heterogeneity of modern life, as these 'other' times are the products of the encounter with modernity itself.

In such a context, Chatterjee argues, the call for universalism – or a nationalism not contaminated by ethnic politics – is often a mask covering the perpetuation of real inequalities. The politics of democratic nationhood can only offer a means for achieving equality by ensuring adequate representation for the underprivileged groups. A strategic politics of groups, classes and ethnicities is thus inevitable. Such a politics of heterogeneity can never claim to yield a general formula valid for all peoples at all times. Its solutions are bound to be contextual, historically specific and thus provisional. But this is the only way forward, at least for the postcolonial theorist, who is born 'only when the mythical time-space of epic modernity has been lost forever'.

Nationalism, universalism and human rights

Fred Halliday takes a quite different approach to the thorny issue of universalism and focuses on the ways in which universal rights conflict with forms of particularism, notably nationalism. He identifies three main obstacles to a compatibility between nationalism and human rights. The first two, sovereignty and culture, do not pose a serious threat to human rights. It is generally agreed that gross violations of human rights forfeit a state's claim to exclusive control over the fate of its citizens. In any case, it is increasingly difficult to talk of domestic processes without international repercussions, and some go so far as to argue that the outer limits of sovereignty have already been reached. As for culture, a modicum of respect for cultural differences will go a long way towards resolving the tension between nationalism and universally held principles. Moreover, we should not forget that culture is not a given, but a site of diversity and change, that there is not much difference between cultures on many key issues (all peoples accept the right to self-determination, for instance) and that the main arguments about rights are not about culture anyway (the critique of universality is often based on a view of the structural inequality in the world system, not on claims that the peoples in question are exempt from universal jurisdiction).

The third obstacle, that of particularism, is much more problematic according to Halliday. At the heart of nationalism lies a contradiction with regard to rights. Nationalism rests upon the assertion of a universal right, namely the right to self-determination, but it has been and remains deeply hostile to the universality of rights. He identifies four areas where this contradiction can be observed: self-determination, laws of war, terrorism and solidarity. In each case, what we see is the assertion of a particular right by nationalist movements, coupled with the denial of that right to others. A graphic illustration of this point is the recent anti-terrorist measures undertaken by the Bush administration, 'The Patriot Act'. According to Halliday, the appeal to patriotism, 'a necessarily partisan and emotional principle', obscures two important issues: first, it absolves the USA and American citizens from any reflection on the past actions of their own states, including support for terrorist groups; secondly, it implicitly exempts the state and its armed forces from respect for universal principles. Patriotism, he argues, is not the first, but the last place to start in legitimating a campaign which may be valid in itself.

In short, nationalism may be compatible with universal principles, alongside other desirable goals, such as democracy, identity, community and international order, but only if a priority of values is established, or if nationalism knows its place. Unfortunately, Halliday concludes, the record of nationalism over the past century does not leave too much room for hope.

Nationalism, globalization and the clash of civilizations

John Hutchinson engages with two influential *fin de siècle* scenarios in his essay, namely the globalization perspective and Samuel Huntington's clash of civilizations thesis which was given a new lease of life after the events of 11 September 2001. Both perspectives forecast the imminent demise of nations as relevant political actors but, Hutchinson argues, they are seriously vitiated by their lack of a long-range historical perspective.

Hutchinson singles out three fundamental problems with the radical version of the globalization perspective. First, globalization is not a modern revolutionary development but a recurring and evolutionary process, with roots, according to some estimates, as far back as the second millennium B.C. Second, globalization should not be equated with Westernization or universalization. Globalization always flows from particular centres and the rise of the 'West' is only the latest of a series of recurring jumps in global power. Third, globalization cannot be defined as a unitary and secular process as globalizing institutions include missionary religions, empires, migrations and long-distance trade as much as secular sciences, technologies and ideologies.

All this should make us reconsider the causal relationship between globalization and nations, argues Hutchinson. If globalization has been in process for a millennia or more, then claims that nations will be supplanted do not make much sense. More importantly, the dynamic and variegated nature of globalization may well produce differentiation, rather than homogenization. At this point, Hutchinson cites four factors that may stimulate ethnic differentiation: the emergence of universalistic scriptural religions, imperial expansions, interstate warfare and the development of long-distance trade. A longer historical perspective, then, will reveal that globalization has gone hand-in-hand with the articulation and crystallization of ethnic and national differences.

The clash of civilizations thesis, on the other hand, does not fare better. There are several obvious criticisms one can make of this thesis, Hutchinson argues. First, the most ferocious conflicts occurred within the European world of Latin Christianity which Huntington defines as a single civilization and not with Muslim, Confucian or Orthodox Christian civilizations. Second, the real enemy for most modern religious movements is 'within', since in most cases what they wish is to morally regenerate a traditional culture that is being eroded by alien cosmopolitan values. And third, the recent religious revivals should not be seen as the domain of 'backward', non-Western countries: religion has been one of the sources of national identity for many avowedly secular states such as Holland and France, and it remains a powerful force in many contemporary 'Western' societies, including the USA, Italy, Ireland and Greece. The current religious revival, Hutchinson concludes, does not offer a serious threat to the contemporary system of nation-states. In many cases, religion becomes ethnicized and much of the current revival is directed against the alleged inauthenticity of secular nationalism. Nation-states continue to be the major actors of world politics and new threats will only intensify nationalism in many parts of the world.

Nationalism and its alternatives

The last two essays of this volume explore some of the alternatives to nationalism. Craig Calhoun begins his critique of 'actually existing' cosmopolitanism by noting the rhetorical advantages of being a 'citizen of the world' in Western academic circles. This image is reinforced by the ubiquity of cosmopolitan diversity (or what he calls 'consumerist cosmopolitanism') in the world's major cities. One can eat Chinese food everywhere now, just like one can buy Kentucky Fried Chicken in Beijing. Yet food, tourism, music, literature and clothes are easy faces of cosmopolitanism and they tell us nothing about the spread of democracy or human rights on a global scale.

There are two fundamental problems with the dominant ways in which cosmopolitanism is conceptualized, according to Calhoun. The first concerns cosmopolitanism's relationship to democracy. Contemporary cosmopolitan theory, with its roots in seventeenth and eighteenth-century rationalism, is deeply suspicious of the local and the traditional, notably religion and ethnicity. This attitude is further reinforced by the spectre of 'bad nationalism' and by a particular construction of ethnicity as the reactionary 'other' to globalization. This, however, makes cosmopolitanism an elite perspective on the world and an agent of the institutional order of power relations and capital.

The second problem with cosmopolitanism relates to the issue of social solidarity. One of the virtues of cosmopolitanism is to challenge the logic of nationalism which holds that the nation has a primacy over any other possible groupings; but it conceptualizes the alternative 'too abstractly and vaguely'. This is dangerous, argues

Calhoun, as an 'attenuated cosmopolitanism is likely to leave us lacking the old sources of solidarity without adequate new ones'. Is there a chance of salvaging cosmopolitan democracy? Calhoun's answer is affirmative, provided that cosmopolitans revise their theories in two ways. First, they should admit that 'immanent struggle for a better world always builds on particular social and cultural bases'. In other words, the construction of viable local communities may be equally central in solving the problems of the nation. Thus cosmopolitans must come to terms with tradition, community, ethnicity, religion and above all nationalism. Second, they should accept that both local community and nationalism have developed remarkable capacities for binding people to each other. 'Cosmopolitan democracy cannot flourish without a comparable basis in social solidarity'. Citizenship, Calhoun notes, must be more than an abstraction. It must be embedded in the practices of everyday life and able to make demands. After all, 'the view from the frequent traveler lounges does not provide an adequate sense of how people in very different circumstances can feel, gain voice and realize their individual and collective projects'.

In the last essay of this volume, Nira Yuval-Davis focuses on the contemporary notion of belonging, with the aim of deconstructing the hegemonic discourses of nationalist politics of belonging 'in which people, territories and states are constructed as immutably connected and the nation is a "natural" extension of one's family to which one should be prepared, if necessary, to sacrifice oneself'.

Yuval-Davis lays special emphasis on the affective dimension of belonging in her essay – belonging not just as being, but also as longing, or yearning – and stresses the differential positionings from which belongings are imagined, in terms of gender, class, stage in the life cycle, and so on. Another crucial intervening factor is the fact that people tend to belong to more than one collectivity and polity. For Yuval-Davis, an examination of the hierarchy and dynamics of power between these collectivities is crucial for a better understanding of belonging.

Yuval-Davis is critical of the dominant versions of identity politics as they do not differentiate between elements of identification and participation in the construction of belonging. She argues that citizenship signifies the participatory dimension of belonging whereas identification relates to the more emotive, affective dimension.

What we need, she claims, is a politics of belonging that encompasses both dimensions.

There are three candidates for such a politics of belonging. The first is Otto Bauer's attempt to revise the conventional principle of national self-determination in terms of what he called 'the personality principle', whereby the members of each national group will govern themselves, no matter where they are located. The second candidate is the discourse of 'indigenousness' with its emphasis on rights and recognition. The aboriginal perception, for instance, that 'they belong to the land', rather than that 'the land belongs to them', may lead to an alternative, non-exclusive, mode of ownership and sovereignty, similar to Bauer's personality principle.

Yuval-Davis is more ambivalent about the third candidate, namely the politics of diaspora. She notes the extent to which the literature on diasporas is characterized by binary, naturalized and essentialist ideas about kinship, nature and territory. Diasporas, she argues, are much more heterogeneous than is generally assumed. Moreover, many theories of diasporas do not take into account the effects diaspora yearning can have on the homeland. Following Anderson she claims that diasporic politics is often reckless politics without accountability and due democratic processes. Yet the development of transport and communicative technologies has intensified the contact between diasporas and homelands. This may lead to new possibilities of getting together and have a positive impact on the discourses of belonging.

Yuval-Davis concludes her essay by highlighting the extent to which belonging is 'multiplex and multi-layered, continuous and shifting, dynamic and attached'. The task ahead of us is to explore the ways of developing a form of political participation 'in which differential belongings and positionings are acknowledged in a non-exclusionary way'.

Concluding remarks

It is customary to end the Introduction to an edited volume by identifying a common theme, or a shared ground, that unites all contributions. This Introduction will be an exception, because it is not the presence of a single thread that runs through all the contributions that gives this volume its value, but the overlapping of 'differing threads, intersecting, entwined, one taking up where another breaks off, all of them posed in effective tensions with one another to form a composite body' (the image of a rope is of course Wittgenstein's; for the particular interpretation adopted here see Geertz 2000). The contributors come from a variety of disciplines, ranging from history to political theory, from sociology to international relations, and espouse different theoretical and normative perspectives. Not surprisingly, their views on a number of important issues diverge.

Take the nature of nationalism. For John A. Hall, nationalism is like the libido, 'essentially labile, characteristically absorbing the flavours of the historical forces with which it interacts'. In an analogous way, Partha Chatterjee stresses 'the ambivalence of the nation as a narrative strategy as well as an apparatus of power'. Craig Calhoun concurs with Hall and Chatterjee with regard to the discursive nature of nationalism, but argues that it is not easily abandoned even if its myths, contents and excesses are debunked. John Hutchinson, on the other hand, accuses what he calls 'modernist interpretations' of downplaying the extent to which states have been shaped by older ethnic identities. For him, nationalism derives much of its power 'from its capacity to overcome contingency by finding "solutions" based on a past believed to be authentic'.

The authors also disagree on the issue of the model of belonging best suited to the needs of the twenty-first century, and the related issue of the tension between universalism and particularism. Chatterjee and Calhoun, for instance, are both critical of ethical universalism, a heritage of Enlightenment thinking, and advocate a politics of heterogeneity which recognizes the needs of various groups and a form of cosmopolitanism more in peace with local democracy and traditional sources of solidarity respectively. By contrast, Halliday remains unabashedly universalist and calls for a hierarchy of values where universal rights will have pride of place. For Hall, the best option seems to be a civil form of nationalism which recognizes diversity, for Yuval-Davis a politics of belonging which will encompass the dimensions of citizenship and identification.

Yet, despite their disagreements on these and a host of other issues, all contributors admit, implicitly or explicitly, that nations and nationalism are not on the verge of extinction. Given this, the challenge for us is to explore ways of reducing, if not eradicating, the pain and suffering nationalism has caused in the past century. This book steers clear from the trap of intellectual complacency and takes up this challenge. 'Regardless of whether we will be living in one, or five, or fifty states', writes Mira Furlan,

> let us not forget the people, each individual, regardless of which side of this Wall of ours the person happens to be on. We were born here by accident, we are this or that by accident, so there must be more than that, mustn't there?[4]

Notes

1 I am grateful to Atsuko Ichijo, Ayhan Kaya and Spyros A. Sofos for their comments on an earlier draft of this essay.
2 These quotations are from 'A Letter to My Co-Citizens', originally published in *Danas*, Zagreb, 5 November 1991 and *Politika*, Belgrade, 10 November 1991, reproduced in The Mira Furlan Information Station, www.geocities.com/~mfinfostation/miraltr.htm.
3 All the quotations in the following four sections are from the essays in this volume, unless otherwise stated.
4 'A Letter to My Co-Citizens', www.geocities.com/~mfinfostation/miraltr.htm.

References

Achenbach, J. (2001) 'The Clash', *Washington Post*, 16 December.
Anderson, B. (1998) *The Spectre of Comparisons: Nationalism, Southeast Asia and the World*, London: Verso.
Bauer, O. (2001) [1924] *The National Question and Social Democracy*, Minnesota: University of Minnesota Press.
Fukuyama, F. (1989) 'The End of History', *The National Interest*, 16, 3–18.
Fukuyama, F. (1992) *The End of History and the Last Man*, London: Hamish Hamilton.
Fukuyama, F. (2001) 'The West Has Won', *The Guardian*, 11 October.
Furlan, M. (1991) 'A Letter to My Co-Citizens', www.geocities.com/~mfinfostation/.
Geertz, C. (2000) *Available Light: Anthropological Reflections on Philosophical Topics*, Princeton: Princeton University Press.
Gellner, E. (1983) *Nations and Nationalism*, Oxford: Blackwell.
Huntington, S. (1997) *The Clash of Civilizations and the Remaking of World Order*, London: Touchstone Books.

2
Conditions for National Homogenizers

John A. Hall

This chapter is an exercise in insubordination. Rather than discussing nationalism in the twenty-first century, I concentrate on the end of the nineteenth century and the first half of the twentieth century. There are two reasons for this. One is that my knowledge base, such as it is, lies within this period. Differently put, I am reluctant to predict the future until I have better understood the past. The second reason is perhaps more meritorious. To discuss homogeneity may be a useful background to some of the points made elsewhere in this volume about the possibility of greater social heterogeneity. Differently put, some historical understanding at the start may be useful if it explains from whence we have come – as an aid to evaluating assertions about our future options. As will be revealed, I am skeptical of claims now being made for social heterogeneity.

Hence it is all the more important for me to stress a preliminary theoretical point, best expressed by saying that Sigmund Freud's contribution to nationalist studies has been sadly neglected. What is in question here is not *Moses and Monotheism*, interesting as is its discussion of Jewish cultural uniqueness. What matters much more is the analogy that can be drawn between the nature of the libido and that of nationalism. The libido is famously promiscuous and perverse, sticky and mobile, prone to gain character from the elements to which it attaches itself. Nationalism is just like that, essentially labile, characteristically absorbing the flavours of the historical forces

15

with which it interacts. Nationalism has, to adapt Sartre, existences rather than any single essence. The point being made can usefully be underscored. Nationalism does change, as will be seen from the analysis proposed here. So any skepticism about the putative increase in heterogeneity should not be ascribed to the author holding any completely essentialist view. Differently put, to admit that change is of the essence of nationalism does not mean that every claim about mutation in this protean force is necessarily veridical. But these are merely warning notes: a good deal more theoretical work will need to be done in the conclusion.

The argument proceeds in four straightforward steps. I begin by out-lining what is meant by the concept of national homogenization. An attempt is then made to outline at least some of the conditions allow-ing national homogenization to become so very powerful and promi-nent between roughly 1890 and 1945. The third section insists that times 'have' changed. More specifically the tone of nationalism in the advanced world has become moderate, even shamefaced – which is not, it should be stressed immediately, to endorse the popular view that the nation-state has somehow lost its salience as the political form for the modern world. The final section will offer some comments about the world outside Europe. These are tentative, limited and exploratory. Nonetheless, these comments can serve to illustrate, by means of sheer contrast, the nature of the European experience.

What is homogenization?

There is nothing complex about the notion of national homogeniza-tion. The definition of nationalism which suggests that each nation should have its own state – and, quite as much, that each state should have its own nation – implies a complete fit between nation-ality and politics, an absolute correspondence that suggests the notion of homogeneity (the definition and the ideas of this para-graph come from Gellner 1983). It takes but a moment's thought to realize that such correspondence has not generally characterized the historical record. Rulers of most pre-modern societies did not share the norms of those over whom they exercised authority – and, typi-cally, there were several sets of norms, given the presence of many and varied ethnic groups within most territories. Hence we must talk about homogenization, that is, the actual nationalizing practices designed to create that perfect fit between nation and state. Accordingly, we need to examine ways in which social differences were obliterated so as to fit people into a common mould. An initial way to get at the matter – with an alternative being proposed in the conclusion – is to imagine a continuum between voluntary practices on the one hand, and very nasty and vicious practices on the other. Let me examine four points along this scale, leaving it to the reader to fit different historical cases within its parameters.

The softer end of the scale is neatly exemplified by the figures of David Hume and Adam Smith, the two great theorists of modernity, of empiricism and of capitalism, who lived and wrote in eighteenth-century Edinburgh. These thinkers characteristically did not think of themselves as Scottish, habitually preferring to refer to themselves as North Britons. They lived in a semi-peripheral zone, and wanted to be part of the action – that is, they looked south of the border at a growing economy and what they felt to be a more civilized world, and sought entry within it. The Scottish education system, then and now, gave such considerable cultural capital that it was possible for Scots to do well within a larger political system. Although there were attempts to block their entry, these basically failed, as can be seen in a myriad of ways – from the leadership of the Labour Party, past and present, to the railways and bridges constructed by Scottish engineers throughout the empire. This relatively easy voluntary assimilation should not lead one into thinking that everything about Britain was liberal. To the contrary, the British state was exceptionally vicious when handling the tribes of the Scottish Highlands after they had provided a military threat in 1745. Nonetheless, the predominant tone of the British experience was liberal: once the Highlands had been subdued, for example, employment for their military capabilities was found within the established forces of the realm.

Somewhat further along the scale stands the complex case of the United States. On the one hand, there are decidedly vicious elements of American history that are all too easily forgotten. First, the United States was created by means of powerful acts, usually directed from below, of cleansing. The ethnic cleansing of the native population has been terribly neglected, perhaps because it was laced with genocidal tendencies. Equally forgotten is the political cleansing that followed secession and revolution. A significant section of the elite – in absolute numerical terms greater than those guillotined during the

French Revolution, and from a smaller population at that – that had supported the Crown was forced to leave (Palmer 1959: 188–202). Canada thereby gained an element of that anti-Americanism which comprises the key part of its national identity. Second, the United States remained unitary only as the result of a very brutal civil war. The Constitution had of course recognized the different interests of the slave-owning southern states, but the division between North and South grew in the early years of the republic. The works of John Calhoun amount to a myth of hierarchy on the basis of which a new nation might have been formed. War destroyed that diversity, with Lincoln trying at the end of the conflict to create unity by means of such new institutions as Thanksgiving. Of course, the South did not lose its cultural autonomy simply as the result of defeat in war, maintaining a key hold on federal politics well into the 1930s. Nonetheless, over time the South has lost its uniqueness, especially in recent years as the result of political change and of population and industrial transfers from North to South. Third, another alternative vision, that of socialism, was defeated in brutal fashion, as is apparent once we note the very large number of deaths in labor disputes in nineteenth-century America (Mann 1993: chapter 18).

On the other hand, the rosier and milder face of American homogeneity can be seen at work in ethnic relations. A warning should be issued before describing what is a remarkable American achievement. Everything that will be said excludes Afro-Americans, whose position inside the United States remains heavily marked by racial discrimination. The hideousness of what is involved can be seen in the desire of the vast majority of Afro-Americans 'to get in', with great bitterness being shown by middle-class blacks who make it economically only to find that integration does not exist in the suburbs to which they move (Hall and Lindholm 1999: chapter 10). But for the majority of Americans, ethnic identity is now – as Mary Waters' superb *Ethnic Options* (1990) makes clear – a choice rather than a destiny imposed from outside. Rates of intermarriage are extremely high, not least for the first generation of Cuban-Americans in Florida, more than 50 per cent of whom marry outside their own group.[1] Ethnic identity has little real content. It is permissible to graduate from kindergarten wearing a sari as long as one does not believe in caste – that is, as long as one is American. There are severe limits to difference because homogenizing forces remain so strong.

A central facet of modern European history, namely the practices of ethnic cleansing and population transfer, marks a point much further along the scale – that is, a point that brings us instantly closer to social viciousness. Cleansing has a long history in Europe. Religious cleansing began with that of the Jews from fifteenth-century Spain, continued with that of the Huguenots from seventeenth-century France, and peaked with the expulsion of perhaps five million Muslims from the Balkans in the years before 1914. Ethnic cleansing *per se* began to be practiced quite generally at the end of the nineteenth century, peaking amid the fog of the two world wars – with the expulsion of Germans from Central Europe in 1945 involving particularly large numbers. This is a world which we know all too well, for its images have been placed on our television screens as the result of the Balkan wars of the last decade.

The extremity of viciousness in the scale is of course reached in genocide, the extirpation of difference through mass murder. Fortunately, there have been relatively few examples of full-scale genocide, although the pattern that began with the Turkish treatment of Armenians and which is forever associated with the Holocaust in fact reached its greatest moment of intensity only recently, in Kampuchea and in Rwanda.[2] This too has been so present on our television screens as to need no further explication.

A first consideration to be stressed about this scale is simple and factual. European history has been the process of national homogenization. At the beginning of the century, perhaps 65 million people lived under alien rule; that is, their states were not led by people co-cultural with themselves. By 1919, this figure had fallen by nearly two-thirds. Today, there are very few countries in all of the advanced world as a whole – Canada, Belgium, the United Kingdom, Switzerland and Spain – which are genuinely multinational, with most of those being, it has to be noted, far from models of social, economic and political stability (Mann 1999). Further, cleansing has been accomplished before our very eyes in the Balkans. Bluntly, this is what European history has been about – making wholly appropriate Mazower's choice of *The Dark Continent* (1998) as the title for his history of twentieth-century Europe.

For this reason, I still believe that the most considerable theorist of nationalism remains Ernest Gellner. This is less because of the way he explained nationalism than because this hideous factual history was absolutely at the center of his vision. He could do no other; his life was given shape by the forces of homogenization. He grew up in Kafka's world of mixed German–Czech–Jewish cultures, bilingual, secular but with a Jewish background. Interwar Prague was made still more remarkable by the presence of a Ukrainian university, and by the entry of scholars such as Einstein, Jakobson and Carnap. He fled aged thirteen only after the Nazis had entered the city, and returned in 1945, after one side of his family had been killed in the Holocaust, to witness the brutal expulsion of the Germans. Convinced that Russian hegemony was inevitable, he became an émigré – only to return again in 1990 in time to see the Czech majority finally get rid of the Slovaks. In the end, the Czech Republic became, as his theory predicted, monotone and unitary. It is important to note that he hated this change, finding the country boring. There was little relation between his descriptive sociology of nationalism and his personal likes and prescriptions.

The maximal point of national homogenization

Softer processes of assimilation to singular standards had taken place, not least for those of Jewish background, since the Enlightenment. Nevertheless, it makes sense to try to understand the maximal point of national homogenization – by which I mean the sudden and general adoption of the more vicious practices described from the end of the nineteenth century. Three factors are considered in turn. It is not claimed that these give a full account, rather that no account can be complete without including them.

The first consideration is that of the difference between two types of nation-building projects in European history. The contrast is that between the sociology of state before nation and that in which nation comes before state. Let me try to spell out the difference.

It is perfectly proper to suggest that England had a centralized polity from 1066. There was a single parliament as conquest had made feudalism centralized, and to this was added an exchequer and, somewhat later, a system of common law. Within this world there was a slow process of homogenization. I do not easily read Chaucer but I understand Shakespeare without any trouble at all – which is to say that a measure of linguistic unification had taken place by the end of the sixteenth century. Such unification took place over a long period, as varied elites found it in their interest to gain the services of the state ruling over them. Care should be taken not to exaggerate. Of course, much of what has been said applies most strongly only to England. Great Britain was a composite affair, with the politics of Ireland resembling those of the world to which we now turn.

The great European land empires of the end of the nineteenth century – those of the Ottomans, the Romanovs and the Habsburgs – were utterly composite, the creation of marriages in which varied territories had different linguistic and administrative systems. States often sought to rationalize their possessions, particularly through linguistic policies, but in a sense their attempt to do so came too late. Nations were able to protect themselves, for they were able to codify their own languages, and to create their own educational systems. This is the classic world of national awakening. This was a movement from below, one element of which – namely the fact that social inequality combined with cultural distinctiveness could encourage secession – stood at the center of Gellner's theory of nationalism. As it happens, his account seems to me incomplete; but that critical point is best made after considering the second explanatory factor. Still, the basic generalization – that when state building occurred late it ran into the problem of nationalities – is certainly correct.

The second factor can be highlighted by reference to Max Weber, the sociologist of the sociologists. His nickname when young among those who knew him well was 'Polish Max'. His first research had been on the Polish laborers on the agricultural estates of East Germany. He warned obsessively that the presence of such aliens represented a threat to the unity of the nation: it might make sense in economic terms, but it would weaken the fabric of a great power. This sort of view – insisting that the seeming homogeneity of Britain had to be copied so as to increase state strength – was the common currency of the time. It is easy to see why. Armed forces would surely be more efficient, as would civil life, if a single language dominated one's complete territory. Further, the removal of difference might allow the sinews of the state to be strengthened by means of an increase in fiscal extraction. Most generally, social energies might be released if all were part of a single project. It should be noted that there were very different options open to the various empires of the time. The British made a distinction between white Greater Britain and the darker colonies. Russia hoped that a nation-state would emerge, but realized that this depended upon the Ukrainians seeing themselves as but 'little Russians' – for without them there simply would not be enough Russian 'ethnics' to build a nation-state. There certainly were not enough Germans to create a nation-state in Austria, while the Ottoman sphere for some time saw a conflict between those who wanted to recreate the empire as Muslim and others who wished to create a new Turkish nation-state.

This is the appropriate moment to comment on the central weakness of Gellner's account. His definition of nationalism binds together the notion of nation and secession; that is, it implies that every national awakening is bound to lead to the creation of a new state. But this was not necessarily so. The awakening of the nations did not – and does not – always mean that every nation sought its own state. Most Slavs were scared, most obviously in 1848, of establishing a series of small states, likely to be so much at odds with one other as to be non-viable, since the resulting power vacuum would almost certainly suck in either Russia or Germany – as indeed happened. Far better would be a liberal Austro-Hungary, that is, a constitutional monarchy, which would respect the historic liberties of the national communities, allowing them linguistic and cultural rights. It is sometimes argued that Masaryk only made up his mind to opt for secession in 1916 when a final imperial plan came out showing that the empire was not going to become liberal.

A similar point can be made about the Ukraine. It was not really necessary for the empire to be so repressive to the Ukrainians. The greatest enemies of the Ukrainians were the Poles – who were distrusted quite as much by the Tsar. Common cause could have been made, and this would anyway have been wise given that the cultural revival of Ukrainian in Austrian Galicia probably meant that the extirpation of Ukrainian culture was impossible. The point being implied here is a simple one. Of course, socio-economic factors, the combination of social inequality with a cultural marker, can undermine social stability. Still, politicized national consciousness – that is, the move from cultural awareness to the demand for a state of one's own – resulted most of all from the behavior of the state with which the nations interacted. Where there was no voice, there was no loyalty – making exit eventually an attractive and rational option. We can adapt a famous image of Gellner's and say: Ruritanias are created by the illiberalism of Megalomanias (for the key concepts in this paragraph see Hirschman 1978; for the more political view of nationalism see Hall 1998 and Lieven 2000). The general point being made can usefully be underscored. Gellner's socio-economic account needs, at the least, to be complemented by an appreciation of political factors. As always, central to politics is the nature of geopolitical relations. In this period such relations were visceral and intense, thereby politicizing nationalism still more. For the third general factor, again brought to mind by Max Weber, is one which very clearly demonstrates the way in which nationalism can be influenced by a particular political context. Weber was famously a 'fleet professor', a member of the Naval League. This biographical detail is mentioned as a symbol of the more general fact that in this period nationalism was associated with imperialism. To be a strong nation-state meant that you needed to have – or, rather more precisely, it was believed that you needed to have – your own secure sources of supply for minerals and food as well as your own markets. Geopolitical autonomy meant, to put the matter differently, that a premium was placed on territorial aggrandizement. There is a famous book written at the time by a friend of Max Weber's, Werner Sombart, suggesting that Germany had a geopolitical choice between heroism and trading (Sombart 1915). The trading option was a real one, given that Germany was prospering within the terms of the established international order – with its economy, for instance, overtaking that of Great Britain in 1913. But Sombart's – and Weber's – preference was for heroism, for Germany to seek to control its destiny in an absolute sense, not least since German 'Kultur' had so much to offer the world. We must recognize and try to understand the rationality behind this position, even though it is not that of our own age. It is worth noting in this connection that if Weber had been privy to the plans of the British Imperial War Cabinet to blockade Germany so as to starve it to death, he would surely have been confirmed in his desire for territory (Offer 1990). The intensity of geopolitical struggle made nationalism aggressive and expansionist.

As it happens, these three factors did not cause the outbreak of war in 1914. The outbreak of hostilities was rather the result of more traditional geopolitical miscalculations. Still, once war had broken out these conditions – the nationalities question, the political drive for national unity, and the link between nationalism and imperialism – made the war exceptionally vicious (*cf.* Kaiser 1990). And without further ado, I will claim that the sheer scale of the First World War led to the Second World War – for a whole series of reasons, including the destruction of social institutions and the presence of disillusioned soldiers drawn to paramilitary activity. The intensity of this great single great conflict exacerbated the variables on which emphasis has been placed. War multiplied fear, and created a fog within which massive ethnic cleansing and genocide took place. No wonder that nationalism gained such an appalling reputation.

Limits to self-satisfaction

It is of course vital to determine whether the world has changed. Gellner felt that his theory of nationalism was general and universal. Could it be, however, that he theorized a particular historical moment? Might it not be the case that a measure of moral development has occurred, with nationalism no longer being the force whose dangers we have seen to be so great? Necessarily ambivalent answers must be given to questions such as these.

The crucial background condition was changed completely. Europe's security dilemma was solved. The arrival of the United States, partly asked to stay by Europeans afraid of themselves as much as of the Russians, meant that with security questions resolved economic recovery could be especially striking. Differently put, NATO was a necessary condition allowing Germany at last to abandon heroism in favour of trading. But the framework of the alliance was complemented by changes within Europe itself. The European Union derives from an earlier agreement about coal and steel. What that agreement testifies to is a measure of genius on the part of French bureaucrats. Having suffered geopolitical disaster three times within a single lifetime, key members of the French elite in effect realized that, since Germany could not be beaten, it was better to seek control through a positive embrace. Differently put, France and Germany agreed to give up their capacity to autonomously produce their own weapons. Interdependence resulted from a geopolitical deal (Milward 1992; Anderson 1997; Moravcsik 1998).

Nationalism changes a great deal in this new context. Crucially, the link between nationalism and imperialism is broken. Two elements are worth distinguishing. First, European states in this period gave up their empires and, against all expectations, entered into the most spectacular period of economic growth in their history. There are not many people who believe that Putin's desire to retain Chechnya is a reliable route by means of which Russia will become wealthy. The recipe for the wealth of nation-states seems to be that of down-sizing territorially, if need be, so as to enter into trading relations of sufficient intensity as to move one's economy up the product cycle. This is an enormous change there. Equally importantly, second, is the fact that many of the new nationalisms – notably those of the Catalans, Scots and Quebecois – have no desire to protect themselves from the market. Very much to the contrary, such nationalisms are often confident that they can best prosper within the larger market once freed from the constricting embrace of the larger polities within which they find themselves. This is modern, free-trade nationalism.

It is important to issue a warning at this point. The points that have been made should not be misinterpreted as lending support to the notion that globalization is undermining the power of the nation-state. Many points can be made against this vastly overstated view (Hall 2000), at least two of which do need to be emphasized here. First, an increase in market exchange – less global than Northern – rests, as we have seen, upon geopolitical calculations. Nation-states have not disappeared: rather, they have learnt that the attempt to be total power containers led to disaster, and that modesty has led to success. In politics, as in architecture, less seems to be more. Secondly, national homogenization has not ceased. Much of the talk about diversity has as a background the blunt fact that ethnic cleansing has in fact largely taken place in this world. Liberal tolerance is easy once there is little actually to tolerate. More generally, the talk about diversity misses key features of modern societies, as can be seen particularly clearly if we return to the case of the United States. On one hand, there remains a good deal of compulsion within the United States. There is no possibility of the United States becoming a multinational society. No one wants a second civil war of visceral intensity. All evidence shows that Americans are overwhelmingly opposed to the idea that Spanish should be recognized as a second official language. The toughness of American civic nationalism is well expressed in the quip used in a Texas gubernatorial election some years ago: 'If English was good enough for Jesus Christ, it is good enough for Texas.' This is surely one element ensuring that Spanish is being lost as a second language as fast as was the case for the languages of other immigrant groups in the nineteenth century. On the other hand, the point about the talk of difference – the point that we noted earlier had been so well emphasized by Waters – is that it is so very common. It is as if everyone has a right to an ethnicity, as long as it is without much content – as is in fact the case, as the high rates of intermarriage so clearly demonstrate. Besides that, of course, stand the great homogenizing pressures of Hollywood and of consumerism. The United States is more united now than it ever was in the past.

Beyond Europe

Most general sociological schemes have at their core the notion of the less developed seeking to catch up with the more advanced. This applied within Europe, with both nationalism and industrialization, and probably socialism, being best seen as late development strategies (for an interpretation of socialism in this light see Szporluk 1988). It is all too easy to imagine that this logic will apply to what is after all best known as the less developed world. The result can only be appalling, as noted, if nationalizing homogenization becomes the norm for the world's polities.

Remarkably, there are grounds for hope that this may not generally be the case: the non-European world may manage its affairs better, by invention rather than by imitation. The clearest and most important example is that of the way in which linguistic diversity has not led to nationalist secessions in India. David Laitin (1992) has argued that this is best explained by the 'three plus or minus one' linguistic repertoire available to – and in fact often possessed by – Indians. The three is reached in the following manner. One official all-India language is Hindi; a second is English. The reason for this dual situation is that Nehru's attempt to linguistically homogenize the newly independent nation failed because of the resistance encountered among his own civil servants – whose cultural capital very largely lay in their mastery of a world language. The third language necessary is that of one's state – that is, of one's province. It is possible to subtract a language if one's state is Hindi-speaking, but necessary to add a language if one is a minority in a non-Hindi-speaking province. There can be no doubt about the diversity of life within India, yet – incredibly to European eyes – the situation is rather stable, a sort of Austro-Hungary that works. None of this is to say that Indian life is without tension. It may yet be the case, for instance, that religious division will tear India apart – although there are reasons to doubt this. Still, what has been achieved does allow hope.

This sort of linguistic repertoire is as present in much of Africa. Since the future of that continent is often seen in wholly negative light, it makes sense at least to note two further factors that may constrain national homogenizers. One is simply the presence of a large number of ethnic groups, perhaps 120 in Tanzania alone, none of which is near demographic dominance. In these circumstances, politics tend to be formed of multiethnic coalitions, none of which dares play the ethnic card. Secondly, Africa has by and large seen little sustained interstate war since decolonization, for all that it has been plagued by low intensity internal strife – and by more recent resource-driven conflicts. A negative side of the absence of sustained geopolitical conflict has been relative failure in state-building – so much so, indeed, that there is not much discussion of 'failed states' in Africa (Herbst 2000). But there is another, more positive side to the picture. One factor that intensified ethnic cleansing in Europe was competing claims to a single piece of territory.[3] Near-absolute endorsement of the principle of non-intervention has meant that this factor has by and large been missing in Africa. Of course, none of this is to say, once again, that sweetness and light can be guaranteed. The darkest alternative is of course represented by the genocide in Rwanda. Further, it looks likely that the war in Congo-Zaire will do nothing but harm – neither state nor nation-building, merely reliable destruction.

Since so much reliance has been placed on Laitin's analysis, it is worth looking at his more recent analysis of the erstwhile Soviet sphere (Laitin 1999). By and large, the optimism found in the developing world (that is, the demonstration that many nations can share political space) is abandoned when dealing with the Baltics. All that the beached Russian diaspora can hope in those countries is genuine civic nationalism – that is, the creation of a homogeneous monolingual society in which the desire to join is accepted. The rather different situation in Kazakhstan seems no more likely to lead to diversity. Perhaps, Laitin argues, the Ukraine will do better. One can hope, but one must also fear given the debilitating failure to undertake political and economic reform.

Conclusion

The theoretical claims of this chapter can be spelt out much more clearly by consideration of the celebrated notions of civic and ethnic nationalism, not least as this suggests a schema differentiating options of belonging. It should be clearly said at the start that we should not accept everything that is implied in the formula ethnic/bad, civic/good. For one thing, there is nothing necessarily terrible about loyalty to one's ethnic group – and this sentiment in fact underlies the supposedly civic nationalism of the French. For another, civic nationalism is not necessarily nice: its injunction can be 'join us or else'. This was certainly true of the early United States, and it is equally true of the way in which Paris treated La Vendée during the early years of the Revolution. Differently put, civic nationalism may be as resolutely homogenizing as is ethnic nationalism. This suggests the following scheme:

Ethnic

   (Non-)Liberal Caging   Civic

Civil

Moving clockwise around this circle allows a series of theoretical points to be made.

Ethnic nationalism is indeed repulsive when it is underwritten by relativist philosophies that insist that one should literally think with one's blood. Much less horrible is the combination of ethnic and civic nationalisms represented by France – that is, a world in which one is taken in or allowed in as long as one absorbs the culture of the dominant ethnic group. Civic nationalism becomes more liberal when it moves towards the pole of civility, best defined in terms of the acceptance of diverse positions or cultures. Whether this move is, so to speak, sociologically real can be measured by asking two questions. First, is the identity to which one is asked to accede relatively thin; that is, does it have at its core political loyalty rather than a collective memory of an ethnic group? Second, are rates of intermarriage high? Differently put, is the claim that one can belong whatever one's background in fact borne out by the facts? All this is obvious. Less so, perhaps, is a tension that lies at the heart of multiculturalism. In the interests of clarity, matters can be put bluntly. Multiculturalism properly understood 'is' civil nationalism, the recognition of diversity. But that diversity is – needs to be, should be – limited by a consensus on shared values. Difference is acceptable only so long as group identities are voluntary; that is, insofar as identities can be changed according to individual desire. What is at issue is neatly encapsulated when we turn to the notion of caging.[4] If multiculturalism means that groups have rights over individuals – if, for example, the leaders of a group have the power to decide to whom young girls should be married – then it becomes repulsive. Such multiculturalism might seem liberal in tolerating difference, but it is in fact the illiberalism of misguided liberalism, diminishing life chances by allowing social caging. Such a position is of course relativist, and it is related to ethnic nationalism in presuming that one must think with one's group. Importantly, the link to ethnic nationalism may be very close indeed. If there are no universal standards, and ethnic groups are held to be in permanent competition, then it is possible, perhaps likely, that one group will seek to dominate another.

These are ideal typical positions. But I have joined 'at the descriptive level' that powerful stream of modern social theory suggesting that some positions have greater viability than others. A series of thinkers, interestingly all liberal, have insisted that homogeneity, whether ethnic or civic, is a 'must' if a society is to function effectively. John Stuart Mill (1862: chapter 16) made this claim when speaking about the workings of democracy, insisting that the nationalities question had to be solved in order for democracy to be viable. The great contemporary theorist of democracy Robert Dahl (1977) has reiterated this idea. The background notion here is straightforward. Human beings cannot take too much conflict, cannot put themselves on the line at all times and in every way. For disagreement to be productive in the way admired by liberalism, it must be contained – that is, it must take place within a frame of common belonging. Very much the same insight underlies David Miller's (1995) view that national homogeneity is a precondition for generous welfare regimes. This is correct: the generosity of Scandinavian countries rests on the willingness to give to people exactly like oneself. But the great theorist of the need for social homogeneity was of course Gellner. As it happens, the explanation he offered for this ever more insistently – that of the necessity of homogeneity so that industrial society can function properly – is rather question begging (for a series of critical reviews on this point see most of the essays in Hall 1998). I have preferred a more political account as an explanation for the maximal period of national homogenization. But the visceral experience underlying his image of political space moving from the world of Kokoschka to that of Modigliani – that is, from a world in which peoples were intermingled to one in which national homogeneity was established – does have very great truth to it (Gellner 1983). It is as well to be clear about what is implied here. Nationalism has no essence, in that modes of integration can vary so much – from the voluntary to the forced, from ethnic to civic, and from imperialistically inclined to enraptured by free trade. Still, my argument does suggest that homogeneity is what nationalism is about, and it is essentialist in that sense. This point can be made differently. Civil nationalism – that is, cultural diversity within a shared commitment to minimal liberal political norms – seems to me 'at the prescriptive' level utterly desirable. But it has been hard to achieve. Europe has little room for self-congratulation in this regard: most countries were cleansed; Belgium, Canada and the United Kingdom do badly, while Switzerland is highly idiosyncratic – leaving only Spain as an exemplar of civil nationalism. The miserable record of Europe relates in part to the intensity of its geopolitical conflicts, in part to the potential for singular groups to dominate whole territories. Such hope as we can realistically allow ourselves rests on the possibility that countries of the developing world – thankfully less engaged in interstate war and often lacking dominant or potentially dominant ethnicities – will build on their institutional accomplishments so as not to imitate those who so blithely consider themselves advanced.

Notes

1 I rely here on the research of Elizabeth Arias of the State University of New York at Stony Brook.
2 I have benefited enormously from reading at a late stage a manuscript by Michael Mann on ethnic cleansing. This powerful analysis is exceptionally rich, and I have not as yet absorbed its full implications for the argument made here.
3 Mann argues this in his manuscript on ethnic cleansing.
4 The notion of 'caging' is of course that of Michael Mann. It is variously used in the first two volumes of his *The Sources of Social Power* (1986, 1993).

References

Anderson, P. (1997) 'Under the Sign of the Interim' and 'The Europe to Come', in P. Gowan and P. Anderson (eds), *The Question of Europe*, London: Verso, 51–76, 126–48.

Dahl, R. (1977) *Polyarchy*, New Haven: Yale University Press.

Gellner, E. (1983) *Nations and Nationalism*, Oxford: Blackwell.

Hall, J.A. (ed.) (1998) *The State of the Nation*, Cambridge: Cambridge University Press.

Hall, J.A. (2000) 'Nationalism and Globalization', *Thesis Eleven*, 63, 63–79.

Hall, J.A. and C. Lindholm (1999) *Is America Breaking Apart?*, Princeton: Princeton University Press.

Herbst, J. (2000) *States and Power in Africa*, Princeton: Princeton University Press.

Hirschman, A. (1978) *Exit, Voice and Loyalty*, Cambridge, Mass.: Harvard University Press.

Kaiser, R. (1990) *Politics and War*, Cambridge, Mass.: Harvard University Press.

Laitin, D. (1992) *Language Repertoires and State Construction in Africa*, Cambridge: Cambridge University Press.

Laitin, D. (1999) *Identity in Formation: The Russian-Speaking Populations in the Near Abroad*, Ithaca: Cornell University Press.

Lieven, D. (2000) *Empire: The Russian Empire and its Rivals*, London: John Murray.

Mann, M. (1986) *The Sources of Social Power*, volume 1, Cambridge: Cambridge University Press.

Mann, M. (1993) *The Sources of Social Power*, volume 2, Cambridge: Cambridge University Press.

Mann, M. (1999) 'The Dark Side of Democracy: The Modern Tradition of Ethnic and Political Cleansing', *New Left Review*, 235, 18–45.

Mazower, M. (1998) *The Dark Continent*, London: Allan Lane.

Mill, J.S. (1862) *Considerations on Representative Government*, New York: Harpers.

Miller, D. (1995) *On Nationality*, Oxford: Oxford University Press.

Milward, A. (1992) *The European Rescue of the Nation-State*, Berkeley: University of California Press.

Moravcsik, A. (1998) *The Choice for Europe: Social Purpose and State Power from Messina to Maastricht*, Ithaca: Cornell University Press.

Offer, A. (1990) *The First World War: An Agrarian Interpretation*, London: Routledge.

Palmer, R. (1959) *The Age of Democratic Revolution*, volume 1, Princeton: Princeton University Press.

Sombart, W. (1915) *Handler und Helden*, Leipzig: Dunckler und Humblot.

Szporluk, R. (1988) *Communism and Nationalism: Karl Marx versus Friedrich List*, Oxford: Oxford University Press.

Waters, M. (1990) *Ethnic Options: Choosing Ethnic Identities in America*, Berkeley: University of California Press.

3

The Nation in Heterogeneous Time

Partha Chatterjee

I

Benedict Anderson, in his now classic *Imagined Communities* (1983), has made famous the argument that the nation lives in homogeneous empty time.[1] In this, he was, in fact, following a dominant strand in modern historical thinking that imagines the social space of modernity as distributed in homogeneous empty time. A Marxist could call this the time of capital. Anderson explicitly adopts the formulation from Walter Benjamin and uses it to brilliant effect to show the material possibilities of large anonymous socialities being formed by the simultaneous experience of reading the daily newspaper or following the private lives of popular fictional characters. It is the same simultaneity experienced in homogeneous empty time that allows us to speak of the reality of such categories of political economy as prices, wages, markets, and so on. Empty homogeneous time is the time of capital. Within its domain, capital allows for no resistance to its free movement. When it encounters an impediment, it thinks it has encountered another time – something out of pre-capital, something that belongs to the pre-modern. Such resistances to capital (or to modernity) are therefore understood as coming out of humanity's past, something people should have left behind but somehow haven't. But by imagining capital (or modernity) as an attribute of time itself, this view succeeds not only in branding the resistances to it as archaic and backward, but also in securing for capital and modernity their ultimate triumph, regardless of what some people may believe or hope, because after all, as everyone knows, time does not stand still.

In his recent book *The Spectre of Comparisons* (1998), Anderson has followed up his analysis in *Imagined Communities* by distinguishing between nationalism and the politics of ethnicity. He does this by identifying two kinds of seriality that are produced by the modern imaginings of community. One is the unbound seriality of the every-day universals of modern social thought: nations, citizens, revolutionaries, bureaucrats, workers, intellectuals, and so on. The other is the bound seriality of governmentality: the finite totals of enumerable classes of population produced by the modern census and the modern electoral systems. Unbound serialities are typically imagined and narrated by means of the classic instruments of print-capitalism, namely, the newspaper and the novel. They afford the opportunity for individuals to imagine themselves as members of larger than face-to-face solidarities, of choosing to act on behalf of those solidarities, of transcending by an act of political imagination the limits imposed by traditional practices. Unbound serialities are potentially liberating. Bound serialities, by contrast, can operate only with integers. This implies that for each category of classification, any individual can count only as one or zero, never as a fraction, which in turn means that all partial or mixed affiliations to a category are ruled out. One can only be black or not black, Muslim or not Muslim, tribal or not tribal, never only partially or contextually so. Bound serialities, Anderson suggests, are constricting and perhaps inherently conflictual: they produce the tools of ethnic politics.

Anderson uses this distinction between bound and unbound serialities to make his argument about the residual goodness of nationalism and the unrelieved nastiness of ethnic politics. Clearly, he is keen to preserve what is genuinely ethical and noble in the universalist critical thought characteristic of the Enlightenment. Faced with the indubitable facts of historical conflict and change, the aspiration here is to affirm an ethical universal that does not deny the variability of human wants and values, or cast them aside as unworthy or ephemeral, but rather encompasses and integrates them as the real historical ground on which that ethical universal must be established. Much philosophical blood was spilt in the nineteenth century over the question of whether there was an idealist and a materialist version of this aspiration and, if so, which was the more truthful. Few take those debates seriously any more. But as the sciences and technologies of governmentality spread their tentacles throughout the populated world in the twentieth century, the critical philosophical mind has been torn by the question of ethical universalism and cultural relativism. The growing strength of anticolonial nationalist politics in the middle decades of that century contributed greatly to the recognition of this problem, even though the very successes of nationalism may also have led to the chimerical hope that the cultural conflicts were merely the superficial signs of the production of a richer, more universal, modernity. Decolonization, however, was soon followed by the crisis of the third-world state, and the culture wars became identified with chauvinism, ethnic hatred and cynically manipulative and corrupt regimes. To all intents and purposes, nationalism became incurably contaminated by ethnic politics.

Ben Anderson is one in a dwindling group of thinkers who have refused to accept this diagnosis. He continues to believe that the politics of nationalism and that of ethnicity arise on different sites, grow on different nutriments, travel through different networks, mobilize on different sentiments, and fight for different causes. But unlike many in the Western academy, he has refused to soothe the liberal bad conscience with the balm of multiculturalism. He has also remained an outspoken critic of the hard-headed developmentalist of the 'realist' school whose recipes for third-world countries flow out of a cynical double standard that says 'ethics for us, economics for them'. Anderson closes *The Spectre of Comparisons* with an evocative listing of some of the ideals and affective moments of nationalism and remarks: 'There is something of value in all this – strange as it may seem … Each in a different but related way shows why, no matter what crimes a nation's government commits and its passing citizenry endorses, My Country is ultimately Good. In these straitened millennial times, can such Goodness be profitably discarded?' (1998: 368) Idealist? I think the question is quite meaningless, especially since we know that Anderson, more than anyone else in recent years, has inspired the study of those material instruments of literary and cultural production that made possible the imagining of modern political communities in virtually every region of the world. Romantic? Perhaps, but then much that is good and noble in modern social thinking has been propelled by romantic impulses. Utopian? Yes. And there lies, I think, a major theoretical and political problem, which is also the chief source of my disagreement with Anderson.

I believe Anderson, in the tradition of much progressive historicist thinking in the twentieth century, sees the politics of universalism as something that belongs to the very character of the time in which we now live. It is futile to participate in, or sympathize with, or even give credence to efforts to resist its sway. In his recent book, Dipesh Chakrabarty has drawn our attention to a remark made by E.P. Thompson, a Marxist historian who was justifiably celebrated for his anti-reductionist view of historical agency. In a famous essay on time and work-discipline in the era of industrial capitalism, Thompson spoke of the inevitability of workers everywhere having to shed their pre-capitalist work habits: 'Without time-discipline we could not have the insistent energies of industrial man; and whether this discipline comes in the form of Methodism, or of Stalinism, or of nationalism, it will come to the developing world' (cited in Chakrabarty 2000). Similarly, Ben Anderson speaks of 'the remarkable planetary spread, not merely of nationalism, but of a profoundly standardized conception of politics, in part by reflecting on the everyday practices, rooted in industrial material civilization, that have displaced the cosmos to make way for the world' (1998: 29). Such a conception of politics requires an understanding of the world as 'one', so that a common activity called politics can be seen to be going on 'everywhere'. Politics, in this sense, inhabits the empty homogeneous time of modernity.

I disagree. I believe this view of modernity, or indeed of capital, is mistaken because it is one-sided. It looks at only one dimension of the time–space of modern life. People can only imagine themselves in empty homogeneous time; they do not live in it. Empty homogeneous time is the utopian time of capital. It linearly connects past, present and future, creating the possibility for all of those historicist imaginings of identity, nationhood, progress, and so on, that Anderson along with others have made familiar to us. But empty homogeneous time is not located anywhere in real space – it is utopian. The real space of modern life consists of heterotopia – my debt to Michel Foucault should be obvious (1998: 175–85). Time here is heterogeneous, unevenly dense. Here, even industrial workers do not all internalize the work-discipline of capitalism, and more curiously, even when they do, they do not do so in the same way. Politics here does not mean the same thing to all people. To ignore this is, I believe, to discard the real for the utopian.

Homi Bhabha, describing the location of the nation in temporality, pointed out a few years ago how the narrative of the nation tended to be split into a double time: in one, the people were an object of national pedagogy because they were always in the making, in a process of historical progress, not yet fully developed to fulfil the nation's destiny, but in the other, the unity of the people, their permanent identification with the nation, had to be continually signified, repeated and performed. Bhabha (1990) also showed how Anderson, in borrowing Walter Benjamin's notion of the homogeneous empty time of the nation's narrative, entirely failed to notice the profound ambivalence that becomes inescapable when one tries to tell the story of the fullness of the nation's life. I will attempt in this chapter to illustrate some of the instances of this ambivalence and argue that they are an inevitable aspect of modern politics itself; to disavow them is either wishful piety or an endorsement of the existing structure of dominance within the nation.

It is possible to cite many examples from the postcolonial world that suggest the presence of a dense and heterogeneous time. In those places, one could show industrial capitalists delaying the closing of a business deal because they hadn't yet had word from their respective astrologers, or industrial workers who would not touch a new machine until it had been consecrated with appropriate religious rites, or voters who would set fire to themselves to mourn the defeat of their favourite leader, or ministers who openly boast of having secured more jobs for people from their own clan and having kept the others out. To call this the co-presence of several times – the time of the modern and the times of the pre-modern – is only to endorse the utopianism of Western modernity. Much recent ethnographic work has established that these 'other' times are not mere survivals of a pre-modern past: they are new products of the encounter with modernity itself. One must therefore call it the heterogeneous time of modernity. And to push my polemical point a little further, I will add that the postcolonial world outside Western Europe and North America actually constitutes 'most' of the populated modern world.

Let me discuss in some detail an example of the continuing tension between the utopian dimension of the homogeneous time of capital and the real space constituted by the heterogeneous time of governmentality and the effects produced by this tension on efforts to narrativize the nation.

II

B.R. Ambedkar (1891–1956) is famous as the foremost political leader in the twentieth century of India's downtrodden Dalit peoples – the former untouchable castes. In this role, he has been both celebrated and vilified for having strenuously fought for the separate political representation of the Dalits, for preferential reservation or affirmative action in their favour in education and government employment, and for constructing their distinct cultural identity going as far as conversion to another religion – Buddhism. At the same time, Ambedkar is also famous as the principal architect of the Indian constitution, a staunch advocate of the interventionist modernizing state and of the legal protection of the modern virtues of equal citizenship and secularism. Seldom has the tension between utopian homogeneity and real heterogeneity been played out more dramatically than in the intellectual and political career of B.R. Ambedkar.

I do not propose to give here a full intellectual biography of Ambedkar, which is a task I am not competent to carry out but on which, I believe, the definitive work still remains to be done. What I will do instead is focus on certain moments in that biography to highlight the contradictions posed for a modern politics by the rival demands of universal citizenship on the one hand and the protection of particularist rights on the other. My task will be to show that there is no available historical narrative of the nation that can resolve those contradictions.

Ambedkar was an unalloyed modernist. He believed in science, history, rationality, secularism and above all in the modern state as the site for the actualization of human reason. But as an intellectual of the Dalit peoples he could not but confront the question: what is the reason for the unique form of social inequality practised within the so-called caste system of India? Being a modernist, he rejected all answers that relied on a faith in mythical religion or the sanctity of the scriptures. He wanted an answer that would stand the tests of science. One such answer commonly given in the early twentieth century was sociological: the caste system, it was said, was the particular form taken in India of the universal principle of the division of labour. Ambedkar submitted this answer to the scrutiny of reason and concluded that, as a system of division of labour, the caste system was utterly irrational, inefficient and a hindrance to the advancement of social production and general prosperity. If a rational principle of division of labour was desired, the caste system should be the first to go (Ambedkar 1936: 47–9).

Another answer that was both respectable and fashionable in the early twentieth century was based on the identification, by the examination of physical features, linguistic affinities and kinship patterns, of racial types. Ambedkar took this evidence seriously, evaluated it at great length in his books *Who Were the Shudras?* (1946) and *The Untouchables* (1948), and came to the unambiguous conclusion that whatever differences there might be of racial types in India, they were distributed regionally and not by caste. That is to say, the upper and lower castes in the social hierarchy of any region of India belonged to the same racial type. As he once put it in a lecture that was never delivered because the upper-caste organizers, on reading an advance copy, chose to cancel the conference rather than have Ambedkar preside over it:

> To hold that distinctions of Castes are really distinctions of race and to treat different Castes as though they were so many different races is a gross perversion of facts.... The Brahmin of the Punjab is racially of the same stock as the Chamar of the Punjab and the Brahmin of Madras is of the same race as the Pariah of Madras. Caste System does not demarcate racial division ... Caste System is a social division of people of the same race. (Ambedkar 1936: 50)

Interestingly, Ambedkar was not questioning the scientific claims of anthropometry or even of eugenics which still enjoyed considerable intellectual prestige. His claim was that biological knowledge proved that caste distinctions were not based on racial distinctions:

> An immense lot of non-sense is talked about heredity and eugenics in defence of the Caste System. Few would object to the Caste System if it was in accord with the basic principle of eugenics because few can object to the improvement of the race by

judicious mating. But one fails to understand how the Caste System secures judicious mating. (*Ibid.*)

He then goes on to show that since the rule of endogamy operates on sub-castes, each sub-caste, of which there were more than a hundred in any regional caste system, would have to be regarded as a pure racial group, which was absurd. It would also leave unexplained why eugenics should prohibit not only marriage with other castes but also dining with other castes (*ibid.*: 50–1). In other words, Ambedkar's argument was that the science of race, using the rational methods of science, could not provide a rational explanation for the caste system. The claim is important because, as Gail Omvedt has pointed out, Ambedkar was here explicitly moving away from racial ideas, such as the division between Aryan and Dravidian peoples, that had become dominant in the Non-Brahmin movements in much of western and southern India (1994: 244–7). Ambedkar was rejecting a narrative that was based on an original racial split between dominant and oppressed castes.

Ambedkar proposed instead a historical narrative of the origins of untouchability. The Shudras, the lowest in the four-fold hierarchy among the Indo-Aryan people, were, he argued, originally part of the warrior caste. There was a long feud between the Shudra kings and the Brahmins in which the latter 'were subjected to many tyrannies and indignities'. Later, when the Brahmins became politically dominant, they satisfied their hatred by imposing ritual degradations on the Shudras and relegating them to the same status as the non-Aryan peoples. Thus, the hierarchical division of caste was produced not by natural or racial or economic reasons but through a political history of conflict (this is the argument in Ambedkar 1946).

But this still does not explain the origin of the untouchables groups who are altogether outside the caste hierarchy. To explain this, Ambedkar considers the historical stage of the transition from nomadic-pastoral society to that of settled cultivation. When different groups among the Indo-Aryan peoples began to form settled villages, they were faced with the problem of defending their settlements against the marauding nomads. They turned to the 'broken men' – those scattered groups of nomadic tribesmen who had been defeated in battle and separated from their original communities. The arrangement was mutually convenient: the villagers needed people to protect them, the broken men needed a means of livelihood. The latter were invited to live outside the village boundaries and guard the village; in return, they were assured of sustenance. But, of course, they lived their own lives in their own ways, outside the social and moral order of the settled village communities. They were separate, but not in any way degraded. That happened much later, when the Brahmins, in their historic struggle against Buddhism, resolved to adopt vegetarianism, and especially the avoidance of beef-eating, as the most potent and morally superior sign of purity. Ambedkar marshalled all sorts of evidence to claim that this could not have happened before the fourth century A.D. The broken men, however, were too poor to forsake the consumption of dead cows, not only for the meat but also for the numerous other uses of the skins and bones in which knowledge, as former pastoralists, they were highly proficient. To this day, argued Ambedkar, they lived outside the village, ate beef, dealt in hides and leather, made things out of cowskins, but of course were treated as untouchables by caste Hindus. But untouchability did not go back to times immemorial; it had a definite history that could be scientifically established to be no longer than about 1500 years (Ambedkar 1948).

It is not necessary for us here to judge the plausibility of Ambedkar's theory of the origin of untouchability, except to say that at least as far as the rise of vegetarianism as a sign of Brahminical purity is concerned, the French sociologist Louis Dumont, a leading modern theorist of the caste system and no subscriber to the Ambedkarist ideology, has connected it explicitly to the reaction to Buddhism (1970: 194). What is more interesting for our purposes is the narrative structure suggested by Ambedkar's historical theory of caste. There was, in the beginning, a state of equality between the Brahmins, the Shudras and the untouchables. The Shudras belonged to the warrior caste and the untouchables were nomadic tribesmen. This equality, moreover, was not in some mythological state of nature but at a definite historical moment in which all Indo-Aryan tribes were nomadic pastoralists. Then came the stage of settled agriculture and the reaction, in the form of Buddhism, to the sacrificial religion of the Vedic tribes. This was followed by the conflict between the Brahmins and the Buddhists, leading to the political defeat of Buddhism, the degradation of the Shudras and the relegation of the beef-eating 'broken men' into untouchability.

The modern struggle for the abolition of caste was thus a quest for a return to that primary equality that was the original historical condition of the nation. The utopian search for homogeneity is thus made historical. It is, as we know, a familiar historicist narrative of modern nationalism.

To show how this narrative is disrupted by the heterogeneous time of colonial governmentality, let me turn to the fiction of nationalism.

III

One of the greatest modernist novels about Indian nationalism is *Dhorai charit manas* (vol. 1, 1949; vol. 2, 1951) by the Bengali writer Satinath Bhaduri (1906–65) (Ghosh and Acharya 1973). The novel is deliberately constructed to fit the form of the *Ramcharitmanas*, the retelling in Hindi by the sixteenth-century saint-poet Tulsidas (1532–1623) of the epic story of Rama, the mythical king who, through his exemplary life and conduct, is supposed to have created the most perfect kingdom on earth. Tulsidas's Ramayana is perhaps the most widely known literary work in the vast Hindi-speaking regions of India, providing an everyday language of moral discourse that cuts across caste, class and sectarian divides. It is also said to have been the most powerful vehicle for the generalization of Brahminical cultural values in northern India. The distinctness of Satinath Bhaduri's modernist retelling of the epic is that its hero, Dhorai, is from one of the backward castes.

Dhorai is a Tatma from northern Bihar (the district is Purnea, but Satinath gives it the fictional name Jirania). It is not an agriculturist group, specializing instead in the thatching of roofs and the digging of wells. When Dhorai is a child, his father dies, and when his mother wants to remarry, she leaves him in the charge of Bauka Bawa, the village holy man. Dhorai grows up going from door to door, accompanying the sadhu with his begging bowl, singing songs, mostly about the legendary king Rama and his perfect kingdom. The mental world of Dhorai is steeped in mythic time. He never goes to school but knows that those who can read the Ramayana are men of great merit and social authority. His elders – those around him – know of the government, of course, and know of the courts and the police, and some in the neighbourhood who worked in the gardens and kitchens of the officials could even tell you when the district magistrate was displeased with the chairman of the district board, or when the new kitchenmaid was spending a little too much time in the evenings in the police officer's bungalow. But their general strategy of survival, perfected over generations of experience, is to stay away from entanglements with government and its procedures. Once, following a feud, the residents of the neighbouring hamlet of Dhangars set fire to Bauka Bawa's hut. The police investigate and Dhorai, the sole eyewitness, is asked to describe what he has seen. As he is about to speak, he notices Bauka Bawa's eyes. 'Don't talk', the Bawa seems to say. 'This is the police, they'll go away in an hour. The Dhangars are our neighbours, we'll have to live with them.' Dhorai understands and tells the police that he had seen nothing and did not know who had set fire to their house.

One day, Dhorai, along with others in the village, hear of Gandhi Bawa who, it was said, was a bigger holy man than their own Bauka Bawa or indeed any Bawa they had known, because he was almost as big as Lord Rama himself. Gandhi Bawa, they heard, ate neither meat nor fish, had never married and roamed around completely naked. Even the Bengali schoolmaster, the most learned man in the area, had become Gandhi Bawa's follower. Soon there is excitement in the village when it is found that an image of Gandhi Bawa appears on a pumpkin. With great festivity, the miraculous pumpkin is installed in the village temple and offerings are made to the greatest holy man in the country. Gandhi Bawa, the Tatmas agreed, was a great soul indeed because even the Muslims promised to stop eating meat and onions, and the village shaman, whom no one had ever seen sober, vowed henceforth to drink only the lightest toddy and to stay away completely from hemp and opium. Some time later, a few villagers went all the way to the district town to see Gandhi Bawa himself, and came back with their enthusiasm somewhat deflated. The huge crowds had prevented them from seeing the great man from close up but what they had seen was incongruous. Gandhi Bawa, they reported, like the fancy lawyers and teachers in town, wore spectacles! Who had ever seen a holy man wear spectacles? One or two even whispered if the man might not, after all, be a fake.

Satinath Bhaduri's intricately crafted account of Dhorai's upbringing among the Tatmas in the early decades of the twentieth century could be easily read as a faithful ethnography of colonial governance and the nationalist movement in northern India. We know, for instance, from Shahid Amin's studies how the stature and authority of Mahatma Gandhi was constructed among India's peasantry through stories of his miraculous powers and rumours about the fate of his followers and detractors, or how the Congress programme and the objectives of the movement were themselves transmitted in the countryside in the language of myth and popular religion (Amin 1984, 1995). If Gandhi and the movements he led in the 1920s and 1930s were a set of common events that connected the lives of millions of people in both the cities and the villages of India, they did not constitute a common experience. Rather, even as they participated in what historians describe as the same great events, their own understandings of those events were narrated in very different languages and inhabited very different life-worlds. The nation, even if it was being constituted through such events, existed only in heterogeneous time.

Of course, it might be objected that the nation is indeed an abstraction, that it is, to use the phrase that Ben Anderson has made famous, only 'an imagined community' and that, therefore, this ideal and empty construct, floating as it were in homogeneous time, can be given a varied content by diverse groups of people, all of whom, remaining different in their concrete locations, can nevertheless become elements in the unbound seriality of national citizens. Without doubt, this is the dream of all nationalists. Satinath Bhaduri, who was himself a leading functionary in the Congress organization in Purnea district, shared the dream. He was acutely aware of the narrowness and particularism of the everyday lives of his characters. They were yet to become national citizens. But he was hopeful of change. He saw that even the lowly Tatmas and Dhangars were stirring. His hero Dhorai leads the Tatmas into defying the local Brahmins and wearing the sacred thread themselves – in a process, occurring all over India at this time, that the sociologist M.N. Srinivas (1966) describes as Sanskritization, but which the historian David Hardiman (1987) has shown to be marked by a bitterly contested and often violent struggle over elite domination and subaltern resistance. The intricate caste and communal grid of governmental classifications is never absent from Satinath's narrative. But in a deliberate allusion to the life-story of the legendary Prince Rama, Satinath throws his hero Dhorai into a cruel conspiracy hatched against him by his kinsmen. He suspects his wife of having a liaison with a

Christian man from the Dhangar hamlet. He leaves the village, goes into exile and resumes his life in another village, among other communities. Dhorai is uprooted from the narrowness of his home and thrown into the world. The new metalled roadway, along which motorcars and trucks now whizz past ponderous bullock-carts, opens up his imagination:

> Where does this road begin? Where does it end? [Dhorai] doesn't know. Perhaps no one knows. Some of the carts are loaded with maize, others bring plaintiffs to the district court, still others carry patients to the hospital. In his mind, Dhorai sees shadows that suggest to him something of the vastness of the country. (Bhaduri 1973: 70)

The nation is coming into shape. Satinath sends his hero off on an epic journey towards the promised goal – not of kingdom, because this is no longer the mythical age of Rama, but of citizenship.

IV

Ambedkar's dream of equal citizenship also had to contend with the fact of governmental classifications. As early as 1920, he had posed the problem of representation faced by untouchables in India: 'The right of representation and the right to hold office under the state are the two most important rights that make up citizenship. But the untouchability of the untouchables puts these rights far beyond their reach ... they [the untouchables] can be represented by the untouchables alone'. The general representation of all citizens would not serve the special requirements of the untouchables, because given the prejudices and entrenched practices among the dominant castes, there was no reason to expect that the latter would use the law to emancipate the untouchables: ' ... a legislature composed of high caste men will not pass a law removing untouchability, sanctioning intermarriages, removing the ban on the use of public streets, public temples, public schools ... This is not because they cannot, but chiefly because they will not' (cited in Omwedt 1994: 146).

But there were several ways in which the special needs of representation of the untouchables could be secured, and many of these had been tried out in colonial India. One was the protection by

colonial officials of the interests of the lower castes against the politically dominant upper castes or the nomination by the colonial
government of distinguished men from the untouchable groups to
serve as their representatives. Another way was to reserve a certain
number of seats in the legislature only for candidates from the lower
castes. Yet another was to have separate electorates of lower-caste
voters who could elect their own representatives. In the immensely
complicated world of late-colonial constitutional politics in India, all
of these methods, with innumerable variations, were debated and
tried out. Besides, caste was not the only contentious issue of ethnic
representation; the even more divisive issue of religious minorities
became inextricably tied up with the politics of citizenship in late
colonial India.

Ambedkar clearly ruled out one of these methods of special
representation – protection by the colonial regime. In 1930, when
the Congress declared independence or Swaraj as its political goal,
Ambedkar declared at a conference of the depressed classes:

> ...the bureaucratic form of Government in India should be
> replaced by a Government which will be a Government of 'the
> people, by the people and for the people... We feel that nobody
> can remove our grievances as well as we can, and we cannot
> remove them unless we get political power in our own hands. No
> share of this political power can evidently come to us so long as
> the British government remains as it is. It is only in a Swaraj con-
> stitution that we stand any chance of getting the political power
> in our own hands, without which we cannot bring salvation to
> our people.... We know that political power is passing from the
> British into the hands of those who wield such tremendous eco-
> nomic, social and religious sway over our existence. We are will-
> ing that it may happen, though the idea of Swaraj recalls to the
> mind of many the tyrannies, oppressions and injustices practiced
> upon us in the past.... (Cited in Omwedt 1994: 168–9)

The dilemma is clearly posed here. The colonial government, for all
its homilies about the need to uplift those oppressed by the religious
tyranny of traditional Hinduism, could only look after the untouchables as its subjects. It could never give them citizenship. Only under
an independent national constitution was citizenship conceivable

for the untouchables. Yet, if independence meant the rule of the upper castes, how could the untouchables expect equal citizenship and the end of the social tyranny from which they had suffered for centuries? Ambedkar's position was clear: the untouchables must support national independence, in the full knowledge that it would lead to the political dominance of the upper castes, but they must press on with the struggle for equality within the framework of the new constitution.

In 1932, the method of achieving equal citizenship for the untouchables became the issue in a dramatic stand-off between Ambedkar and Gandhi. In the course of negotiations between the British government and Indian political leaders on constitutional reforms, Ambedkar, representing the so-called depressed classes, had argued that they must be allowed to constitute a separate electorate and elect their own representatives to the central and provincial legislatures. The Congress, which had by this time conceded a similar demand for separate electorates for the Muslims, refused to accept that the untouchables were a community separate from the Hindus and was prepared instead to have reserved seats for them to be elected by the general electorate. Ambedkar clarified that he would be prepared to accept this formula if there was any hope that the British would grant universal adult suffrage to all Indians. But since the suffrage was severely limited by property and education qualifications, the depressed castes, dispersed as a thin minority within the general population and, unlike the Muslim minority, lacking any significant territorial concentrations, were unlikely to have any influence at all over the elections. The only way to ensure that the legislature contained at least some who were the true representatives of the untouchables was to allow them to be elected by a separate electorate of the depressed classes.

Gandhi reacted fiercely to Ambedkar's suggestion that upper-caste Congress leaders could never properly represent the untouchables, calling it 'the unkindest cut of all'. Indulging in a rather un-mahatma-like boasting, he declared:

> I claim myself in my own person to represent the vast mass of the Untouchables. Here I speak not merely on behalf of the Congress, but I speak on my own behalf, and I claim that I would get, if there was a referendum of the Untouchables, their vote, and that I would top the poll.

He insisted that unlike the question of the religious minorities, the issue of untouchability was a matter internal to Hinduism and had to be resolved within it.

> I do not mind Untouchables, if they so desire, being converted to Islam or Christianity. I should tolerate that, but I cannot possibly tolerate what is in store for Hinduism if there are two divisions set forth in the villages. Those who speak of the political rights of Untouchables do not know their India, do not know how Indian society is today constructed, and therefore I want to say with all the emphasis that I can command that if I was the only person to resist this thing I would resist it with my life.

True to his word, Gandhi threatened to go on a fast rather than concede the demand for separate electorates for the depressed classes. Put under enormous pressure, Ambedkar conceded and, after negotiations, signed with Gandhi what is known as the Poona Pact by which the Dalits were given a substantial number of reserved seats but within the Hindu electorate (for accounts of the Poona Pact and the relevant citations, see Kumar 1987; Omvedt 1994: 161–89). As it happened, this remained the basic form for the representation of the former untouchable castes in the constitution of independent India, but of course, by this time the country had been divided into two sovereign nation-states.

The problem of national homogeneity and minority citizenship was posed and temporarily resolved in India in the early 1930s; but the form of the resolution is instructive. It graphically illustrates that ambivalence of the nation as a narrative strategy as well as an apparatus of power which, as Homi Bhabha (1990) has pointed out, 'produces a continual slippage into analogous, even metonymic, categories, like the people, minorities, or "cultural difference" that continually overlap in the act of writing the nation'. Ambedkar, as we have seen, had no quarrel with the idea of the homogeneous nation as a pedagogical category – the nation as progress, the nation in the process of becoming – except that he would have insisted with Gandhi and the other Congress leaders that it was not just the ignorant masses that needed training in proper citizenship but the upper-caste elites as well who had still not accepted that democratic equality was incompatible with caste inequality. But Ambedkar refused to join Gandhi in performing that homogeneity in constitutional negotiations over citizenship. The untouchables, he insisted, were a minority within the nation and needed special representation in the political body. On the other hand, Gandhi and the Congress, while asserting that the nation was one and indivisible, had already conceded that the Muslims were a minority within the nation. The untouchables? They represented a problem internal to Hinduism. Imperceptibly, the homogeneity of India slides into the homogeneity of the Hindus. The removal of untouchability remains a pedagogical task, to be accomplished by social reform, if necessary by law, but caste inequality among the Hindus is not to be performed before the British rulers or the Muslim minority. Homogeneity breaks down on one plane only to be reasserted on another. Heterogeneity, unstoppable at one point, is forcibly suppressed at another.

In the meantime, our fictional hero Dhorai continues, in the 1930s, to receive his education in nationalism. Loosed from his moorings, he drifts to another village and starts life afresh among the Koeri, a backward caste of sharecroppers and labourers. Dhorai begins to learn the realities of peasant life – of Rajput landlords and Koeri adhiars and Santal labourers, of growing paddy and jute and tobacco and maize, of moneylenders and traders. In January 1934, Bihar is ripped apart by the most violent earthquake in its recorded history. Government officers come to survey the damage; so do the nationalist volunteers from the Congress. For more than a year, the Koeris hear vaguely that they are going to be given 'relief'. And then they are told that the survey had found that the Koeri huts, being made of mud walls and thatched roofs, had been easily repaired by the Koeris themselves, but the brick houses of the Rajput landlords had suffered severe damage. The report had recommended, therefore, that the bulk of the relief should be given to the Rajputs.

Thus begins a new chapter in Dhorai's education – his discovery that the Bengali lawyers and Rajput landlords were fast becoming the principal followers of the Mahatma. But even as the old exploiters become the new messengers of Swaraj, the mystique of the Mahatma remains untarnished. One day, a volunteer arrives in the village with letters from the Mahatma. He tells the Koeris that they in turn must send a letter each to the Mahatma. No, no, they don't have to pay for the postage stamp. All they have to do is walk up to the officer who would give them a letter which they must put in Mahatmaji's postbox – the white one, remember, not the coloured ones. This was called the 'vote'. The volunteer instructs Dhorai: 'Your name is Dhorai Koeri, your father is Kirtu Koeri. Remember to say that to the officer. Your father is Kirtu Koeri.' Dhorai does as he is told.

> Inside the voting booth, Dhorai stood with folded hands in front of the white box and dropped the letter into it. Praise to Mahatmaji, praise to the Congress volunteer, they had given Dhorai the little role of the squirrel in the great task of building the kingdom of Rama. But his heart broke with sorrow – if only he could write, he would have written the letter himself to the Mahatma. Just imagine, all these people writing letters to the Mahatma, from one end of the country to the other, all together, at the same time. Tatmatuli, Jirania,...Dhorai,...the volunteer,...they all wanted the same thing. They had all sent the same letter to the Mahatma. The government, the officers, the police, the landlords,...all were against them. They belonged to many different castes, and yet they had come so close ... They were linked as though by a spider's web; the fibre was so thin that if you tried to grab it, it would break. Indeed, you couldn't always tell if it was there or not. When it swayed gently in the breeze, or the morning dewdrops clung to it, or when a sudden ray of the sun fell on it, you saw it, and even then only for a moment. This was the land of Ramji over which his avatar Mahatmaji was weaving his thin web...'Hey, what are you doing inside the booth?' The officer's voice broke his reverie. Dhorai came out quickly. (Bhaduri 1973: 222–3)

The vote is the great anonymous performance of citizenship, which is why it probably did not matter too much that Dhorai's introduction to this ritual was through an act of impersonation. But it only concealed the question of who represents whom within the nation. Although the Koeris voted faithfully for the Mahatma, they were dismayed to find that the Rajput landlord with whom they had fought for years was elected chairman of the district board with support from the Congress. Mahatmaji's men, they heard, were now ministers in the government, but when a new road was built, sure enough, it went right past the Rajput houses.

But Dhorai bought himself a copy of the Ramayana. One day, he promised himself, he would learn to read it. The passage to the kingdom of Rama, however, was suddenly disrupted when news arrived that the Mahatma had been arrested by the British. This was the final struggle, the Mahatma had announced. Every true follower of Mahatmaji must now join his army. Yes, the army; they must act against the tyrants, not wait to be arrested. Dhorai is mobilized into the Quit India movement of 1942. This was a war unlike any other; it was, the volunteers said, a revolution. Together, they stormed the police station, setting fire to it. By the morning, the district magistrate, the police superintendent, and all senior officers had fled. Victory to Mahatmaji, victory to the revolution! The district had won Swaraj, they were free.

It didn't last long. Weeks later, the troops moved in, with trucks and guns. Along with the volunteers, Dhorai left for the forests. He was now a wanted man, a rebel. But they were all wanted men – they were Mahatmaji's soldiers. There was a strange equality among them in the forest. They had dropped their original names and called each other Gandhi, Jawahar, Patel, Azad – they were so many anonymous replicas of the representatives of the nation. Except they had been driven away from its everyday life. Sometime later, the word came that the British had won the war with the Germans and the Japanese, the Congress leaders were about to be released and all revolutionaries must surrender. Surrender? And be tried and jailed? Who knows, may be even hanged? Dhorai's unit resolves not to surrender.

V

On the national stage, the Muslim League resolved in March 1940 that any constitutional plan for devolution of power in India must include an arrangement by which geographically contiguous areas with Muslim majorities could be grouped into independent states, autonomous and sovereign. This became known as the Pakistan resolution. The Congress opposed the plan. A few months later, in December 1940, Ambedkar wrote a long book entitled *Pakistan or Partition of India* (1945) in which he discussed in detail the pros and cons of the proposal. It is a book that is, surprisingly, seldom mentioned, even today when there is such a great Ambedkar revival (except by such exemplars of politically sanctioned ignorance and prejudice as Shourie 1997). Apart from the fact that it shows his superb skills as a political analyst and a truly astonishing prescience, I think it is a text in which Ambedkar grappled most productively with the twofold demand on his politics – one, to further the struggle for universal and equal citizenship within the nation, and two, to secure special representation for the depressed castes in the body politic.

The book is almost Socratic in its dialogical structure, presenting first, in the strongest possible terms, the Muslim case for Pakistan, and then the Hindu case against Pakistan, and then considering the alternatives available to the Muslims and the Hindus if there were no partition. What is striking is the way in which Ambedkar, as the unstated representative of the untouchables, adopts a position of perfect neutrality in the debate, with no stake at all in how the matter is resolved – he belongs neither to the Muslim nor to the Hindu side. All he is concerned with is to judge the rival arguments and recommend what seems to him the most realistic solution. But, of course, this is only a narrative strategy. We know that Ambedkar did have a great stake in the question: the most important issue for him was whether or not partition would be better for the untouchables of India. The significance of *Pakistan or Partition of India* is that Ambedkar is judging here the utopian claims of nationhood in the concrete terms of realist politics.

After dissecting the arguments of both sides, Ambedkar comes to the conclusion that, on balance, partition would be better for both Muslims and Hindus. The clinching arguments come when he considers the alternative to partition: how was a united and independent India, free from British rule, likely to be governed? Given the hostility of Muslims to the idea of a single central government, inevitably dominated by the Hindu majority, it was certain that if there were no partition, India would have to live with a weak central government, with most powers devolved to the provinces. It would be 'an anaemic and sickly state'. The animosities and mutual suspicions would remain: 'burying Pakistan is not the same thing as burying the ghost of Pakistan' (Ambedkar 1945: vii). Moreover, there was the question of the armed forces of independent India. In a long chapter, Ambedkar goes straight to the heart of colonial governance and discusses the communal composition of the British Indian army, a subject on which there was a virtual conspiracy of silence. He points out that almost

60 per cent of the Indian army consisted of men from the Punjab, the North-West Frontier and Kashmir, and of them more than half were Muslims. Would a weak central government, regarded with suspicion by the Muslim population, command the loyalty of these troops? On the other hand, should the new government attempt to change the communal composition of the army, would that be accepted without protest by the Muslims of the north-west? (*ibid.*: 55–87).

Judged positively, the new state of Pakistan would be a homogeneous state. The boundaries of Punjab and Bengal could be redrawn to form relatively homogeneous Muslim and Hindu regions to be integrated with Pakistan and India, respectively. Long before anyone had demanded the partition of the two provinces, Ambedkar foresaw that the Hindus and Sikhs would not agree to live in a country specifically created for Muslims and would want to join India. For the North-West Frontier Province and Sind, where the Hindu population was thinly distributed, the only realistic solution was an officially supervised transfer of population, as had happened in Turkey, Greece and Bulgaria. The India or Hindustan that would be created would be composite, not homogeneous. But the minority question could then be handled more reasonably. 'To me, it seems that if Pakistan does not solve the communal problem within Hindustan, it substantially reduces its proportion and makes it of minor significance and much easier of peaceful solution' (*ibid.*: 105).

And then, in a string of brilliant moves of realpolitik logic, Ambedkar shows that only in united India, in which more than a third of the population is Muslim, could Hindu dominance be a serious threat. In such a state, the Muslims, fearing the tyranny of the majority, would organize themselves into a Muslim party such as the Muslim League, provoking in turn the rise of Hindu parties calling for Hindu Raj. Following partition, the Muslims in Hindustan would be a small and widely scattered minority. They would inevitably join this or that political party, pursuing different social and economic programmes. Similarly, there would be little ground left for a party like the Hindu Mahasabha which would wither away. And as for the lower orders of Hindu society, they would make common cause with the Muslim minority to fight the Hindu high castes for their rights of citizenship and social dignity (*ibid.*: 352–8).

Once again, we need not spend time here trying to assess the intrinsic merits of Ambedkar's arguments for and against the partition of India, although in the discursive context of the early 1940s they are remarkably perspicacious. What I have tried to emphasize is the ground on which he lays his arguments. He is fully aware of the value of universal and equal citizenship and wholly endorses the ethical significance of unbound serialities. On the other hand, he realizes that the slogan of universality is often a mask to cover the perpetuation of real inequalities. The politics of democratic nationhood offers a means for achieving a more substantive equality, but only by ensuring adequate representation for the underprivileged groups within the body politic. A strategic politics of groups, classes, communities, ethnicities – bound serialities of all sorts – is thus inevitable. Homogeneity is not thereby forsaken; on the contrary, in specific contexts, it can often supply the clue to a strategic solution, such as partition, to a problem of intractable heterogeneity. On the other hand, unlike the utopian claims of universalist nationalism, the politics of heterogeneity can never claim to yield a general formula for all peoples at all times: its solutions are always strategic, contextual, historically specific and, inevitably, provisional.

Let me then finally return to Anderson's distinction between nationalism and the politics of ethnicity. He agrees that the 'bound serialities' of governmentality can create a sense of community, which is precisely what the politics of ethnic identity feeds on. But this sense of community, Anderson thinks, is illusory. In these real and imagined censuses, 'thanks to capitalism, state machineries and mathematics, integral bodies become identical, and thus serially aggregable as phantom communities' (Anderson 1998: 44). By contrast, the 'unbound serialities' of nationalism do not, one presumes, need to turn the free individual members of the national community into integers. It can imagine the nation as having existed in identical form from the dawn of historical time to the present without requiring a census-like verification of its identity. It can also experience the simultaneity of the imagined collective life of the nation without imposing rigid and arbitrary criteria of membership. Can such 'unbound serialities' exist anywhere except in utopian space?

To endorse these 'unbound serialities' while rejecting the 'bound' ones is, in fact, to imagine nationalism without modern governmentality. What modern politics can we have that has no truck with capitalism, state machineries or mathematics? The historical moment Anderson seems keen to preserve is the mythical moment when classical nationalism merges with modernity. I believe it is no longer productive to reassert the utopian politics of classical nationalism. Or rather, I do not believe it is an option that is available for a theorist from the postcolonial world. Such a theorist must chart a course that steers away from global cosmopolitanism on the one hand and ethnic chauvinism on the other. It means necessarily to dirty one's hands in the complicated business of the politics of governmentality. The assymmetries produced and legitimized by the universalisms of modern nationalism have not left room for any ethically neat choice here. For the postcolonial theorist, like the postcolonial novelist, is born only when the mythical time-space of epic modernity has been lost for ever. Let me end by describing the fate of our fictional hero Dhorai.

Living in the forests with his band of fugitive rebels, Dhorai is brought face to face with the limits to his dreams of equality and freedom. It is not the bound serialities of caste and community that prove illusory, but rather the promise of equal citizenship. The harshness of fugitive life scrapes the veneer off the shell of comradeship and the old hierarchies reappear. Suspicion, intrigue, revenge and recrimination become the ruling sentiments. Dhorai's copy of the Ramayana lies tied up in its bundle, unopened, unread. In the middle of all this, a young boy joins the band. He is a Christian Dhangar, he says, from the hamlet next to Tatmatuli. Dhorai feels a strange bond with the boy. Might he be, he imagines, the son he has never seen? Dhorai looks after the boy and asks him many questions. The more he talks to him, the more he is convinced that this indeed is his son. The boy falls ill, and Dhorai decides to take him to his mother. As he approaches Tatmatuli, he can hardly control his excitement. Was this going to be the epic denouement of the latter-day untouchable Rama? Was he going to be united with his banished wife and son? The mother appears, takes her son in, comes out again and invites the kind stranger to sit down. She talks about her son, about her dead husband. Dhorai listens to her. She is someone else, not his wife. The boy is someone else, not his son. Dhorai makes polite conversation for a few minutes and then goes, we don't know where. But he leaves behind his bundle, along with the copy of the Ramayana for which he has no further need. Dhorai has lost for ever his promised place in prophetic time.

Or has he? Following independence, B.R. Ambedkar became chairman of the drafting committee of the Indian constitution and later the minister of law. In these capacities, he was instrumental in putting together one of the most progressive democratic constitutions in the world, guaranteeing the fundamental rights of freedom and equality irrespective of religion or caste and at the same providing for special representation in the legislatures for the formerly untouchable castes (for the story of the legal provision of opportunities for the depressed castes in independent India, see Galanter 1984). But changing the law was one thing; changing social practices was another matter. Frustrated by the ineffectiveness of the state in putting an end to caste discrimination in Hindu society, Ambedkar decided in 1956 to convert to Buddhism. It was an act of separatism, to be sure, but at the same time, it was also, as Ambedkar pointed out, affiliating with a religion that was far more universalist than Hinduism in its endorsement of social equality (for a recent discussion on Ambedkar's conversion, see Viswanathan 1998: 211–39). Ambedkar died only a few weeks after his conversion, only to be reborn some twenty years later as the prophet of Dalit liberation. That is his status today – a source of both realist wisdom and emancipatory dreams for India's oppressed castes.

At a recent meeting in an Indian research institute, after a distinguished panel of academics and policymakers had bemoaned the decline of universalist ideals and moral values in national life, a Dalit activist from the audience asked why it was the case that liberal and leftist intellectuals were so pessimistic about where history was moving at the turn of the millennium. As far as he could see, the latter half of the twentieth century had been the brightest period in the entire history of the Dalits, since they had got rid of the worst forms of untouchability, mobilized themselves politically as a community, and were now making strategic alliances with other oppressed groups in order to get a share of governmental power. All this could happen because the conditions of mass democracy had thrown open the bastions of caste privilege to attack from the representatives of oppressed groups organized into electoral majorities. The panelists were silenced by this impassioned intervention. I came away persuaded once more that it is morally illegitimate to uphold the universalist ideals of nationalism without simultaneously demanding that the politics spawned by governmentality be recognized as an equally legitimate part of the real time-space of the modern political life of the nation. Without it, governmental technologies will continue to proliferate and serve, much as they did in the colonial era, as manipulable instruments of class rule in a global capitalist order. By seeking to find real ethical spaces for their operation in heterogeneous time, the incipient resistances to that order may succeed in inventing new terms of political justice.

Note

1 This paper was written in 2001 when I was Fellow of the Wissenschaftskolleg in Berlin. I have discussed it at meetings in London, Istanbul and Calcutta. I am grateful to all those who commented on earlier drafts.

References

Ambedkar, B.R. [1936] (1990) *Annihilation of Caste: An Undelivered Speech*, New Delhi: Arnold Publishers.

Ambedkar, B.R. (1945) *Pakistan or the Partition of India* (2nd edn), Bombay: Thacker.

Ambedkar, B.R. [1946] (1970) *Who Were the Shudras? How They Came to be the Fourth Varna in the Indo-Aryan Society*, Bombay: Thacker.

Ambedkar, B.R. (1948) *The Untouchables: Who Were They and Why They Became Untouchables?*, New Delhi: Amrit Book Company.

Amin, S. (1984) 'Gandhi as Mahatma', in R. Guha (ed.), *Subaltern Studies III*, Delhi: Oxford University Press, 1–61.

Amin, S. (1995) *Event, Metaphor, Memory: Chauri Chaura 1922–1992*, Delhi: Oxford University Press.

Anderson, B. (1983) *Imagined Communities: Reflections on the Origins and Spread of Nationalism*, London: Verso.

Anderson, B. (1998) *The Spectre of Comparisons: Nationalism, Southeast Asia and the World*, London and New York: Verso.

Bhabha, H. (1990) 'DissemiNation', in H. Bhabha (ed.), *Nation and Narration*, London: Routledge, 291–322.

Bhaduri, S. (1973) *Dhorai carit manas* (vol. 1, 1949 and vol. 2, 1951), in S. Ghosh and N. Acharya (eds), *Satinath granthabali*, vol. 2, Calcutta: Signet, 1–296.

Chakrabarty, D. (2000) *Provincializing Europe: Postcolonial Thought and Historical Difference*, Princeton: Princeton University Press.

Dumont, L. (1970) *Homo Hierarchicus: An Essay on the Caste System* (translated by Mark Sainsbury), Chicago: University of Chicago Press.

Foucault, M. (1998) 'Different Spaces', in James D. Faubion (ed.), *Aesthetics, Method and Epistemology*, New York: New Press, 175–85.

Galanter, M. (1984) *Competing Equalities: Law and the Backward Classes in India*, Delhi: Oxford University Press.

Hardiman, D. (1987) *The Coming of the Devi: Adivasi Assertion in Western India*, Delhi: Oxford University Press.

Kumar, R. (1987) 'Gandhi, Ambedkar and the Poona Pact, 1932', in J. Masselos (ed.), *Struggling and Ruling: The Indian National Congress, 1885–1985*, New Delhi: Sterling.

Omvedt, G. (1994) *Dalits and the Democratic Revolution: Dr. Ambedkar and the Dalit Movement in Colonial India*, New Delhi: Sage.

Shourie, A. (1997) *Worshipping False Gods: Ambedkar and the Facts Which Have Been Erased*, New Delhi: ASA Publications.

Srinivas, M.N. (1966) *Social Change in Modern India*, Berkeley: University of California Press.

Viswanathan, G. (1998) *Outside the Fold: Conversion, Modernity, and Belief*, Princeton: Princeton University Press.

4

Universality and Rights: the Challenges to Nationalism

Fred Halliday

Introduction

It is conventional in much current discussion to approach the question of human rights and nationalism by questioning the assumptions underlying human rights.[1] In this chapter I would like to do the opposite, to interrogate nationalism in the light of the assumptions of universality and reason that underpin the modern conception of rights, individual and group. In essence what I want to suggest is that it is time, after two centuries of nationalism, and of the prospect of many more decades when this ideology will hold sway, to conduct an assessment of the record of nationalism, in effect an audit, in the light of general rights criteria. Such a broad form of audit has been conducted with regard to other forms of collective human endeavour – war on the one hand, democracy on the other. Nationalism recognizes no higher authority than the nation and the claims of those who are said to speak for the nation. But that is nationalism's problem, not ours. Our challenge is to place on record, in a necessarily imperfect but unyielding way, what the record of nationalism has been with regard to rights and to identify where elements of complementarity, and also of contradiction, may lie. Nationalism need not, and should not, be exempt from such an assessment. Hence the title of my paper, the 'challenges to nationalism' rather than the modish challenges of nationalism, with regard to human rights.

Sovereignty and culture

This argument involves a choice as to the starting point of such an assessment. In much of contemporary discussion there is, rightly, examination of the ways in which a claim to universal rights conflicts with forms of particularism. Within international relations this revolves around the question of sovereignty, the right of states to order their internal affairs as they see fit provided they do not threaten others or, in the language of the UN, 'threaten international peace and security'. Much discussion in the 1990s revolved around this question, the argument for universality moving between, or encompassing, two broad claims: one that some forms of violation of rights within a state are of such enormity that the state forfeits its claim to exclusive control, that it (in the honoured phrase) 'shocks the conscience of mankind'; the other argument has been that it is no longer possible, or decreasingly so, to talk of domestic processes that are without international repercussions – much of the argument for recent interventions has claimed this, for example, that Kosovo produces refugees, that Afghanistan generates terrorists and drugs. This argument allows for no easy resolution and it may be that the outer limits of sovereignty have been reached. Others would claim that it was only alive as a universal principle for a brief period between the end of colonialism, which was a massive, internationally sanctioned violation of sovereignty, and the 1990s, in other words as a function of the Cold War. Yalta, not Westphalia, was the founding moment of sovereignty.

The other issue, debated within international relations, and in political theory, anthropology and sociology is that of what is termed culture, that different peoples have different practices, that others, the hypostatized 'we', should respect this and this is a necessary qualification, and in legal terms reservation, as far as human rights codes and enforcement are concerned. At its mildest, this involves a respect of cultural difference, provided it does not involve gross sovereignty-challenging violation of human rights. At its less mild, it involves a robust denial that all peoples can be incorporated into a rights regime, 'cannibalism for the cannibals' as it has been termed. This is a serious and necessary issue, but I would question how far it affects the international relations of rights and indeed how far it is the main obstacle, historically or today, to a reconciliation of nationalism with human rights. I would argue this qualification for four reasons, all of which are well known. First, culture is in many cases not a given, a set of timeless and universally held principles within a society, but an idiom and a site of diversity, as well as of change: what is presented as the unitary, and timeless, culture of a people or religion may now be so. There is diversity, and change, and it is up to each generation, and each member of that community, to interpret the culture, as they speak the language. Second, when it comes to modern times, there is on many key issues no difference between cultures: the right that lies at the centre of nationalism, the principle of self-determination, is held by all peoples, religions and states. Similarly, views as to the equality of peoples, the need for a just international economic system, a right to development, now the need to protect the environment are propounded universally: indeed much of the third-world, non-hegemonic critique of the west is not that it is thrusting inappropriate principles onto them, but that it is failing to live up to its own proclaimed, universal, standards. Third, in international relations the main arguments about rights are not about culture: some states do introduce reservations, for example, about women, or children, but the main authority for resistance to external pressure, or internal demands, is not the claim that 'we' are different, but rather that those pressing for changes have other, hostile, agenda, that they are a product of imperialism and domination. In other words, the critique of universality is based on a view of the structural inequality in the world system, not on claims that the peoples in question are somehow exempt from universal jurisdiction. Finally, the argument about culture has, in much academic writing, an apologetic, or even evasive, form; I would cite as an example of this the work of John Rawls, on a 'law of peoples'. Rawls calls for a liberal acceptance of difference on a global scale, and posits the existence of peoples who are not governed by liberal principles but who meet certain minimal criteria for being 'well-ordered'. Universal principles should not be imposed on them. But this model is deeply flawed: it assumes unitary cultures, where none exists; it posits an ideal type of society for which there are almost no real world examples – perhaps Singapore on a good day; it is wholly innocent of a sociology of values, be this in terms of who within that society determines and enforces the interpretation of culture, and in terms of how the values, indeed the state and community concerned, are themselves formed and sustained by international factors. It of course provides no answer, other than robust interventionism, to the challenge of states that are not well-ordered of which there would appear to be some few dozen in the world of today.

Particularism

At the core of nationalism lies a contradiction with regard to rights: nationalism rests upon the assertion of a right, and a right deemed to be universal, namely the right of nations to self-determination, and, with this, a set of subordinate claims about territory, sovereignty *vis-à-vis* other states, and the obligation of members of that nation to it; yet nationalism, while asserting a right and claim universality, has been and remains deeply hostile to rights. My argument here is, in essence, that it is this tension, or contradiction, as much as and possibly more than the issues of sovereignty and culture, which besets the relation of nationalism to human rights. We need to recognize that over its two centuries, and stretching into the future, nationalism is profoundly hostile to the universality of rights, and it is this hostility that an audit will bring out. Rather than engage in general, and necessarily inconclusive, discussion of this, I would like to focus on four areas where this contradiction may be examined.

Self-determination

The basis of nationalism is the principle of self-determination: this need not necessarily mean independence, but could involve various forms of autonomy, federalism, protection of minority rights or even equality within a general provision for citizenship. This principle is deemed to be the basis of international peace and security, is inscribed in the UN Charter and in Article 1 of both the ICPPR and the ICESCR. The problem is that this right operates without any legal or internationally recognized criteria: it is not justifiable, and in practice has been resolved on an ad hoc basis.

There are, moreover, several practical problems with it that go to the heart of nationalism's uses and misuses of universality and to the record of nationalism these two centuries past. One such problem is that the claim to self-determination rests as nationalism ordains on a claim to history, and where territory is concerned to prior title. If divine sanction can be brought in, all the better. Yet even the smallest dose of modernism, or historiographic and legal scepticism, would show that the history, and supposedly historical titles, used to justify claims to self-determination are bogus. Modernism would indeed suggest that we take the world as it is, that whatever peoples of sufficient weight and noise who show up should be given states, but it does so on an ad hoc basis, not by accepting the history, title and usually expansionist territorial claims associated with nationalism.

The basis on which a right is recognized within a modernist perspective is, therefore, in contrast to the kinds of historic right claimed by nationalism. Indeed the more intense the nationalist claim, the less reason or defensible legal claim there is. The supreme conflict revolves around a right much proclaimed by nationalism but, to my knowledge, without any legal, or moral, basis at all, a 'birthright'. In the canon of normative claims this must have a special place. So must claims based on invocation of the divine: God, or gods, do not have legal or constitutional status under international law, yet they have come to be used as authority in the adjudication of territorial disputes. They have also been used to invest places of alleged sacred character with special national significance. Modernism can, and should, have short shrift with this: the divine is a projection, indeed an invention, of the human, and should be subject to human control.

The most serious problem with the right to self-determination is however something that is inherent, not necessarily but recurrently, within nationalism, namely its selective application. This, not imprecision of criteria, or cultural difference as to the interpretation of rights, has been the central problem over the past century. What we see in conflict after conflict is the declamatory assertion of one right coupled with the denial of that right to others. The right itself is not denied, indeed it is celebrated, with much historical, expansionist and emotional baggage. But when it comes to the other people, their nationality is suspect. To take three obvious examples: Northern Ireland, former Yugoslavia, and the Palestine–Israel conflict. In each case the argument is not about territory as such, or about the applicability of the modernist principle of self-determination, but about whether the 'others' are really a nation at all. Thus for Catholic nationalists the Protestants are colonizers, fascists, British agents, when not 'soupers', Catholics who converted to Protestantism at times of famine in order to get fed. For Protestants the Catholic minority are treasonable, agents of a foreign power (the Irish

Republic, when not the Vatican) and without legitimate national rights. In former Yugoslavia Serbs and Croats, colluding when not killing each other, deny the right of either Bosnians or Kosovo Albanians to self-determination, just as the Greeks long denied the same to the Macedonians. In the Arab-Israeli context arguments long dormant have now been revived as to the illegitimacy of one or other ethnic group: the Israelis are just foreign colonizers, or a religious minority, or all, down to the last child, agents of a fascist power. The Palestinians have no right to self-determination because they are just Arabs, with lots of other places to go, or Syrians, or plain inhuman terrorists. I repeat: this denial of the rights of others is not logically part of the principle of self-determination, it is a recurrent part of its proclamation and application.

Laws of war

This contradictory relation to human rights, at once proclamation and denial, is evident in a second area of rights, that of the laws of war. Here the record of nationalism is one, over many decades, of the most bloodthirsty, cruel, and vindictive behaviour and of partisan, manipulative use of these principles when charges of violation of the rules of war are made. Far too much time is devoted to using the rules of war as an instrument of conflict, too little to holding all parties, irrespective of their affiliation, responsible before the principles themselves. Humanitarian principles relating to the conduct of war, the Geneva Conventions of 1949, the Additional Protocols of 1977, the Genocide Convention and a range of agreements relating to the use of particular weapons may not be part of the core principles of human rights as they have evolved since the Second World War, but they are taken to be part of the broader code, not only because they relate to principles of conduct by states and opposition movements alike, but because they rest in large measure on conceptions of rights, be they of combatants or civilians.

Here as much as in regard to self-determination problems arise, but they are not primarily a function of culture. This is so in the obvious sense that all states, whatever their culture are bound by these conventions, and all have, in modern times, committed acts that, in the opinion of most, violate them. There may be cultural issues to do with the specifics of war, for example, how exactly to cut the throat, break the bones, or humiliate the women, or mutilate the corpses of those from the other side, but the broad picture is universal: nationalist movements eager to proclaim their own self-righteousness have been responsible and are today responsible for gross violations of the rules of war. Again, the examples are evident. The history of the Balkans from the mid-nineteenth century through to 1914 reveals a pattern of repeated, universal crimes involving the killing of civilians, the destruction of villages, the raping of women, the starving of civil populations, the abuse of prisoners of war, not to mention the expulsion of populations. Here Serbs, Croats, Montenegrans, Greeks, Bulgarians, Turks, Albanians all participated in a frenzy of cruelty and vindictiveness. In the 1970s and 1980s we saw in the Lebanese civil war terrible massacres, bombardment of cities, mass kidnapping and killing of prisoners, destruction of villages and much else besides by the participants in that conflict, one on which the external states involved, Syria and Israel, also played their part. In the past 10 years we have the massacres in Rwanda, repeated violations of rights of combatants and civilians in Bosnia and Kosovo, atrocities in the Armeno-Azeri conflict. In Northern Ireland militant nationalists on both sides have, over three decades, engaged in sectarian murder. We have seen in the Palestine–Israel conflict of the past three years evidence of repeated violations of the rules of war by both the Israeli armed forces and Palestinian militants, actions that, if not ordered, were in general terms sanctioned and retrospectively condoned by their political leaders and by their own peoples. To all of this has been added an orgy of unrestrained hate-speech, the cult of violence, violation of the rights of children and reckless disregard for the security of their own peoples.

Faced with this record, and one which I repeat is far from over, there is little space for attributing this to a culture problem with human rights. Both sides are quick to acknowledge the relevance of universal principles by dint of their denunciation of the other side. Equally, their ferocious denial that they have been involved in any such actions is testimony to a recognition that these principles do apply and that they would suffer, in terms of reputation and morale, if the claims of violation are proven. Yet too quickly the partisans in such conflicts, and outside observers, get drawn into a trap, one set by nationalism, which is not only that of partisan assertion or denial, but one of comparison. Hence the argument becomes as to which side is worse, has, as it were, piled up more corpses, burnt villages or rape victims, or, alternatively, which side bears prime responsibility, be this in terms of origination, 'who started it?', or in terms of continued obstruction of a settlement. Yet the very fact of comparison is itself a mistake, a concession to nationalism. For the issue is not which side is worse, or who started it, but whether the principles that are generally held to apply in armed conflict, and which are binding on all parties, have been respected or not. When within a domestic juridical process two people are charged with theft, or murder, we do not ask which of them committed a worse theft, or a more horrible murder. We ask whether they have broken the law. The same can, and should, apply in ethnic conflicts and in nationalist wars. We need to wrench the argument back to that of universality, and away from the nationalist traps, of denial/denunciation, or comparison, that are too easily set.

Terrorism

This issue of the relation of nationalism to human rights in the context of war has been most forcefully posed with regard to the question much discussed in the contemporary context of terrorism. 'Terrorism' is a word that allows for much polemical usage, but it also allows for more precise political and legal usage, and for some historical context. This context allows us to remember that the use of violence against a civilian population in violation of the rules of war is a practice that is resorted to by all parties in conflict, states as well as their opponents, and that it is part of a political conflict in which the subjugation of the political will of the opposing side is a central goal.

That nationalists have throughout much of the past century resorted to terrorism, against their national oppressors and against subject peoples, needs little underlining. It is the prerogative of no nation, or religion. There are, moreover, three ways in which it is particularly central to any question of the relation of nationalism to human rights. First of all, terrorism, by states or their opponents, is an extreme example of the denial of universality: it involves a rejection not only of the claims of an opposing group, but of their very humanity. This is a more extreme version of the denial of universality involved in rejecting the claims of rival groups for self-determination. Second, the use of terrorist practices touches on one of the central problems and occasions for abuse inherent in discussion of the rules of war, the relation between *jus as bellum* and *jus in bello*. These are

two separate categories, and supposedly discrete: there is no necessary relation, no rate of exchange, between the issue of just authority, and that of the methods used in war, or, in this case between the
claims of a nationalist movement, to territory or independence, and
the methods used. But in practice such a rate of exchange is understood: all parties to a conflict know that if they are held to account
on the methods used this will prejudice their overall case. Those who
wish to discredit the authority of another side are liable to accuse it
of atrocities. Those accused of atrocities deny these, in order to project the overall claim of just case. Hence a group that carries out terrorist acts, beyond the political aim involved, must also be concerned
to prevent its practice of abhorrent actions from prejudicing its
broader political goal. Those who are generally held to be terrorists
forfeit, or at least to a considerable degree prejudice, their general
political goals. All of this both obscures, but also implicitly endorses,
the relevance of universal principles to the conduct of nationalist,
inter-ethnic and communal conflicts.

There is a third aspect of the relation between terrorism and
human rights that needs examination here, and which is much in
evidence, namely the relation of anti- or counter-terrorist policies to
human rights. States and peoples have a right to their own security,
to protect themselves against attack. This applies to terrorism as
much as any other more conventional form of attack. But nationalism is a very dubious basis on which to base such a response: it opens
the possibility of discrimination in the treatment of detainees and
suspects, and, implicitly, places those suspected of terrorism, or by
extension from communities suspected of harbouring terrorists, on
an inferior basis. Alternatively, and often at the same time, it places
those opposed to terrorism outside the jurisdiction of law and norms
of combat. It is a denial of universality, and on both sides. Herein and
beyond any special suspension of human rights provisions in deten-
tion and trial lies the danger of the response taken by the US administration to the events of 11 September 2001. The set of anti-terrorism
measures put through in October was termed 'The Patriot Act' as if
patriotism, a necessarily partisan and emotional principle, provides
any basis for law. Equally, the appeal to patriotism and to a national,
as distinct from universal, legitimation obscures two other aspects of
the issue: one, it absolves the USA and American citizens, the
presumed beneficiaries of this patriotism, from any reflection on the

past actions of their own states including support for groups practising terrorism in the 1980s; second, it implicitly exempts the state, armed forces and citizens from respect for universal principles. Patriotism is not the first, but the last, place to start in legitimating a campaign, which may be valid in itself, in terms of a universally defensible goal of security, against terrorism.

Solidarity

So far I have discussed three areas where a conflict between universal principles of human rights and nationalism may be observed and where a connection, iterative if not necessary, between the two has been evident these many decades past. I want in conclusion to examine another area where the partisan character of nationalism may affect concern with human rights, and this is an area that might at first sight appear to be one suited to a more positive relation with nationalism, namely solidarity. In essence I want to argue, in the face of much liberal and internationally concerned involvement in conflicts and nationalism the world over, that solidarity on the basis of ethnic or national affiliation, far from being desirable, is in contradiction with universal principles and is inherently undesirable as far as human rights are concerned.

We are all familiar with the role of solidarity in inter-ethnic conflicts. Some of this is based on well-intentioned, but often politically motivated, campaigns of support for particular ethnic or national groups – Tibetans, Bosnians, Palestinians, Israelis, Kurds, Irish, Chechens and so forth. Much of the activity of support for such groups is promoted by groups who claim an ethnic affiliation with the oppressed people, diasporas. In both of these cases there is a conflict with universality. To proclaim support for a particular people, on the basis of 'universal' principles, is inherently contradictory: be it in regard to self-determination or the conduct of war it necessarily ascribes moral import to one side and denies it, to a greater or lesser extent, to the other. We can see this very clearly in discussion of the Palestine–Israel conflict. The great majority of public comment, let alone letters to the press, is simply one-sided, invoking moral principles but in a partisan spirit. Even more questionable is the advocacy of support on the basis of national, diaspora, or particularist affiliation. This of necessity establishes a split between moral and legal systems, let alone one in which history, religious claims and nationalist invention of all kinds have free rein. To take an obvious example: the arguments advocated by diaspora Jews on the one side and by Muslims on the other with regard to the status or division of Jerusalem. Here I would suggest that such affiliation is by dint of the very basis on which it is proclaimed suspect. Not only are they partisan, but they are, on universal grounds, simply invalid: given their starting point, neither side, and the religious or other spokespersons for them, have not just qualified, but no moral standing in the matter. To spell it out: in terms of discussion in Britain, I would argue that neither the Jewish community nor the Muslim community, both of these rather flexible terms, have any special standing in the matter of the Israel–Palestine conflict. Indeed, given the partisan positions they uphold, and which are not conducive to peace, they lay a negative role. The same would go for other diaspora solidarities, be these disaporas comprising peoples who themselves migrated, or whose relatives recently migrated, or on historic links of affect and invention that are so often invoked. This would apply to campaigns about the rights of religious minorities: campaigns about the rights of Christians were in the late nineteenth century the stock-in-trade of European states, and this issue has come up again through agitation in the USA. The rights of believers are an important part of human rights concern, but on the basis of universal principles. Those who proclaim by dint of some affinity of faith or superstition solidarity with kindred believers also, by the logic of their position, exclude others from the same rights.

Conclusion

Much has been said, by political theorists and by the academic friends of nationalism, about how nationalism, as an ideology and as a practice, may be compatible with other, desirable, modern goals: democracy, identity, community and international order. As I have argued elsewhere, this may all be so, but it will only be so if a priority of values is established; if, in other words, nationalism knows its place. Faced with the record over the past century and more of nationalists in regard to universal principles and with the ongoing abuse of human rights by nationalists in several continents, we should be uneasy about accepting too readily that such a compatibility will occur. Of the three major obstacles to such a compatibility, sovereignty, culture and particularity of interpretation, it is the last, I would suggest, which poses the greatest problem and danger for human rights.

Note

1   An earlier version of this paper was presented to ASEN Conference on 'Human Rights and Nationalism', London School of Economics, 26 April 2002.

5

Nationalism, Globalism, and the Conflict of Civilizations

John Hutchinson

In recent decades the apparent global resurgence of nationalism has been accompanied by a questioning of the future of nationalism, the nation, and the nation-state. Will nationalism and national identities continue to exist in the next century? If so, will their characteristics change, and in what manner? If not, what are the alternatives to the nation and the nation-state? Since the nation is generally regarded as the hegemonic cultural and political unit of the modern period and the nation-state remains the major institution (via its membership of international organizations) through which the great planetary problems such as environmental change, nuclear proliferation, world trade, human rights, and the rights of minorities are addressed, these questions are naturally of great import.

Social scientists cannot pretend to offer authoritative answers to any of these questions. Karl Popper (1960) has rightly argued that since the action of human beings is shaped by their knowledge of the world, a successful long-range historical prediction would require us to know now what people will know in the future and the conditions they will encounter. Forecasting, then, in any strict sense is impossible. All we can do is examine if there are 'current' indications that the nation is being superseded by other forms of collective identification.

This in itself is a potentially vast topic, and I can offer here only a very limited analysis. What I wish to do in this chapter is to expose the lack of a long-range historical perspective displayed by many social scientists (including historians) who assert the declining salience of nations. I address two influential positions, which have

71

raised doubts about the continued existence of the secular nation. The first argues that globalization is leading to the supersession of nations as relevant political actors, whereas the second, rapidly gaining audibility since the recent extraordinary terrorist attacks on the USA, suggests that the major battleground in the world is not between nations and their states but between civilizations, in particular religious civilizations.

I reject both arguments. Each suffers, I argue, from a characteristic of social scientists to (in John Breuilly's words) 'eternalise the present' (Breuilly 2000). A longer historical perspective indicates that globalization rather than eroding may engender ethnic- and nation-formation, and that the current religious resurgence is as likely to contribute to nation-formation as to undermine it. Both positions exaggerate the uniqueness of the nation to the modern world and the potency of nations and nation-states within modernity. We require a much more calibrated discussion that charts the changing characteristics of ethnic communities and nations over historical periods and their fluctuating strength *vis-à-vis* other collective attachments.

Some rough definitions are necessary to make this discussion intelligible. Many differences between scholars arise out of competing definitions of the nation. I follow Anthony Smith in agreeing with the modernist school that nations, as entities based on conceptions of popular sovereignty and common citizenship, a consolidated territory and economy, are generally post-eighteenth-century formations (Smith 2001: 19). But as Smith observes, nations are also communities of sentiment that in large part rest on ethnic cultures which predate the modern period. These provide the nation with a collective name, myths of unique origins, a sense of belonging to a homeland, of shared history and culture, and common political fate. It is the sense of belonging to an ancient 'timeless' community that gives the ideology of nationalism such potency in the modern world. Nationalism is able to bind individuals into a society (the nation) through which they can overcome contingency and death, achieving immortality by adding their story to that of an eternal unit (Smith 1999: 88).

This is not to say that populations without pre-modern ethnic traditions cannot become nations. Ethnogenesis continues in the contemporary period. It is also true that many states or populations claiming to be nations do not fit the above definition, but I suggest that part of the disruption in world politics is an attempt to create substantive nations. In contrast with those commentators who suggest we are reaching the end of the nation period, I would argue that many areas of Eurasia, Africa, and Latin America are still in the early processes of nation-formation, and that this will be accompanied by social and political upheavals.

The globalization perspective

In 1919 John Maynard Keynes recalled the liberal Europe before the First World War:

> What an extraordinary episode in the economic progress of man that age was which came to an end in August 1914! The inhabitant of London could order by telephone, sipping his morning tea in bed, the various products of the whole earth, in such quantity as he might see fit, and reasonably expect their delivery early upon his doorstep; he could at the same moment and by the same means adventure his wealth in the natural resources and new enterprises of any quarter of the world...he could secure forthwith, if he wished it, cheap and comfortable means of transit to any country or climate without passport or other formality...But, most important of all, he regarded this state of affairs as normal, certain, and permanent, except in the direction of further improvement.... (Keynes 1920: 9–10)

What is interesting in this passage is Keynes' nostalgic presentation of this cosmopolitan free-trading civilization as a golden 'past', destroyed by war and fanatical ideologies (nationalism and Bolshevism).

This should make us cautious about the claims of scholars (Giddens 1990; Albrow 1996; Castells 1996) who argue that we have recently entered a new period of world history, a shift from modernity to postmodernity, engendered by globalization. 'Globalization' is defined here as an intensification of interconnectedness between the populations of the world. As Montserrat Guibernau (2001) ably outlines, theorists of globalization argue that advances in technology and communications have intensified the contacts between the world's populations to such an extent that time and space have been compressed to form human populations into one world, transforming our sense of the 'local'. The later twentieth century saw the rise of multinational corporations and transnational non-governmental agencies, international codes to protect human rights, world bodies such as GATT, the World Bank and the UN, regional associations of states such as the EU and NAFTA, and the new world language of English. As we seek in our everyday life to adjust to these new horizons and organizations, the nation and nation-state, those primary institutions of modernity, cease to be our primary political and cultural reference points, unable by themselves to manage the problems of nuclear proliferation, international terrorism, long-distance economic migrations and refugee flows, and climatic change.

There are various versions of this thesis, but they all presume the existence during the nineteenth and early twentieth centuries of a 'classical' nation-state that was politically sovereign, militarily autonomous, territorially bounded, culturally homogeneous, and economically integrated. A radical version, heavily economistic in its assumptions, predicts the erosion of such nation-states by institutions of global and regional governance as they become increasingly powerless to regulate the new borderless world of economic transactions (Wriston 1992; Ohmae 1996).

A more qualified version argues for a transformation not a destruction of the classical nation-state (Giddens 1990; Held *et al.* 1999). According to this, the autonomy of the nation-'state' is qualified by the growth of transnational institutions that has resulted in a pooling or loss of sovereignty. The 'identity of nations' has also been recast. The unlikelihood of large-scale war between great powers means that 'internal' others such as immigrants and refugees are substituting for 'external' enemies for purposes of collective differentiation; the greater visibility of immigrants and national minorities means that homogeneous national cultures are being pluralized and hybridized; and the rise of English as the world language, carried by transnational media channels, encourages a global consciousness and culture at elite and popular levels. There are countervailing tendencies that are strengthening nations and nation-states, but the overall effect is weakened nation-states having to come to terms with multiculturalism (on all this, see Guibernau 2001).

My chief target will be the radical thesis, for there is some substance in the transformationalist thesis (though not as much as is being claimed). I wish, however, to take issue with the assumptions of both positions, because of the confusion of globalization with modernity or postmodernity; an over-technological conception of globalization that underplays the importance of religious and military factors; and a unitary conception of globalization. Globalization has a long historical duration and brings with it a sense of unpredictable threat that in turn has often resulted in the crystallization and articulation of ethnic and national differences.

World history, nation-states, and the global

The starting ground for a discussion of the global in history is W.H. McNeill's *The Rise of the West* (1963), one of the great works of twentieth-century historical scholarship. McNeill argued that a global society formed in the twentieth century out of the world dominance of the West (first in the guise of European nation-states, then the USA). The origins of globalism lay in fifteenth-century Europe when militaristic nation-states, forming out of a competitive continental state system, expanded overseas, given impetus by revolutions in science and communications, until by the twentieth century, Europe had overthrown all other civilizations. The revolutionary disruptions engendered by the scientific ideas could be seen as either a threat to all older religiously-based civilizations, including the West, or a continuation of the West's revolutionary potential. What was undeniable was the emergence for the first time of a single cosmopolitan humanity, one that would make obsolete the European nation-states themselves. In McNeill's treatment nation-states precede and indirectly engender globalism only to be superseded by it, and globalism is defined as Western in origin and secular in character (McNeill 1963: chapters 11–13).

McNeill's account is richly illuminating but it can be criticized on three grounds. First, globalization should not be seen as a modern revolutionary development but a recurring and evolutionary process. Adshead, in dating the origins of world history in the Mongolian 'explosion' of the thirteenth century in Central Asia, argues that the contemporary world system was built on successive layers of interlocking networks: information, microbial and military circuits, religious internationals, the republic of letters, the global armoury, the world commodity market, the world technological bank, and a common consciousness expressed through the use of English (Adshead 1993: 3–4). McNeill, himself, influenced by the work of Janet Abu-Lughod (1989) revised his earlier views to admit precursors of the Western capitalist world system on which the latter built. The first developed in the Middle East from the second millennium B.C. until its decay by A.D. 200. A second, again centred on the Middle East, accompanied the rise of Islam from A.D. 600–1000. Between A.D. 1000–1500 China, borrowing from the Middle East, became the centre of gravity of a third world economic network, stretching through the Middle East to Western Europe (McNeill 1990).

A second criticism made by Marshall Hodgson (1993) among others (see Eaton 1990) is that globalization cannot be equated with either 'Western' characteristics or 'universalization'. Islam preceded Christian Europe as a global civilization, and in Marshall Hodgson's interregional model world history is the story of the interactions of four major culture zones over a period of 2 800 years, each of which emerged for a time as the leading edge before its innovations were (over a period of 500 years) assimilated by the other zones. This inter-action was both peaceful (e.g. trade and the diffusion of ideas) and violent (warfare and imperial conquest). Among the implications are that 'globalization' always flows from 'particular' centres and that the rise of the 'West' (an area previously on the margins) is not unprecedented or final but is only the latest manifestation of a series of 'jumps' in global social power, which is at present being absorbed world-wide. In the 1980s it appeared that Japan as the second indus-trial economy had developed an alternative mode of development to the (American-dominated) 'West', leading some to speak of its impact on Asian societies as a form of 'Easternization'. Although the Japanese model is now discredited, it is not inconceivable that China now rapidly developing its vast population could eventually become the global centre of gravity. Indeed, there is a Chinese project to present a Neo-Confucianism, emphasizing harmonious co-operation, as a non-exclusionary principle of world order superior to the competitive messianism of the West (Zheng 1999: chapter 4). A growing sense of threat from China has evoked at times an ethnocentric response from the supposedly global civilization of the USA. In short, global currents come with their own ethnocultural assumptions and can provoke countervailing visions, and in turn rival nationalisms.

Third, globalization cannot be defined as a unitary and secular process, since 'globalizing' institutions include as well as the rise of secular sciences, technologies, and ideologies, those of missionary religious expansion, imperial conquest and colonization, migrations, and long-distance trade that often cut across each other. Michael Mann (1986) has analysed patterns of world history through the interaction of four overlapping and competing networks of power, the economic, the political, the ideological, and the military, each of which has its own technologies and boundaries.

This should make us reconsider the causal relationship between globalization, ethnic formations and nations in three ways. If globalization has been in process for a millennium or more, then claims that it will result in the supersession of nations become problematic, to say the least. It is at least possible that ethnic- and nation-formation accompanies and is engendered by globalization. Secondly, globalization is an inherently dynamic process which produces differentiation rather than homogenization, since it always comes laden with the assumptions from an originating region and is transformed into the specificities of the 'receiving' culture as it seeks to absorb it and compete with its challenger. Thirdly, because there are multiple agents and processes of interconnectedness, there is an inherent unpredictability in world history. Indeed, McNeill's own 'contact model' acknowledges this. He argues that world history is marked by a shift from isolated individuals to increasing social interdependence, and through this an enhancement of human power over nature. But he noted that such contacts have engendered conflict as well as harmony (Costello 1993: 197–9). Indeed, I would argue that such contacts (and conflicts) are one of the catalysts of ethnic crystallizations and that the resulting ethnic communities have often provided the basis of the modern nation.

The emergence of universalistic scriptural religions has been critical in binding dispersed populations together into the major civilizations. Although such religions have traditionally been regarded as undercutting ethnic affiliations to territory and culture, they have often been catalysts of ethnic and, some would argue, national formation. Adrian Hastings (1997) claims a special role for Christianity. Evangelism, inspired by the Biblical recognition of linguistic diversity, effected the translation of the scriptures into local languages and the proliferation of written vernaculars. In presenting Israel as the exemplary political community of unified kingdom, sacred territory, and holy people bound by their distinctive culture, the Old Testament diffused the model of the nation first in Western Europe, then, as Christianity expanded with European imperialism, world-wide.

In fact, all religions have stimulated ethnic formation for several reasons. This includes their need to accommodate (in different degrees) to the ethos and practices of the previous cultures in order to reach the people; their proclivity to schism and internal differentiation which can transform ethnic categories into rival ethnic communities; their employment by rulers seeking to build culturally cohesive populations, differentiated from neighbouring groups; and, above all, by wars between political communities of different faiths.

The tendency of all great religions including Islam, Buddhism and Hinduism to fissure into different traditions endowed them with ethnogenetic potential, particularly whenever rival traditions took root in adjacent and competing populations or states. The rooting of Shiite Islam in Iran was accompanied by a rejection of Arab dominance and a Persian ethnocultural revival, given intensity by the wars of the Sunni Ottoman Empire against the Safavid Empire of Persia. As John Armstrong (1982: chapter 3) demonstrates, the global ambitions of rival proselytizing religions brought them into military conflict, and states on the fault lines developed *antemurale* myths that depicted them as elect polities, destined to be the border guards of their civilization. The conflict between Islam and Christianity over 1 000 years saw several polities claim such a status: on the Christian side, Byzantium, Castilian Spain, and Tsarist Russia, and on the Muslim, Mameluke Egypt. The *antemurale* myth also re-emerged from wars between Catholic and Orthodox, and between Catholic and Protestant states. Poland regarded itself as the defender of Catholicism in Western Europe against 'Eastern' Russian Orthodoxy, and England and Holland saw themselves as the bulwarks of Protestant liberty against Catholic despotism. All of these identifications have strongly shaped the trajectories of the modern nations.

Although imperial expansions would seem to be the enemy of ethnic communities, ethnic consciousness can also arise as one of their unintended consequences. A powerful ethnicity was forged in Armenian and Jewish populations, caught on exposed trade and communication routes between rival Roman and Persian empires, who suffered collective subjugation and 'exile' (Armstrong

1982: chapter 7). Empires, Michael Hechter (2000: chapters 2 and 3) claims, can consolidate ethnic communities through systems of indirect rule that reinforce indigenous leaderships, as under the Ottoman *millet* system.

In general, interstate warfare intensifies ethnic difference. Of course, the psychological and social effects of warfare on the populations are more limited when warfare is conducted by mercenaries or a small aristocratic stratum, or, indeed, limited corps of professionals. But recurring and protracted interstate wars even between feudal states such as the Hundred Years War between the French and English kingdoms have resulted in a social penetration of ethnic sentiments. Warfare, as Anthony Smith (1981) has argued, has had ethnicizing effects by mobilizing localized groups into a state army and thereby creating an identification with a larger territory which becomes a homeland, by generating propaganda through which mutually opposing ethnic stereotypes are constructed between opposing populations, and by throwing up heroes and epochal battles. These epic events when celebrated subsequently by poets, historians and artists, such as Froissart and Shakespeare, popular legends and commemorative rituals become institutionalized in the group consciousness. Even when wars result in the overthrow of a state, an ethnic consciousness may persist, especially where groups define themselves in religious terms, interpreting their defeat like the Serbs at the battle of Kosovo as a test of their commitment to the true religion. A religious sense of election thus explains away defeat, indeed instils a reinforced drive to defend collective traditions as a means of eventually regaining divine favour.

The development of long-distance trade also excited the ambitions of groups to control it and brought far distant cultures into contact and often conflict. Waves of steppe migrations or invasions of the wealthy agrarian civilizations of Asia and Europe followed the land-based silk and spice routes, including the Mongol drive for world Empire of the thirteenth century, which has been explained by historians as motivated in part by a desire to seize control of the silk route. The memory of the 'Mongol/Tatar yoke' had a deep impact on Muscovite Russia and Magyar Hungary. The Mongol unification of the silk route created an information circuit linking Asia and Europe, inspiring Western states to discover a seaward route to the riches of the East (Adshead 1993: chapter 3). Out of this came the European 'discovery' of America, large-scale colonizations which led to new ethnic crystallizations, and wars between the European great powers in their struggle for overseas wealth and empire. This together with the religio-dynastic wars of the Reformation on the European subcontinent was the context from which early modern national identities developed, articulating a universal religious mission.

In short, globalization when conceived in *la longue durée* has gone hand-in-hand with an intensification, if not creation, of differentiation that often takes ethnic forms. To adopt McNeill's contact perspective, of course, is to accept a 'non-essentialist' concept of ethnicity that acknowledges that ethnic formations even when strongly institutionalized are subject to recurring challenges of different kinds. These may result in internally generated innovation, imposed syncretization through conquest, and possibly dissolution through voluntary or coerced assimilation and ethnocidal programmes. The multiple and disaggregated nature of these challenges entails that ethnic formations may often contain a repertoire of many different pasts, cultural heritages, and hence models of individual and collective identity to which they can turn in order to negotiate change. Ireland had such heritages as the pagan era of Celtic aristocratic warriors, encoded in epic poetry; the Catholic *insula sacra* of St Patrick and Irish saints; the Anglo-Norman heritage of parliamentary autonomy. In Magyar lands, the early pagan warrior ethos of the steppes was overlaid (though never obliterated) after conversion to Christianity by an *antemurale* identity as a European Christian bulwark against the Asiatic Islam.

Yet, globalization theorists build on modernist interpretations such as Ernest Gellner (1983), Eric Hobsbawm (1990), and John Breuilly (1993) who view the nationalism and nations of the late eighteenth century onwards as discontinuous with older ethnic communities, and as products of a secular modernization process that is tending to the organization of humanity into ever more extensive units. Although the nation may profess a commitment to an ethnic past, it is essentially novel, even invented, for the nation is a political unit, territorially more extensive and bounded, economically integrated and culturally homogeneous than earlier entities. It is the bearer of the scientific and technological revolutions of the eighteenth century that made possible a single and uniform culture based on a new rational and political concept of humanity, which is transmitted by a centralized territorial state and an industrial economy over much of the planet. Nations are directed to modernity and modern nation-states are much alike: they share an industrial culture, and the very rise of a nation-state system is the harbinger of increasing interdependence and of a new system of governance by transnational institutions.

One can agree with modernists that modern nations contain many novel features, but they underplay the degree to which states have been shaped in their modernizing policies by older ethnic identities. For example, the emancipatory mission of revolutionary France to Europe fed off medieval conceptions of the kingdom of France as the 'elected' defender of European Catholic Christianity. Modernists also exaggerate the autonomy of states when facing the greater intensity of contacts since the eighteenth century, and fail to observe the recurring crises that states encounter and that compel them to 'return' to the resources of ethnocommunal heritages. Throughout the modern period, states, whether they were long established empires or indeed avowedly nation-states, have periodically been shaken or even destroyed by unforeseen events such as warfare, economic crises, migrations and demographic shifts, ecological changes, and ideological challenges.

Although modern states can exert immense power by mobilizing their populations through efficient administrations, educational systems, and their economic alliances, as only one of many power actors, they are tested to their limits by unpredictable political, military, economic, and ideological challengers. Warfare in the modern period (particularly in the Napoleonic era and the First World War) has required a continuous redefinition of political communities with respect to each other. An ethnic nationalism has been fanned as a result of the overthrow and rise of states, the shifting of states into new geopolitical spaces, the turning of dominant groups into national minorities and vice versa, and large-scale transfers of population. Even aggressive secular French republican nationalists found it necessary to appropriate traditional religious symbols such as St Joan to mobilize support for the state after France's defeat in 1871 and during the German invasion of the First World War (Gildea 1994: 154–65).

Waves of transnational economic revolutions have also upset the power of states *vis-à-vis* each other and the status of regions and classes within nation-states. Prussia's deployment of railways in its crushing defeat of France in 1871, and German leadership of the 'second industrial revolution' of iron and steel seemed to presage the rise of new hegemon in Europe. This resulted in intensified nationalist rivalries in the early twentieth century with France and also with Britain, which felt its traditional naval superiority threatened by Tirpitz's development of an armoured battleship fleet. The economic depressions of the 1870s, the threat to the traditional European landed order from an emerging world agrarian market, together with the growth in rapidly expanding cities of a large and politicized unskilled working class attracted to militant socialist parties, large migrations of Jews resulted in racial and anti-semitic nationalism across Western Europe.

Competing ideological movements arising from the heritage of the Enlightenment and religious counter challenges, transmitted through transnational institutions such as churches, revolutionary internationals, diaspora groups, and printed media have fed national antagonisms. In the late eighteenth century fears of the imposition of radical republican ideas by France and its ideological supporters resulted in the crystallization of a modern conservative British identity focused on the crown (Colley 1992: 216–20). Papal 'ultramontane' rejection of secular nationalist principles, culminating in the Syllabus of Errors (1863) and the declaration of the doctrine of Infallibility (1870), intensified tensions within Britain, France, Germany, and Italy. The Bolshevik revolution, in similar vein, created a nationalist panic in Western and Eastern Europe, particularly among conservative middle-class groups, fearful not just of a large external enemy but also an enemy within in the form of an internationalist working class.

Furthermore, closer contact between peoples has intensified and widened the impact of 'natural' disturbances – diseases, famines, ecological disturbances, shifts in fertility patterns – heightening national tensions and conflict. The inability of the British government to avoid the Great Famine in mid-nineteenth-century Ireland permanently alienated the Catholic Irish from the union with Britain, and the flight of many thousands of diseased emigrants to the cities of the USA and Britain stoked nativist reactions in both these countries. A racial nationalism in the European imperial nations in the early twentieth century was fuelled by fears of demographic decline in the face of the superior fertility of the 'yellow races'.

Under such circumstances the motifs of cultural nationalism of communal self-help and of the recreation of social and political institutions from below have resonated, often reviving and allying with older ethnoreligious sentiments. At the same time, the plurality of ethnic heritages allows nationalists to reject 'failed' traditions, and to justify necessary social innovations by appealing to 'forgotten' golden ages. Although Greek national identity was defined by reference to Orthodox and the Byzantine Empire, it was the 'rediscovery' of secular Hellenism that inspired early nineteenth Greek nationalists to secede from the Ottoman Empire and 'rejoin' the European 'West'. In short, nationalism does have modernizing objectives, but even more its power derives from its capacity to overcome contingency by finding 'solutions' based on a past believed to be authentic.

The idea of a sovereign nineteenth-century nation-state, then, is something of a myth since there have long been oscillations between national and imperial, class, regional, and religious identities throughout the modern period. Eugene Weber's analysis (1976) of the strength of regionalism in the 1870s implies a decline in the pervasiveness of French nationalism since the period of the revolutionary wars. Modern nations and nation-states from their very beginnings have operated in alliance or contest with the transnational institutions of empires, the great religions, revolutionary internationals, and capitalism. Acknowledging this, states have historically developed different strategies to overcome such challenges. The British adherence to liberal economics until the early twentieth century accorded with their self-interest, whereas late nineteenth-century Germany espoused a protectionist approach. For much of the nineteenth century Britain remained a world power, in part because of its skill in mustering coalitions of states against the dominant great power on the European sub-continent.

In short, the crisis of the contemporary nation-'state' which is allegedly no longer capable of exercising sovereignty in military, economic, and religious matters has a long history, and has been addressed by a recurring mobilizing of social networks. As one would expect, national identities have evolved in response to new contingencies, dramatically so in the case of (West) Germany after 1945 when there was an attempt to reject the national past and redefine the German future in terms of a commitment to 'European' democracy; but even here recent debates since unification indicate a desire to 'legitimize' Germany by discovering an acceptable national heritage (Fulbrook 1999: chapters 4, 6, and 7).

What then of the claims that these problems are intensifying and that globalization has reached a new stage in the post-1945 world with the rise of transnational and international alliances that indicate that the era of the nation-state is passing? Do we not see emerging a new planetary consciousness expressed through the rise of English (or American) as the world language, global institutions – political (the UN) and economic (GATT), a world civil society represented by NGOs, international covenants recognizing human rights: and this in response to a recognition of problems such as nuclear proliferation, environmental threats, large-scale international migratory and refugee flows, and terrorism that cannot be managed by existing nation-states? Does not the rise of regional associations (the European Union, ASEAN) represent a fundamental revulsion against the national principle in the name of wider cultural loyalties? The European Union, for example, is expropriating many of the traditional powers of the nation-state in monetary management, defence, border controls, and sub-continental elite networks are forming which suggest to some that a new European identity will develop, possibly based on an extension of civic models of the nation to a European scale (see Wallace 1990).

From the world historical perspective it is much too soon to evaluate the long-term viability of the European Union as a 'federal' enterprise, when in recent years we have seen the collapse or destabilization of multinational states whether federal or unitary from the USSR to Indonesia, in response to global military and economic pressures. It is more plausible to explain the development of the EU as a new strategy of national elites to maximize their sovereignty in an increasingly globalized world, rather than a rejection of the nation-state principle. The negotiating games of national elites are complex. Nation-states in Europe can use the 'global' as an instrument in their struggle for autonomy against the 'regional' pressures of 'Europe' or against the dominant regional power. The global military reach of the USA, made possible by technology, is welcomed by many states, both in Europe and Asia.

In any case, the very growth of transnational institutions has provoked a widespread reaction against the 'Western' values that they seem to embody in Asia, Africa, and Latin America. What many scholars have failed to acknowledge is that in the contemporary period religious organizations such as Islam and Evangelical Protestantism, their reach extended by modern communication systems, remain among the most potent globalizing agents and offer a rival vision to that of secular modernity. The recent attacks on the USA have shown the unpredictable threats that can arise out of a global interconnectedness. Globalization has engendered a hostility against the West on the part of large sections of the Muslim world, and whereas the prophets of multiculturalism have viewed the migration of millions of peoples from a historically rival civilization into the heartlands of the West as encouraging a creative hybridization of identities, many now fear instead an intensified and perhaps even religio-racial nationalism in the West triggered by conflicts between host community and 'immigrants'. Already there are calls to 'nationalize' immigrant minorities by teaching them the values of citizenship defined in terms of the dominant ethnic culture.

The clash of civilizations

If a single world united by secular cosmopolitan values seems some way off, is secular nationalism not threatened in a new way by a resurgence of religious movements in much of the Middle East, Asia, and Latin America? Mark Juergensmeyer (1993) has likened this religious revival to that of a new Cold War against the West. The current Islamist revival against Western secularism, highlighted in the Iranian revolution, has not only reshaped the politics of states with a Muslim majority, but also fanned a widely-based ethnocentric reaction in European nation-states against Muslim immigrants, including France where politicians of the left and right have expressed fears of the erosion of secular republican traditions by militant Islam. Samuel Huntington (1997), while rejecting this notion of a binary conflict between a secular West and a religious non-West, offers a vision of a future as one of a battle of civilizations, underpinned for the most part (though not exclusively) by antagonistic religious heritages. In such a vision, states (and nation-states) play a secondary role as leading political actors within their civilization. Civilizations without leading states will be politically flaccid, but equally states that seek to escape their historic civilization such as Greece (which should be in the Orthodox camp but seeks to be Western) or Turkey (which also seeks to be European but is within Islam) will be perpetually torn. Those interstate or intrastate conflicts that coincide with religio-civilizational fault-lines are likely to be the most intense. There are several obvious criticisms one can make of Huntington's associations of conflict with civilizational difference. He defines the European world of Latin Christianity as a single civilization, but from the sixteenth century the most ferocious conflicts conducted by Europeans were not with Muslim, Confucian or Orthodox Christian civilizations but were rather directed against themselves, first in the wars of Reformation and Counter Reformation in the sixteenth and seventeenth centuries, and then in the world wars of the twentieth century. One should note also that the modern religious movements can be directed against the West, but even where they critique external 'others' the 'real' enemy is generally not global but particular. Their main target is 'within', since they wish to morally regenerate a traditional culture being eroded by secular forces allied to alien cosmopolitan principles. The Islamist movements of the Middle East, Africa, and Pakistan and the Hindu revivalism in India were born out of the failures of secular (socialist) nationalism: to deliver the promises of social and economic emancipation of the masses, to free them from Western 'neo-colonialist' capitalism, and to provide law and order (see Hutchinson 1994: chapter 3). In each case there has been a 'return' to older religious traditions and institutions long predating the Western colonial impact in order to provide more 'authentic' models of development and to deliver basic social services, including justice (hence the appeal for Muslims of shariah law), education and the relief of poverty. The objective of these movements has been focused on a transformation of an existing political community rather than a world-wide crusade. Where the target is also external this is often a neighbouring country (Israel for Middle Eastern Muslims, India for Pakistanis), and, just as in the case of medieval *antemurale* kingdoms, this has reinforced a sense of nationality by defining the community as a unique custodian of spiritual values now under threat.

It is also a mistake to view such religious revivals as the domain of 'backward' non-western countries. Religion has been one of the sources of national identity for many avowedly secular states such as Holland and France, and it remains a powerful force in many contemporary 'Western' societies, including the USA, Germany, Italy,

Ireland and Greece. The current religious revival can be viewed as the most recent manifestation of a long-recurring conflict between secular and religious concepts 'within' nations. Internal conflicts erupted in Europe at the very beginning of modern nationalism as part of a general reaction against the secularism of the French revolution, and they continued through the nineteenth century, including within France itself. Such contestations are often an integral part of the nation-building process, for even many 'established' nations are riven by embedded cultural differences that generate rival symbolic and political projects. As I have already argued, nations are culturally plural, and the assumptions that there is a trend towards homogenization means that the centrality of cultural struggles in nation-formation has been neglected.

In Russia competition between Slavophiles and Westerners, the first defenders of Russia's distinctive Orthodox traditions and the second looking to Western European models, originated in the early nineteenth century and continues. In France the struggle between republicans, and clerico-legitimists since the French Revolution recurs in various forms, most visibly in the campaigns of Le Pen's National Front against the Fifth Republic. British society was riven in the nineteenth and twentieth centuries between the nonconformist Protestantism of the urban regions of Wales, Scotland and the north of England, which supported, first the Liberal, then the Labour parties, and the rural Anglican Protestantism of the more rural South, which supported Conservatism.

These divisions may reflect radically different views of the structure of politics, the status of social groups, relations between regions, the countryside and the city, economic and social policies and foreign policy. They also reflect the diverse heritages of populations whose geopolitical setting continues to expose them to unpredictable changes from several directions. Modern Russia has been shaped by interaction with Western and Central Europe, Byzantium, and the Asian steppes, and both Westerners and Slavophiles have recognized the validity of the other. The westerner Alexander Herzen, uneasy at wholesale importation of European ideas, especially after the failure of 1848, declared that Westerners would be cut off from the people as long as they ignored the questions posed by the Slavophiles (Neumann 1996: 170). Similarly, Dostoyevsky, advocate of Russia's Orthodox mission and its Eastern destiny, reveals the ambivalence of neo-Slavophiles, and how they had internalized assumptions of the Westerners: 'In Europe we were Asiatics, whereas in Asia we too are Europeans ... We shall go to Asia as masters' (*ibid.*: 64). These visions have alternated in power both at the level of state and of 'educated society', with groups, at times switching positions, in part affected by the sense of place and security of the national territory.

In short, nations are pluralist not culturally homogenous and the survival of these diverse heritages, it may be argued, is explained by geopolitical setting and because they enhance options for societies when faced with unpredictable challenges. But it may be fairly asked what prevents internal conflicts, especially when religiously based, leading to social breakdown and civil wars.

This is a pressing question today in many parts of the world. The short answer might be that breakdown is avoided when there is a potent common live ethnic heritage to which both secular and religious movements can appeal. In the Middle East and Asia the secular nation-state is threatened because it is seen to be a derivative of Western colonial rule, compared to 'rooted' religious traditions which can claim at times of crisis to be more 'authentic'. Often the nationalist initiative is taken by religious minorities, the Copts in Egypt, Maronite Christians in Lebanon, and Protestants in Korea who find in a secular nationalism, based on an historic period before that of the dominant religion, an instrument that allows them participation as (at least) equals in the political community. The danger is that this ethnic vision has little resonance with the majority. In those nation-states or state-nations with multiple ethnoreligious or religious communities, an ethnoreligious rejection of secular nationalism by the dominant group threatens to dissolve the state.

Even in contemporary Israel, a nation-state that claims legitimation by reference to an ancient kingdom and a myth of chosenness, the conflict with Palestinians only partly keeps the lid on an internal battle between secular Zionists and Orthodox Jews and religious nationalists about the character of the state. The Israeli example illustrates two important points. First, it reveals the inability of secular nationalism to override an old and institutionalized ethnoreligious heritage, in spite of its association with a powerful modern foundation story and victorious wars that have generated a pantheon of heroes and legends of sacrifice. Second, accommodation, although at times precarious, has been possible between bitter secular and religious rivals because of a common ethnoreligious heritage to which they refer, but there has been an increasing shift in the identity of the state from its original more secular and socialist orientation.

One may wonder whether the 'success' (however qualified) of the Israeli case is something of an exception in those non-European countries where secular nationalism aggressively confronts a traditional religious heritage. States, however, remain the potent instruments of politics, with the result that religious movements tend to become particularized, and the world religions lack transnational institutions and foci capable of mobilizing the faithful in alternative political formations. That said, the emergence of a world of 'real' nation-states if it ever comes is likely to be a protracted, uncertain, and potentially reversible process.

Conclusions

We can agree with some points of proponents of 'globalization' and of 'the clash of civilizations'. Undoubtedly, the strategies and perhaps the forms of nation-state are changing to face the new international environment. States in many contexts pool sovereignty; international institutions and doctrines have emerged restricting sovereignty, though uncertainly; and trends in democracy and human rights have enabled ethnoregional movements to become more visible and salient. But much discussion of the post-national state remains West Eurocentric, and the conditions that allow such national 'weakening' may remain temporary; even in Europe. I have expressed scepticism about the erosion of national principle in the face of globalization.

Globalization is a much longer phenomenon than most theorists of the subject are willing to acknowledge, and the agents and processes are not simply secular but include religion and warfare, both of which encourage difference. Before the modern world such factors resulted in large-scale ethnic phenomena, much of which has shaped the way that modern societies have evolved. More intense forms of interaction, engendered by scientific and technological revolutions, in turn produced the rise of nationalism but the empowerment of states came through not an eradication but a transformation of older ethnic heritages. Although the state principle is widely regarded as obsolete in the face of transnational entities, the major world actors remain nation-states or would-be nation-states. New threats will intensify nationalism in many parts of the world, such as climatic changes on states already locked in conflict over such natural resources as water, a major issue between Israel and Jordan and between India and Bangladesh.

We can also agree that religions remain potent global agents in the contemporary world, and that in much of the world nationalism is thinly based, statist and bearing little relation to ethnic and other traditional realities. Nonetheless, the current religious revival does not offer a significant threat to the system of nation states; as we have seen religion has, in many cases, become ethnicized. Religious conflict has as often as not intensified ethnic or national identity between neighbouring states (India and Pakistan). Much of the current religious revival is directed internally at the supposed inauthenticity of secular nationalism in relation to native heritages. Its effects vary according to context. In some cases it is a reflection of the multiple heritage of communities and can provide an alternative option to populations at times of crisis when established ideas and institutions have failed. In other cases where there are no common ethnic memories to which the rival projects can appeal, internal conflict becomes problematic. The result is likely to be long-term instability and state paralysis.

References

Abu-Lughod, J. (1989) *Before European Hegemony: The World System A.D. 1250–1350*, Oxford: Oxford University Press.

Adshead, C.M. (1993) *Central Asia in World History*, London: Macmillan (now Palgrave Macmillan).

Albrow, M. (1996) *The Global Age: State and Society Beyond Modernity*, Cambridge: Cambridge University Press.

Armstrong, J. (1982) *Nations Before Nationalism*, Chapel Hill: University of North Carolina Press.

Breuilly, J. (1982) *Nationalism and the State*, Manchester: Manchester University Press.

Breuilly, J. (1993) *Nationalism and the State*, 2nd edn, Manchester: Manchester University Press.

Breuilly, J. (2000) Panel Discussion 'Nationalism and the State', Annual Association for the Study of Ethnicity and Nationalism Conference, London, March.

Castells, M. (1996) *The Rise of the Network Society*, Oxford: Blackwell.

Colley, L. (1992) *Britons: Forging the Nation 1707–1837*, New Haven: Yale University Press.

Costello, P. (1993) *World Historians and Their Goals: Twentieth-Century Answers to Modernism*, DeKalb: Northern Illinois University Press.

Eaton, R.M. (1990) *Islam in World History*, Washington, D.C.: American Historical Association.

Fulbrook, M. (1999) *German National Identity After the Holocaust*, Cambridge: Polity.

Gellner, E. (1983) *Nations and Nationalism*, Oxford: Blackwell.

Giddens, A. (1990) *The Consequences of Modernity*, Cambridge: Polity.

Gildea, R. (1994) *The Past in French History*, New Haven, CT: Yale University Press.

Guibernau, M. (2001) 'Globalization and the Nation-state', in M. Guibernau and J. Hutchinson (eds), *Understanding Nationalism*, Cambridge: Polity, 242–68.

Hastings, A. (1997) *The Construction of Nationhood: Ethnicity, Religion, and Nationalism*, Cambridge: Cambridge University Press.

Hechter, M. (2000) *Containing Nationalism*, Oxford: Oxford University Press.

Held, D., A. McGrew, D. Goldblatt and J. Perraton (1999) *Global Transformations: Politics, Economics and Culture*, Cambridge: Polity.

Hobsbawm, E.J. (1990) *Nations and Nationalism Since 1780*, Cambridge: Cambridge University Press.

Hodgson, M. (1993) *Rethinking World History: Essays on Europe, Islam and World History* (edited by E. Burke III), Cambridge: Cambridge University Press.

Huntington, S. (1997) *The Clash of Civilizations and the Remaking of World Order*, London: Touchstone Books.

Hutchinson, J. (1994) *Modern Nationalism*, London: Fontana Press.

Juergensmeyer, M. (1993) *The New Cold War? Religious Nationalism Confronts the Secular State*, Berkeley: University of California Press.

Keynes, J.M. (1920) *The Economic Consequences of the Peace*, London: Macmillan and Co.

Mann, M. (1986) *The Sources of Social Power*, volume 1, Cambridge: Cambridge University Press.

McNeill, W.H. (1963) *The Rise of the West: A History of the Human Community with a Retrospective Essay*, Chicago: University of Chicago Press.

McNeill, W.H. (1990) 'The Rise of the West after Twenty-Five Years', *Journal of World History*, 1 (1), 1–21.

Neumann, I.B. (1996) *Russia and the Idea of Europe: A Study in Identity and International Relations*, London: Routledge.

Ohmae, K. (1996) *The End of the Nation State: The Rise of Regional Economies*, New York: Free Press.

Popper, K.R. (1960) *The Poverty of Historicism*, London: Routledge.

Smith, A.D. (1981) 'War and Ethnicity: The Role of Warfare in the Formation of Self-Images and Cohesion of Ethnic Communities', *Ethnic and Racial Studies*, 4 (4), 375–97.

Smith, A.D. (1999) *Myths and Memories of the Nation*, Oxford: Oxford University Press.

Smith, A.D. (2001) 'Nations and History', in M. Guibernau and J. Hutchinson (eds), *Understanding Nationalism*, Cambridge: Polity, 9–31.

Wallace, W. (1990) *The Transformation of Western Europe*, London: Pinter.

Weber, E. (1976) *Peasants into Frenchmen: the Modernization of Rural France (1870–1914)*, Stanford: Stanford University Press.

Wriston, W. (1992) *The Twilight of Sovereignty: How the Information Revolution is Transforming Our World*, New York: Charles Scribner's Sons.

Zheng, Y. (1999) *Discovering Nationalism in China: Modernisation, Identity, and International Relations*, Cambridge: Cambridge University Press.

6

Nationalism and Cosmopolitanism

Craig Calhoun

> A certain attenuated cosmopolitanism had taken place of the old home feeling.
>
> Thomas Carlyle, 1857
>
> Among the great struggles of man – good/evil, reason/ unreason, etc. – there is also this mighty conflict between the fantasy of Home and the fantasy of Away, the dream of roots and the mirage of the journey.
>
> Salman Rushdie, 2000

Nineteenth-century thinkers, like Thomas Carlyle, were often ambivalent about cosmopolitanism.[1] They worried that it was somehow an 'attenuated' solidarity by comparison with those rooted in more specific local cultures and communities. Today cosmopolitanism has considerable rhetorical advantage. It seems hard not to want to be a 'citizen of the world'. Certainly, at least in Western academic circles, it is hard to imagine preferring to be known as parochial. But what does it mean to be a 'citizen of the world'? Through what institutions is this 'citizenship' effectively expressed? Is it mediated through various particular, more local solidarities or is it direct? Does it present a new, expanded category of identification as better than older, narrower ones (as the nation has frequently been as opposed to the province or village) or does it pursue better relations among a diverse range of traditions and communities? How does this citizenship contend with global capitalism and with non-cosmopolitan dimensions of globalization?

My questioning is meant not as an attack on cosmopolitanism but as a challenge to the dominant ways in which it has been conceptualized. First, I want to question the social bases for cosmopolitanism. What experiences make this an intuitively appealing approach to the world? Which are reflected poorly or not at all? What issues are obscured from view? Second, I want to ask how much the political theory of cosmopolitanism is shaped by the spectre of bad nationalism and a poorly drawn fight with communitarianism, how well it defends social achievements against neoliberal capitalism, and to what extent it substitutes ethics for politics. Finally, I wish to offer a plea for the importance of the local and particular – not least as a basis for democracy, no less important for being necessarily incomplete. Whatever its failings, 'the old home feeling' helped to produce a sense of mutual obligations and what Edward Thompson (1971) echoed an old tradition in calling a 'moral economy'.

A thoroughgoing cosmopolitanism might indeed bring concern for the fate of all humanity to the fore, but a more attenuated cosmopolitanism is likely to leave us lacking the old sources of solidarity without adequate new ones. Much cosmopolitanism focuses on the development of world government or at least global political institutions. These, advocates argue, must be strengthened if democracy is to have much future in a world where nation-states are challenged by global capitalism, cross-border flows and international media and accordingly less able to manage collective affairs (Held 1995; Archibugi and Held 1995; Archibugi *et al.* 1998).[2] At the same time, most of these advocates see growing domestic heterogeneity and newly divisive subnational politics as reducing the efficacy of nation-states from within. While most embrace diversity as a basic value, they simultaneously see multiculturalism as a political problem. In the dominant cosmopolitan theories, it is the global advance of democracy that receives most attention and in which most hopes are vested. But cosmopolitanism without the strengthening of local democracy is likely to be a very elite affair.

The political theory of cosmopolitan democracy

Generally speaking, to say 'cosmopolitan' has been to say anything *but* 'democratic'. Cosmopolitanism was the project of empires, and as an intellectual and a personal style – and indeed a legal arrangement – it flourished in imperial capitals and trading cities. The tolerance of diversity in cosmopolitan imperial cities reflects among other things precisely the absence of a need to organize self-rule (I have argued this at greater length in Calhoun 1993 and 1995).

Liberalism and belonging

Contemporary cosmopolitanism is the latest effort to revitalize liberalism.[3] It has much to recommend it. Aside from world peace and more diverse ethnic restaurants, there is the promise to attend to one of the great lacunae of more traditional liberalism. This is the assumption of nationality as the basis for membership in states, even though this implies a seemingly illiberal reliance on inheritance and ascription rather than choice, and an exclusiveness hard to justify on liberal terms.

Political theory has surprisingly often avoided addressing the problems of political belonging in a serious, analytic way by presuming that nations exist as the prepolitical bases of state-level politics. I do not mean that political theorists are nationalists in their political preferences, but rather that their way of framing analytic problems is shaped by the rhetoric of nationalism and the ways in which this has become basic to the modern social imaginary (on the predominance of nationalist understandings in conceptions of 'society', see Calhoun 1999). 'Let us imagine a society', theoretical deliberations characteristically begin, 'and then consider what form of government would be just for it'. Nationalism provides this singular and bounded notion of society with its intuitive meaning.

Even so Kantian, methodologically individualistic and generally non-nationalist a theorist as Rawls exemplifies the standard procedure, seeking in *A Theory of Justice* to understand what kind of society individuals behind the veil of ignorance would choose – but presuming that they would imagine this society on the model of a nation-state. Rawls modifies his arguments in considering international affairs in *Political Liberalism* and *The Law of Peoples*, but continues to assume something like the nation-state as the natural form of society. As he unhelpfully and unrealistically writes:

...we have assumed that a democratic society, like any political society, is to be viewed as a complete and closed social system.

It is complete in that it is self-sufficient and has a place for all the
main purposes of human life. It is also closed, in that entry into it
is only by birth and exit from it is only by death. (Rawls 1993: 41)

Rawls is aware of migration, war and global media, of course. But he
imagines questions of international justice to be just as that phrase and
much diplomatic practice implies: questions 'between peoples', each of
which should be understood as unitary. Note also the absence of attention to local or other constituent communities within this conception
of society. Individuals and the whole society have a kind of primacy
over any other possible groupings. This is the logic of nationalism.

   This is precisely what cosmopolitanism contests – at least at its
best – and rightly so. Indeed, one of the reasons given for the very
term is that it is less likely than 'international' to be confused with
exclusively intergovernmental relations (Archibugi 1998: 216).
Advocates of cosmopolitanism argue that people belong to a range of
polities of which nation-states are only one, and that the range of
significant relationships formed across state borders is growing. Their
goal is to extend citizenship rights and responsibilities to the full
range of associations thus created. In David Held's words,

   people would come, thus, to enjoy multiple citizenships – political
   membership in the diverse political communities which significantly affected them. They would be citizens of their immediate
   political communities, and of the wider regional and global networks which impacted upon their lives. (1995: 233)

Though it is unclear how this might work out in practice, this challenge to the presumption of nationality as the basis for citizenship is
one of the most important contributions of cosmopolitanism.

   The cosmopolitan tension with the assumption of nation as the
prepolitical basis for citizenship is domestic as well as international.
As Jurgen Habermas puts it,

   the nation-state owes its historical success to the fact that it sub-
   stituted relations of solidarity between the citizens for the disinte-
   grating corporate ties of early modern society. But this republican
   achievement is endangered when, conversely, the integrative force
   of the nation of citizens is traced back to the prepolitical fact of a

quasi-natural people, that is, to something independent of and prior to the political opinion- and will-formation of the citizens themselves. (2000: 115)

But pause here and notice the temporal order implied in this passage. *First* there were local communities, guilds, religious bodies and other 'corporative bonds'. *Then* there was republican citizenship with its emphasis on the civic identity of each. *Then* this was undermined by ethnonationalism. What this misses is the extent to which each of these ways of organizing social life existed simultaneously with the others, sometimes in struggle and sometimes symbiotically. New 'corporative ties' have been created, for example, notably in the labour movement and in religious communities. Conversely, there was no 'pure republican' moment when ideas of nationality did not inform the image of the republic and the constitution of its boundaries.

As Habermas goes on, however, 'the question arises of whether there exists a functional equivalent for the fusion of the nation of citizens with the ethnic nation' (*ibid.*: 117). (Note that Habermas tends to equate 'nation' with 'ethnic nation'.) We need not accept his idealized history or entire theoretical framework to see that this raises a basic issue. That is, for polities not constructed as ethnic nations, what makes membership compelling? This is a question for the European Union, certainly, but also arguably for the United States itself, and for most projects of cosmopolitan citizenship. Democracy requires a sense of mutual commitment among citizens that goes beyond mere legal classification, holding a passport, or even respect for particular institutions. As Charles Taylor (2002) has argued forcefully, 'self-governing societies', have need 'of a high degree of cohesion'.

One of the challenges for cosmopolitanism is to account for how social solidarity and public discourse might develop in these various wider networks such that they could become the basis for an active citizenship. So far, most versions of cosmopolitanism share with traditional liberalism a thin conception of social life, commitment and belonging. Actually existing cosmopolitanism exemplifies this deficit in its 'social imaginary'. That is, it conceives of society – and issues of social belonging and social participation – in too thin and casual a manner.

The result is an emerging theory of transnational politics that suffers from an inadequate sociological foundation. As Bellamy and

Castiglione (1998) write, hoping to bridge the opposition between cosmopolitanism and communitarianism, 'a pure cosmopolitanism cannot generate the full range of obligations its advocates generally wish to ascribe to it. For the proper acknowledgement of "thin" basic rights rests on their being specified and overlaid by a "thicker" web of special obligations'. Held agrees: persons inhabit not only rights and obligations, but relationships and commitments within and across groups of all sorts including the nation.

This image of multiple, layered citizenship directly challenges the tendency of many communitarians to suggest not only that community is necessary and/or good, but that people normally inhabit one and only one community.[4] It also points to the possibility – so far not realized – of a rapprochement between cosmopolitanism and communitarianism. More often, cosmopolitans have treated communitarianism as an enemy, or at least used it as a foil. At the same time, though, advocates of cosmopolitan democracy find themselves falling back on notions of 'peoples' as though these exist naturally and prepolitically. They appeal, for example, for the representation of peoples rather than or in addition to states in various global institutions including an eventual world parliament (Archibugi 2000: 146). This poses deeper problems than is commonly realized. Not only is the definition of 'people' problematic, the idea of representation is extremely complex. Representing peoples has been one of the primary functions of modern states – however great the problems with how they do it. Absent state-like forms of explicit self-governance, it is not clear how the representation of peoples escapes arbitrariness.

Cosmopolitan democracy would appear to require not only a stronger account of representation, but also a stronger account of social solidarity and the formation and transformation of social groups. If one of its virtues is challenging the idea that nationality (or ethnic or other identities understood as analogous to nationality) provides people with an unambiguous and singular collective membership, one of its faults is to conceptualize the alternative too abstractly and vaguely. Equally, most cosmopolitanism seems to underestimate the positive side of nationalism, the virtues of identification with a larger whole, which as a polity or potential polity is more open to democratization than religions or some other kinds of larger groupings.

Part of the problem is that cosmopolitanism relies heavily on a purely political conception of human beings. Such a conception has two weak points. First, it does not attend enough to all the ways in which solidarity is achieved outside of political organization, and does not adequately appreciate the bearing of these on questions of political legitimacy. Second, it does not consider the extent to which high political ideals founder on the shoals of everyday needs and desires – including quite legitimate ones. The ideal of civil society has sometimes been expressed in recent years as though it should refer to a constant mobilization of all of us all the time in various sorts of voluntary organizations.[5] But in fact one of the things people quite reasonably want from a good political order is to be left alone some of the time – to enjoy a non-political life in civil society. In something of the same sense, Oscar Wilde famously said of socialism that it requires too many evenings. We could say of cosmopolitanism that it requires too much travel, too many dinners out at ethnic restaurants, too much volunteering with *Médecins Sans Frontières*. Perhaps not too much or too many for academics (though I wouldn't leap to that presumption) but too much and too many to base a political order on the expectation that everyone will choose to participate.

A key issue is simply what people choose to do with their time. In addition, actually existing politics have developed a less engaging face than they might have. But surely scale is a third factor. Participation rates are low in local and national politics; it is not clear that the spread of global social movements offers evidence enough for a possible reversal on the supranational scale. On the contrary, there is good reason to think that the very scale of the global ecumene will make participation even narrower and more a province of elites than in national politics. Not only does Michels's law of oligarchy apply, if perhaps not with the iron force he imagined, but the capacities to engage cosmopolitan politics – from literacy to computer literacy to familiarity with the range of acronyms – are apt to continue to be unevenly distributed. Indeed, there are less commonly noted but significant inequalities directly tied to locality. Within almost any social movement or activist NGO, as one moves from the local to the national and global in either public actions or levels of internal organization one sees a reduction in women's participation. Largely because so much labour of social reproduction – child care, for instance – is carried out by women, women find it harder to work outside of their localities. This is true even for social movements in which women predominate at the local level.

Rationalism and difference

Contemporary cosmopolitan theory is attentive to the diversity of people's social engagements and connections. But this cosmopolitanism is also rooted in seventeenth- and eighteenth-century rationalism with its ethical universalism counterposed specifically to traditional religion and more generally to deeply-rooted political identities. Against the force of universal reason, the claims of traditional culture and communities were deemed to have little standing. These were at best particularistic, local understandings that grasped universal truths only inaccurately and partially. At worst, they were outright errors, the darkness to which Enlightenment was opposed. Rationalism challenged more than just the mysticism of faith. The sixteenth- and seventeenth-century wars of faith seemed to cry out for universalistic reason and a cosmopolitan outlook. Yet, this rationalism was also rich with contractarian metaphors and embedded in the social imaginary of a nascent commercial culture. It approached social life on the basis of a proto-utilitarian calculus, an idea of individual interests as the basis of judgement, and a search for the one right solution. Its emphasis on individual autonomy, whatever its other merits, was deployed with a blind eye to the differences and distortions of private property. The claims of community appeared often as hindrances on individuals. They were justified mainly when community was abstracted to the level of nation, and the wealth of nations made the focus of political as well as economic attention.

Like this earlier vision of cosmopolis, the current one responds to international conflict and crisis.[6] It offers an attractive sense of shared responsibility for developing a better society and transcending both the interests and intolerance that have often lain behind war and other crimes against humanity. However, this appears primarily in the guise of ethical obligation, an account of what would be good actions and how institutions and loyalties ought to be rearranged. Connection is seldom established to any idea of political action rooted in immanent contradictions of the social order. From the liberal rationalist tradition, contemporary cosmopolitanism also inherits suspicion of religion and rooted traditions; a powerful language of rights that is also sometimes a blinder against recognition of the embeddedness of individuals in culture and social relations; and an opposition of reason and rights to community. This last has appeared in various guises through 300 years of contrast between allegedly inherited and constraining local community life and the ostensibly freely chosen social relationships of modern cities, markets, associational life and more generally cosmopolis.

Confronting similar concerns in the mid-twentieth century, Theodore Adorno wrote:

> An emancipated society... would not be a unitary state, but the realization of universality in the reconciliation of differences. Politics that are still seriously concerned with such a society ought not, therefore, propound the abstract equality of men even as an idea. Instead, they should point to the bad equality today... and conceive the better state as one in which people could be different without fear. (1974: 103)

The tension between abstract accounts of equality and rooted accounts of difference has been renewed in the recent professional quarrels between liberal and communitarian political theorists. For the most part, cosmopolitans model political life on a fairly abstract notion of person as a bearer of rights and obligations.[7] This is readily addressed in rationalist and indeed proceduralist terms. And, however widely challenged in recent years, rationalism retains at least in intellectual circles a certain presumptive superiority. It is easy to paint communitarian claims for the importance of particular cultures as irrational, arbitrary and only a shade less relativist than the worst sort of postmodernism. But immanent struggle for a better world always builds on particular social and cultural bases.[8] Moreover, rationalist universalism is liable not only to shift into the mode of 'pure ought' but to approach human diversity as an inherited obstacle not a resource or a basic result of creativity.

Entering this quarrel on the liberal side, but with care for diversity, Held suggests that national communities cease to be treated as primary political communities. He does not go so far as some and claim that they should (or naturally will) cease to exist, but rather imagines them as one sort of relevant unit of political organization among many. What he favours is a cosmopolitan democratic community:

> a community of all democratic communities must become an obligation for democrats, an obligation to build a transnational,

common structure of political action which alone, ultimately, can support the politics of self-determination. (Held 1995: 232)

In such a cosmopolitan community, 'people would come ... to enjoy multiple citizenships – political membership in the diverse political communities which significantly affected them' (*ibid.*: 233). Sovereignty would then be 'stripped away from the idea of fixed borders and territories and thought of as, in principle, malleable time–space clusters. ... it could be entrenched and drawn upon in diverse self-regulating associations, from cities to states to corporations' (*ibid.*: 234). Indeed, so strong is Held's commitment to the notion that there are a variety of kinds of associations within which people might exercise their democratic rights, that he imagines 'the formation of an authoritative assembly of all democratic states and agencies, a reformed General Assembly of the United Nations ... ' with its operating rules to be worked out in 'an international constitutional convention involving states, IGOs, NGOs, citizen groups and social movements' (*ibid.*: 273–4). The deep question is whether this all-embracing unity comes at the expense of cultural particularity – a reduction to liberal individualism – or provides the best hope of sustaining particular achievements and openings for creativity in the face of neoliberal capitalism.

The very idea of democracy suggests that it cannot be imposed from above, simply as a matter of rational plan. It is inherently a matter of differences – of values, perceptions, interests and understandings. The power of states and global corporations and the systemic imperatives of global markets suggest that advancing democracy will require struggle. This means not only struggle against states or corporations, but struggle within them to determine the way they work as institutions, how they distribute benefits, what kinds of participation they invite. The struggle for democracy, accordingly, cannot be only a cosmopolitan struggle from social locations that transcend these domains, it must be also a local struggle within them. Moreover, it would be a mistake to imagine that cosmopolitan ethics – universally applied – could somehow substitute for a multiplicity of political, economic and cultural struggles. Indeed, the very struggle may be an occasion and source for solidarity.

Moreover, it is important that democracy grows out of the lifeworld, that theories of democracy seek to empower people not in the abstract but in the actual conditions of their lives. To empower people where they are means to empower them within communities and traditions, not in spite of them, and as members of groups not only as individuals. This does not mean accepting old definitions of all groups; there may be struggle over how groups are constituted. For example, appeals to aboriginal rights need not negate the possibility of struggle within Native American or other groups over such issues as gender bias in leadership (this is a central issue in debates over group rights; see for example Kymlicka 1995). It is important that we recognize that legitimacy is not the same as motivation. We need to pay attention to the social contexts in which people are moved by commitments to each other. A cosmopolitanism that does so will be variously articulated with locality, community and tradition, not simply a matter of common denominators. It will depend to a very large extent on local and particularistic border crossings and pluralisms, not universalism.

Such a cosmopolitanism would challenge the abandonment of globalization to neoliberalism (whether with enthusiasm or a sense of helpless pessimism) and challenge the impulse to respond simply by defending nations or communities that experience globalization as a threat. It is unclear, however, just what social life is like in 'malleable time–space clusters' and what it would mean for global politics to be a matter of cross-cutting membership in a host of different 'agencies' from communities to corporations. Multiplicity is one issue; scale is another. It is clear, moreover, that cosmopolitanism has yet to come to terms with tradition, community, ethnicity, religion and above all nationalism. In offering a seeming 'view from nowhere', cosmopolitans commonly offer a view from Brussels (where the postnational is identified with the strength of the European Union rather than the weakness of, say, African states), or from Davos (where the postnational is corporate) or from the university (where the illusion of a free-floating intelligentsia is supported by relatively fluid exchange of ideas across national borders).

The spectre of bad nationalism

Acknowledging diversity is basic to the political theory of cosmopolitan democracy. But the theory is nonetheless ambivalent. Cosmopolitanism seems to be more about transcending cultural specificity and differences of local institutions than about defending them. The claims of ethnicity and nationhood appear primarily as problems, and are analysed in terms of the prejudicial opposition of cosmopolitan liberalism to communitarianism and nationalism. Cosmopolitan thought has a hard time with cultural particularity, local commitments, and even emotional attachments. This comes partly from its Enlightenment liberal heritage of rationalist challenge to religious and communal solidarities as 'backward'. It is reinforced powerfully by the image of 'bad nationalism'. For many advocates of cosmopolitanism, this image of the 'other' is definitive. Nazi Germany is paradigmatic, but more recent examples, such as Milosevic's Serbia and ethnic war in Rwanda and Burundi, also inform the theories. At the core of each instance, as generally understood, is an ethnic solidarity triumphant over civility and liberal values and ultimately turning to horrific violence.

Advocates of a postnational or transnational cosmopolitanism, however, do themselves and theory no favours by equating nationalism with ethnonationalism and understanding this primarily through its most distasteful examples. Nations have often had ethnic pedigrees and employed ethnic rhetorics, but they are modern products of shared political, culture and social participation, not mere inheritances. To treat nationalism as a relic of an earlier order, a sort of irrational expression, or a kind of moral mistake is to fail to see both the continuing power of nationalism as a discursive formation and the work – sometimes positive – that nationalist solidarities continue to do in the world. As a result, nationalism is not easily abandoned even if its myths, contents and excesses are easily debunked (I have discussed nationalism as a discursive formation in Calhoun 1997). The way in which it still informs notions of the representation of 'peoples' is a case in point.

Not only this, the attempt to equate nationalism with problematic ethnonationalism sometimes ends up leading cosmopolitans to place all 'thick' understandings of culture and the cultural constitution of political practices, forms and identities on the nationalist side of the classification. Only quite thin notions of 'political culture' are retained on the attractive cosmopolitan side (see, for example, Habermas 1994; Taylor 1994; on the cosmopolitan side, see Thompson 1998). Yet republicanism and democracy depend on more than narrowly political culture; they depend on richer ways of constituting life together.

Democracy and cosmopolitanism have not always been close fellow travellers. The current pursuit of cosmopolitan democracy flies in the face of a long history in which the cosmopolitan has thrived in market cities, imperial capitals, and court society. Historically, cosmopolitanism often flourished precisely where democracy was not an option. It thrived in Ottoman Istanbul, for example, and old regime Paris, and both ancient and later colonial Alexandria, because in none of these were members of different cultures and communities invited to organize government for themselves. It was precisely when democracy became a popular passion and a political project that nationalism flourished. Nationalism – not cosmopolitanism – has been the social imaginary most compatible – one might say complicit – with democracy. Democracy, in particular, has depended on strong notions of who 'the people' behind phrases like 'we the people' might be, and who might make legitimate the performative declarations of constitution-making and the less verbal performances of revolution (Taylor 2002). In this respect, its seventeenth-century ancestors are less the liberal individualists of social contract theory than early English nationalists.

It is not only nationalism that figures as a defining 'other' to cosmopolitanism. It is also community, ethnicity and religion. Indeed, part of the problem is that the 'bad nationalist' image informs the whole reading of tradition and community. Religion is a particular issue in this. Communitarians generally acknowledge the importance of religion as a basis for community, whether they personally embrace faith or not. Liberals may advocate tolerance, but partly as believers in tolerance they are troubled by the deep prejudices against other ways of life implicit in many religious faiths. But attitudes towards Catholicism and Islam remain litmus tests for the distinction, not least when it is extended into international affairs. Are these potentially sources for alternative and possibly better visions of modernity? Or are they illiberal challenges to a modernity that is necessarily rational-individualist in character?

Cosmopolitanism is in this sense a latent bad conscience to liberalism, a reminder that most liberals had become tacit nationalists, allowing their universalism to extend only to the borders of the countries. Implicitly, liberals had fallen into accepting the illiberal idea that inheritance – birth – rather than choice should be the basis of political identity. A liberal internationalism developed, to be sure, but it was itself rooted in liberal nationalism. Assistance offered to 'less developed countries' was never extended on the basis of the same universalism as that conditioned on domestic citizenship (even if the latter too allowed great inequality and often reduced what should have been universal entitlements to acts of charity). But for the most part, liberalism simply accepted national identities as framing the boundaries of political communities and didn't push the point very hard.

While the cosmopolitan challenge to deeply-rooted traditional identities was often deployed against claims to ground national identity in ethnicity throughout the eighteenth, nineteenth and twentieth centuries, liberals also seized on the state apparatus to promote national integration and homogenization within nation-states. Projects of rational planning and liberal modernization were developed within the boundaries offered by nation-states – even though liberal theory could offer no good account of why those boundaries should be defended against immigrants. It is perhaps paradoxical that in their struggle against benighted local prejudice, against provincialism, that liberals were the advocates of homogenizing nationalism – for example, in education policy – that now helps to underwrite the idea of the nation as a primary and self-sufficient solidarity.

Tradition and self-determination

The idea of approaching autonomy in terms of national self-determination is especially troubling to cosmopolitans. First, it privileges an unchosen whole over individual choice. Second, the idea of nation typically involves a strong claim to stand alone as politically self-sufficient. Third, national self-determination may even be impossible given the contemporary geopolitical challenges to national autonomy.

As David Held writes in what remains the best-developed, most thorough and thoughtful account of cosmopolitan democracy:

> The idea of a community which rightly governs itself and determines its own future – an idea at the very heart of the democratic polity itself – is ... today deeply problematic. (1995: 17)

Held goes on to note the importance of the fact that 'nations' are not today strong containers of the social connections of individuals – if indeed they ever were.

... in a highly interconnected world, 'others' include not just those found in the immediate community, but all those whose fates are interlocked in networks of economic, political and environmental interaction. (*ibid.*: 228)

It is worth pausing to note that 'immediate community' refers here to nation more than to any actual networks of local or other directly interpersonal relationships. The nation is indeed in a sense a 'direct access' construct – individuals are members immediately rather than through their membership of smaller groupings (see Calhoun 1997). But this is in fact part of what is novel about the modern nation by comparison with empires or feudal kingdoms or a variety of other forms of polity. If we are to take seriously people's different forms of belonging to social groups, then we need to avoid using the term 'community' in such an elastic sense that nations, religious confessions, cities and neighbourhoods all appear as exemplars.

This is important for thinking about ethnicity. Too easily, ethnicity is rendered the 'other' to globalization. It is treated as static, or at best grudgingly resistant to modernization and cosmopolitan virtues. It is described as a matter of 'tradition' in a usage that resembles Bagehot's notion of 'the hard cake of culture' rather than emphasizing the importance of passing on creations, sharing ideas and values, reproducing meanings, learning culture in directly interpersonal relations. Like all forms of traditional culture, ethnicity is changed dramatically by the introduction of mass literacy, reliance on fixed texts and authorized interpreters – not to mention newer communications technologies. In efforts to fix and stabilize tradition, the contents of ethnicity are sometimes hardened – though it is almost always the case that if ethnic cultures remain alive this hardening is challenged by new generations and new creativity.

Moreover, ethnicity is not simply an inheritance from the past of small, kin-organized communities. It developed in the context of cities, states and migrations as a distinctive way of constructing identities and solidarities on relatively large scales to which kinship and similar relational structures of very local life were inadequate. It exists not as a simple carry-over from an earlier world of 'pure' local identities, thus, but as a means of managing the interrelationship of the local and the translocal, the interpersonally communicated and the impersonally communicated, the social organization constructed by markets and bureaucracies and that built out of direct relationships. It combines abstract categories of identity with concrete identification within social networks. It is a way of participating in globalization – and other large-scale processes – not their opposite.

Community has always been stronger at local levels than national ones, and necessarily so. This is obscured by use of the same term to refer to the national 'political community' and to neighbourhoods, towns and villages. Accounts of local democracy are strikingly underdeveloped in cosmopolitan theory. It is as though theorists assume that the problems of the nation are to be solved entirely by its transcendence in a welter of border crossings. In fact, the construction of viable local communities – and more democratic local communities – may be equally central. The nation has no monopoly on being a 'community of fate'. At the same time, the existence of communities of fate is not simply conservative. It is also, and often at the same time, in the sort of tension with dominant trends that makes it a basis for radical struggle (on this point, see Calhoun 1983). This struggle, it is true, may be resistance more than proactive construction. Capitalist globalization has spawned a variety of movements seeking exemption from its dictates. But the existence of deep roots for struggle, deep roots to community, does not mean simply resistance. It means also a foundation for serious and radical struggle. This depends on roots and bonds that cannot be simply matters of immediate choice, and thus often on local community. Indeed, one of the oddities of the cosmopolitan hostility to communitarianism is neglect of the extent to which communitarian arguments are actually about sub-national communities, not nations.

For the most part, contemporary public discourse is conceived within a social imaginary in which the idea of nation is still basic, defining not only a new sense of local which is not local at all but national, but also defining the global often as the international. This takes attention away from the extent to which transnational corporations organize apparently international relations. These corporations themselves, like nations, depend on this social imaginary to be construed as natural, normal. In this social imaginary, cosmopolitanism appears mainly in the guise of adaptation to the institutional order of power relations and capital.

Social foundations of cosmopolitanism

Cosmopolitanism presents itself simply as global citizenship. It offers a claim to being without determinate social basis that is reminiscent of Mannheim's idea of the free-floating intellectual. But the view from nowhere or everywhere is more located than this. Cosmopolitanism reflects an elite perspective on the world (although what academic theory does not?).

It is worth recalling the extent to which the top ranks of capitalist corporations and the NGOs that support them – from the World Bank to organizations setting accountancy standards – provide exemplars of cosmopolitanism. Even the ideas of cosmopolitan democracy and humanitarian activism reflect the kind of awareness of the world that is made possible by the proliferation of non-governmental organizations working to solve environmental and humanitarian problems, and by the growth of media attention to those problems. These are important – indeed vital – concerns. Nonetheless, the concerns, the media and the NGOs need to be grasped reflexively as the basis for an intellectual perspective. It is a perspective, for example, that makes nationalism appear one-sidedly as negative. This is determined first perhaps by the prominence of ethnonationalist violence in recent humanitarian crises, but also by the tensions between states and international NGOs. It is also shaped by specifically European versions of transnationalism. Both nationalism and questions of whether states should be strengthened would look different from an African vantage point. Similarly, the development of the 'emergency' as a basic category for understanding the world opens ours eyes to important issues, but also structures the way we see them.

Various crises of the nation-state set the stage for the revitalization of cosmopolitanism. The crises were occasioned by the acceleration of global economic restructuring in the 1990s, new transnational communications media, new flows of migrants, and proliferation of civil wars and humanitarian crises in the wake of the Cold War. The last could no longer be comprehended in terms of the Cold War, which is one reason why they often appeared in the language of ethnicity and nationalism. Among their many implications, these crises all challenged liberalism's established understandings of (or perhaps wilful blind spot towards) the issues of political membership and sovereignty. They presented several problems simultaneously: (1) Why should the benefits of membership in any one polity not be available to all people? (2) On what bases might some polities legitimately intervene in the affairs of others? (3) What standing should organizations have that operate across borders without being the agents of any single state (this problem, I might add, applies as much to business corporations as to NGOs and social movements) and conversely how might states appropriately regulate them?

Enter cosmopolitanism. Borders should be abandoned as much as possible and left porous where they must be maintained. Intervention on behalf of human rights is good. NGOs and transnational social movements offer models for the future of the world. These are not bad ideas, but they are limited ideas.

Cosmopolitanism is a discourse centred in a Western view of the world.[9] It sets itself up commonly as a 'Third Way' between rampant corporate globalization and reactionary traditionalism or nationalism. If Giddens' account of the Third Way is most familiar, Barber's notion of a path beyond 'Jihad vs. McWorld' is equally typical.[10] Such oppositions are faulty, though, and get in the way of actually achieving some of the goals of cosmopolitan democracy. In the first place, they reflect a problematic denigration of tradition, including ethnicity and religion. This can be misleading in even a sheer factual sense, as for example Barber describes Islamism as the reaction of small and relatively homogeneous countries to capitalist globalization. The oppositions are also prejudicial. Note, for example, the tendency to treat the West as the site of both capitalist globalization and cosmopolitanism but to approach the non-West through the category of tradition.

It is worth noting that cosmopolitanism is itself a tradition, with roots in the ancient world (perhaps especially in Hellenism), in early modern Humanism, and in empire. More generally, the opposition to tradition (and with it to community, religion, ethnicity and the like) is based on a limited and static view. This does damage especially to the notion of ethnicity as living, creative culture. In this connection, we should also recall how recent, temporary, and never complete the apparent autonomy and closure of 'nation' was. Looked at from the standpoint of India, say, or Ethiopia, it is not at all clear whether 'nation' belongs on the side of tradition or on that of developing cosmopolitanism. Or is it perhaps distinct from both – a novel form of solidarity and a basis for political claims on the state, one which presumes and to some extent demands performance of internal unity and external boundedness?

This way of ordering the world does not simply reflect pre-existing internal histories of nations; it reflects also their development amid the struggles of religious wars, imperialism and capitalist competition. Nation-states offered powerful means for disciplining their members – not least in demanding loyalty across class lines – as well as attempting to control transborder flows. But imperialism is intrinsic not only to the story of nation-states but of cosmopolitanism. It shapes the very claim to difference within unity that is basic to most modern cosmopolitan theories. As Brennan puts it, cosmopolitanism 'designates an enthusiasm for customary differences, but as ethical or aesthetic material for a unified polychromatic culture – a new singularity born of a blending and merging of multiple local constituents' (2001: 76).[11] But this very claim to unity – echoing on a grander scale that of great empires and great religions – underwrites the cosmopolitan's appeal for all-encompassing world government.

Consumerist cosmopolitanism

Even while the internal homogeneity of national cultures was being promoted by linguistic and educational standardization (among other means), the great imperial and trading cities stood as centres of diversity. Enjoying this diversity was one of the marks of the sophisticated modern urbanite by contrast to the 'traditional' hick. To be a cosmopolitan was to be comfortable in heterogeneous public space.[12] This diversity (and to some extent this elitist attitude) remains a support for cosmopolitanism today.

The notion of cosmopolitanism gains currency from the flourishing of multiculturalism – and the opposition of those who consider themselves multiculturally modern feel to those rooted in monocultural traditions. The latter, say the former, are locals with limited perspective, if not outright racists. It is easier to sneer at the far right, but too much claiming of ethnic solidarity by minorities also falls foul of cosmopolitanism. It is no accident either that the case against Salman Rushdie began to be formulated among diasporic Asians in Britain, or that cosmopolitan theory is notably ambivalent towards them. Integrationist white liberals in the United States are similarly unsure what to make of what some of them see as 'reverse racism' on the part of blacks striving to maintain local communities. Debates over English as a common language reveal related ambivalence towards Hispanics and others.

In the world's global cities, though, and even in a good many of its small towns, certain forms of cosmopolitan diversity appear ubiquitous. Certainly Chinese food is now a global cuisine – both in a generic form that exists especially as a global cuisine and in more 'authentic' regional versions prepared for more cultivated global palates. And one can buy Kentucky Fried Chicken in Beijing. Local taste cultures that were once more closed and insular have indeed opened up. Samosas are now English food just as pizza is American and Indonesian curry is Dutch. Even where the hint of the exotic (and the uniformity of the local) is stronger, one can eat internationally – Mexican food in Norway, Ethiopian in Italy. This is not all 'MacDonaldization' and it is not to be decried in the name of cultural survival. Nonetheless, this tells us little about whether to expect democracy on global scale, successful accommodation of immigrants at home, or respect for human rights across the board. Food, tourism, music, literature and clothes are all easy faces of cosmopolitanism. They are indeed broadening, literally after a fashion, but they are not hard tests for the relationship between local solidarity and international civil society.

The spread of consumerist cosmopolitanism is, rather, a reassurance to liberals that in fact globalization is bringing a happy cultural pluralism. This encourages the idea that a cosmopolitan, post-national politics is a potential path to democracy.[13] It becomes identified with the slogan of a Third Way between unbridled global capitalism and reactionary nationalism. Too many states still wage war or take on projects like ethnic cleansing that an international public might constrain or at least condemn. Transnational flows of people, weapons, drugs and diseases all suggest need for regulation, or at least better design of global institutions, backed up by recognition of human rights. This lends itself to technocratic approaches more than democracy, largely because it neglects the question of solidarity, and it cedes a good deal to neoliberal capitalism. In many versions, the appeal to cosmopolitanism also substitutes ethics for politics, appeals to abstract human rights for efforts to construct a more just and democratic social order.

Cosmopolitanism – though not necessarily cosmopolitan democracy – is now largely the project of capitalism, and it flourishes in the top management of multinational corporations and even more in the consulting firms that serve them. Such cosmopolitanism often joins elites across national borders while ordinary people live in local communities. This is not simply because common folk are less sympathetic to diversity – a self-serving notion of elites. It is also because the class structuring of public life excludes many workers and others. This is not an entirely new story. One of the striking changes of the nineteenth and especially twentieth centuries was a displacement of cosmopolitanism from cities to international travel and mass media. International travel, moreover, meant something different to those who travelled for business or diplomacy and those who served in armies fighting wars to expand or control the cosmopolis. If diplomacy was war by other means, it was also war by other classes who paid less dearly for it.

Yet there are also non-cosmopolitan facets and forms of globalization. There are mass migrations of non-elites whose experiences abroad make them newly conservative about the culture of 'home'. There are projects for alternative modernities – some in the name of revitalizing ancient religions. An effectively democratic future must be built in a world in which these are powerful, not simply a world of diverse individuals.

Capitalism

The current enthusiasm for global citizenship and cosmopolitanism reflects not just a sense of its inherent moral worth but the challenge of an increasingly global capitalism. It is perhaps no accident that the first cited usage under 'cosmopolitan' in the Oxford English Dictionary comes from John Stuart Mill's *Political Economy* in 1848: 'Capital is becoming more and more cosmopolitan' (this is a point made also by Robbins 1993: 182; see also Robbins 2001). Cosmopolitan, after all, means 'belonging to all parts of the world; not restricted to any one country or its inhabitants'. As the quotation from Mill reminds us, the latest .wave of globalization was not required to demonstrate that capital fits this bill. Indeed, as Marx and Engels wrote in the *Communist Manifesto*,

> ... the bourgeoisie has through its exploitation of the world market given a cosmopolitan character to production and consumption in every country. ... All old-established national industries have been destroyed or are daily being destroyed. ... In place of the

old local and national seclusion and self-sufficiency, we have intercourse in every direction, universal inter-dependence of nations. And as in material so also in intellectual production. The intellectual creations of individual nations become common property. National one-sidedness and narrow-mindedness become more and more impossible, and from the numerous national and local literatures, there arises a world literature. (1976: 488)

This is progress, of a sort, but not an altogether happy story. 'The bourgeoisie', Marx and Engels go on,

by the rapid improvement of all instruments of production, by the immensely facilitated means of communication, draws all, even the most barbarian, nations into civilisation. ... It compels all nations, on pain of extinction, to adopt the bourgeois mode of production; it compels them to introduce what it calls civilization into their midst, i.e. to become bourgeois themselves. In one word, it creates a world after its own image. (*Ibid.*)

My purpose here is not to celebrate Marx and Engels for their insight, remarkable as it is. They were, after all, fallible prognosticators. Not much later in the *Communist Manifesto* they reported that modern subjection to capital had already stripped workers of 'every trace of national character' (*ibid.*: 494). The First World War came as a cruel lesson to their followers and nationalism remains an issue today. My point, rather, is to take a little of the shine of novelty off the idea of cosmopolitanism.

Deep inequalities in the political economy of empire and of cap-italism mean that some people laboured and labour to support others in the pursuit of global relations and acquisition. Cosmopolitanism did not and does not in itself speak to these systemic inequalities, any more than did the rights of bourgeois man that Marx criticized in the 1840s. If there is to be a major redistribution of wealth, or a challenge to the way the means of production are controlled in global capitalism, it is not likely to be guided by cosmopolitanism as such. Of course, it may well depend on transnational – even cosmopolitan – solidarities among workers or other groups.

The juxtaposition of empire and capitalism should remind us, moreover, that the rise of the modern world system marked a historical turn against empire. Capitalist globalization has been married to the dominance of nation-states in politics (this is a central point of Wallerstein 1974). Capitalist cosmopolitans indeed have traversed the globe, from early modern merchants to today's World Bank officials and venture capitalists. They forged relations that crossed the borders of nation-states. But they relied on states and a global order of states to maintain property rights and other conditions of production and trade. Their passports bore stamps of many countries, but they were still passports and good cosmopolitans knew which ones got them past inspectors at borders and airports. Not least of all, cosmopolitanism offered only weak defense against reactionary nationalism. This was clearly *declassé* so far as most cosmopolitans were concerned. But Berlin in the 1930s was a very cosmopolitan city. If having cosmopolitan elites were a guarantee of respect for civil or human rights, then Hitler would never have ruled Germany, Chile would have been spared Pinochet, and neither the Guomindang nor the Communists would have come to power in China.

I don't want to paint too strong a picture. Cosmopolitanism is not responsible for empire or capitalism or fascism or communism. Nor does any of this make cosmopolitanism in itself a bad thing. On the contrary, in many ways it is a good and attractive approach to life in a globally interconnected world. The point is that we need to be clear about what work we can reasonably expect cosmopolitanism to do and what is beyond it. In fighting reactionary rightist racism and nationalism, for example, local democracy may be more important than global cosmopolitanism. The two are not contradictory; I hope they can be mutually reinforcing. But they are not the same thing. And in order for them to flourish together it is important not only that local democrats recognize the importance of globalization and the virtues of other cultures, but that cosmopolitans recognize the value of local communities and traditions.[14] The 'catch' to proposing this last recognition is that it flies in the face of capitalist destruction of those communities and violation of those traditions. It is also impeded by the affinity of cosmopolitanism to rationalist liberal individualism.

This has blinded many cosmopolitans to some of the destructions neoliberalism has wrought and the damage it portends to hard-won social achievements. Pierre Bourdieu has rightly called attention to the enormous investment of struggle that has made possible relatively autonomous social fields and at least partial rights of open access to them (see the essays in Bourdieu 1999 and 2001). Such fields are organized largely on national bases, at present, though they include transnational linkages and could become far more global. This might be aided by the 'new internationalism' (especially of intellectuals) that Bourdieu proposes in opposition to the globalization of neoliberal capitalism. The latter imposes a reduction to market forces that undermines both the specific values and autonomy of distinctive fields – including higher education and science – and many rights won from nation-states by workers and others. In this context, defense of existing institutions including parts of national states is not merely reactionary. Yet it is commonly presented this way, and cosmopolitan discourse too easily encourages the equation of the global with the modern and the national or local with the backwardly traditional.

Neoliberalism – the cosmopolitanism of capital – presents one international agenda as simply a force of necessity to which all people, organizations and states have no choice but to adapt. Much of the specific form of integration of the European Union, for example, has been sold as the necessary and indeed all but inevitable response to global competition (Calhoun 2002). This obscures the reality that transnational relations might be built in a variety of ways, and indeed that the shifting forces bringing globalization can also be made the objects of collective choice.

Re-imagining social solidarity

What is needed here is a theory of social solidarity. This would give an account of why mutual obligations should be compelling. But it would also reveal that not all forms of solidarity can with equal ease be made matters of choice. Collective choice about the terms and nature of social institutions and shared life is distinctively a matter of the public sphere. But both public life itself, and society more generally, also depend on systems, categorical identities, and networks of social relations including communities.

Lacking time to develop such a theory in any fullness, let me simply sketch some distinctions among kinds of solidarity. By invoking this term, I mean to recall both Durkheim and the labour movement. That is, I mean to recall both the sociological problem of explaining different sources and forms of social cohesion and the practical problem of developing the kinds of mutual commitments that enable collective action. Solidarity, thus, should not be identified solely with either the unchosen, inherited or systemic forces that bind people to each other or the choice to identify certain others as brothers or sisters. Rather, the question of how much choice different forms of solidarity offer should arise alongside that of how strongly they join people together. Solidarity will always be constraining as well as enabling; it is falsely theorized if we imagine it can offer the latter without the former. Moreover, we should not assume that being bound together is always a matter of harmony and consensus. It is often a matter of argument and struggle; it is organized by competition as much as cooperation; it is marred by ethnic jokes as well as honoured in ritual celebrations. What is key is that people treat the others to whom they are connected as necessary to their lives, not optional.

First, there are systemic or functional forms of integration, such as those of markets. These are powerful, probably the most powerful in the world today. But they present themselves as forces of necessity to which people adapt. One of the challenges of critical theory is to reduce the reification of such forces, but it remains the case that part of their power stems from the fact that they organize social life without requiring collective choices as to their overall form. International civil society can challenge and shape but not replace systemic integration. Much of international civil society actually exists to serve it: NGOs are not all activist or philanthropic organizations; they include professional associations, arbitrators and groups seeking to standardize accountancy rules.

Secondly, there is power, especially as organized in states, but also as deployed inside business corporations – which, as Coase (1937) showed years ago, are not creatures of markets but of hierarchies (Williamson 1975, 1991). It is important to distinguish between corporations and markets, because the former are not simply forces of necessity, dictated by efficiency or the invisible hand. Corporations are institutions that people create and inhabit. They are not an automatic response to the market but a way of organizing work and investment that is shaped by culture and choice as well as power, and potentially a setting for important solidarities that do not reduce to the economy as such. Organizations and movements in international civil society focus largely on trying to influence states and corporations. The influence may come through voting, public opinion or boycotts and other market tactics. We should be clear, though, that the protesters outside WTO meetings do not wield comparable power to the officials of states and corporations represented inside.

Third, there are categorical identities, cultural framings of similarity among people. These include race, ethnicity and nation but also gender and class. Their key feature is to represent people in series, as tokens of a type, as equivalents in respect of some common attribute. International social movements and NGOs rely heavily on categorical identities representing either interests or affinities. Often dispersed members provide financial support to causes with which they identify. Nation has proved the most influential categorical identity in the modern world. Religions often join adherents in a sense of categorical identity. Religion, however, usually involves the combination of categorical identity with embeddedness in specific institutions, practices and relationships.

Social relationships offer a distinct and fourth kind of solidarity. There is no necessary reason for categorical identities to become communities – that is, for similarities to be matched with dense webs of interpersonal relationships. On the contrary, local communities are often precisely the settings in which these categorical identities are combined, in which social relationships establish bridges across race, religion or other lines of categorical difference. In some cases, to be sure, categorical identities are paired with a relatively high density of network relations; they become what Harrison White (1992) calls CATNETS (White's formulation builds on Nadel (1957)). This is part of what gives religious groups force in international civil society. Paying attention to the distinction is important in thinking about community, though, because the word is often used in an ambiguous way. It draws much of its emotional force and attraction from the image of a village or a neighbourhood in which direct ties among people are close. It is used, sometimes ideologically, to refer to nations or other groupings on a very large scale. But the sense of unity that unites millions of people through similarity is importantly distinct from networks of direct interpersonal relationships. Nations are no more communities in this sense than they are families, however often nationalists use either term for its rhetorical value, to promote an illusion of greater closeness than exists.

Fifth, solidarity is created in the production – and continued reproduction and modification – of common culture. This is a matter of shared practices as well as artifacts. In LeRoi Jones's memorable phrase, 'hunting is not those heads on the wall'. Tradition, likewise, is not the result of cultural creativity, it is the process. Living tradition is never simply inheritance, it is also creative reproduction. To be a speaker of a language is to share in this, though of course some are more influential than others. But to be a speaker of a language is also to be joined to other speakers, and not merely by a sense of categorical similarity. Common language is a basis for shared arguments, for identification and even celebration of difference. More generally, the production of shared culture offers people in local settings, and people in subordinate positions, the occasion to resist the domination of authoritative culture from above, whether this is a class-based construction of the nation, or the culture of a dominant ethnic group or mass consumer culture.

Finally, for this list, public discourse itself is potentially a form of solidarity. It is usually treated simply as a source of opinions, and often an occasion for expressing opinions already formed in less public settings. But engaging in common arguments involves forming relationships of a sort. These are marked by the creation or modification of culture as well as the making of more or less rational decisions. That is, people's identities and understandings of the world are changed by participation in public discourse. Commonalities with others are established, not just found, and common interests are explored. But the importance of public discourse is not simply a matter of finding or developing common interests; it is also in and of itself a form of solidarity. The women's movement offers a prominent example; it transformed identities, it did not just express the interests of women whose identities were set in advance. It created both an arena of discourse among women and a stronger voice for women in discourses that were male dominated (even when they were ostensibly gender neutral). The solidarity formed among women had to do with the capacity of this discourse meaningfully to bridge concerns of private life and large-scale institutions and culture. We can also see the converse, the extent to which this gendered production of solidarity is changed as feminist public discourse is replaced by mass-marketing to women and the production of feminism's successor as a gendered consumer identity in which liberation is reduced to freedom to purchase.

In short, there are a variety of ways in which people are joined to each other, within and across the boundaries of states and other polities. Theorists of cosmopolitan democracy are right to stress the multiplicity of connections. But we need to complement the liberal idea of rights with a stronger sense of what binds people to each other. One of the peculiarities of nation-states has been the extent to which they were able to combine elements of each of these different sorts of solidarity. They did not do so perfectly, of course. Markets flowed over their borders from the beginning, and some states were weak containers of either economic organization or power. Not all states had a populace with a strong national identity, or pursued policies able to shape a common identity among citizens. Indeed, those that repressed public discourse suffered a particular liability to fissure along the lines of ethnicity or older national identities weakly amalgamated into the new whole; the Soviet Union is a notable case. Conversely, though, the opportunity to participate in a public sphere and seek to influence the state was an important source of solidarity within it. Coupled with a strong and open public sphere, national levels of solidarity remain still among the most important units within which ordinary people can defend the gains of previous social struggles against reductions to global market forces.

Actually existing international civil society includes some level of each of the different forms of solidarity I listed. In very few cases, however, are these joined strongly to each other at a transnational level. There is community among the expatriate staffs of NGOs; there is public discourse on the Internet. But few of the categorical identities that express people's sense of themselves are matched to strong organizations of either power or community at a transnational level. What this means is that international civil society offers a weak counterweight to systemic integration and power. If hopes for cosmopolitan democracy are to be realized, they depend on developing more social solidarity.

Conclusion

One way of looking at modern history is as a race in which popular forces and solidarities are always running behind. It is a race to achieve social integration, to structure the connections among people and organize the world. Capital is out in front. States come close to catching up and state power is clearly a force to be reckoned with in its own right. Workers and ordinary citizens are always in the position of trying to catch up. As they get organized on local levels, capital and power integrate on larger scales. The integration of nation-states is an ambivalent step in this process. On the one hand, this represents a flow of organizing capacity away from local communities. On the other hand, democracy at a national level constitutes the greatest success that ordinary people have had in catching up to capital and power. Because markets and corporations increasingly transcend states, there is new catching up to do. This is why cosmopolitan democracy is appealing. But it would be a misunderstanding to see nationalism as simply a tradition to overcome, rather than a central moment in the process of expanding scale of social integration, and one with a democratic as well as an authoritarian side.

Even in Europe, it has proved hard to achieve comparable democracy, or public discourse, or labour organization on the scale of the EU than on that of member states. European transnationalism has been driven – and represented publicly – more by the claims of economic necessity (global competition) than by the pursuit of cosmopolitan democracy. The example does not suggest that cosmopolitan democracy should not be pursued – quite the contrary; it only points to how far behind it lags even in a setting where it has considerable advantages. The example of Europe should also remind us that the characteristic oppositions of global to local, universal to particularistic, cosmopolitan to traditional obscure a host of scales of social life between the village and the globe. Not only is nation rendered as local, but the importance of region is obscured. In fact, globalization produces and reproduces regionalization. Much transnationalism – and, indeed, growing cosmopolitanism – is organized on a regional not a global level.

In different ways, both local community and nationalism have developed remarkable capacities for binding people to each other. In the former case this grows out of directly interpersonal relationships; in the latter case it is more a matter of representation. But in both cases this is reproduced in the concrete experiences of everyday life as well as in extraordinary moments. The solidarity of community and nation also offer individuals a sense of location and context vital to a strong sense of self. But community and nation also require commitments and can be limiting. This is one of the reasons for a paradox found especially among second generation immigrants (but not unique to them): the desire to preserve a community one does not wish to be bound by. The tension is real, and community survives only to the extent that some commitments are binding. Nationalism also makes demands on citizens – not least for military service. Cosmopolitan democracy cannot flourish without a comparable basis in social solidarity. Citizenship must be more than an abstraction; to flourish it must be embedded in the practices of everyday life, of civil society. It must be able to make demands. Transnational solidarity can only be based on community to a small extent – though, in fact, diversity of local communities may predispose people to it. UN peace-keeping missions are only a very distant analogue to national service. But humanitarian missions and volunteer service of various sorts do give people a compelling sense of transnational solidarity. These are woven into everyday life over the long term for only a small minority of people, however. Employment in global NGOs affects more, and employment in global corporations still more. But what form of solidarity they produce remains to be studied.

Feeling at home cannot be enough of a basis for life in modern global society (and in its sense of exclusive localism cannot readily be recovered). Attenuated cosmopolitanism won't ground mutual commitment and responsibility. Some relationship between roots – local or other – and broader relationships and awareness needs to be found to provide the solidarity on which cosmopolitan democracy must depend. The view from the frequent traveler lounges does not provide an adequate sense of how people in very different circumstances can feel, gain voice and realize their individual and collective projects.

Notes

1   This chapter draws on presentations to conferences at the International Studies Association in February 2001; Istanbul Bilgi University in May 2001; the University of North Carolina in March 2001; and at Candido Mendes University in May 2001. I am grateful for comments on all these occasions and especially from Michael Kennedy, Laura MacDonald, Thomas McCarthy, Umut Özkırımlı and Kathryn Sikkink. Parts of the chapter were published in *South Atlantic Quarterly* (101) 4: 869–97, and are reprinted with permission.
2   Held, Archibugi and colleagues conceptualize democratic cosmopolitan politics as a matter of several layers of participation in discourse and decision-making, including especially the strengthening of institutions of global civil society, rather than an international politics dominated by nation-states.

3 Liberalism of course embraces a wide spectrum of views in which emphases may fall more on property rights or more on democracy. So too cosmopolitanism can imply a global view that is liberal not specifically democratic. Archibugi prefers 'cosmopolitics' to 'cosmopolitan' in order to signal just this departure from a more general image of liberal global unity (see Archibugi 2000).

4 It is this last tendency which invites liberal rationalists occasionally to ascribe to communitarians and advocates of local culture complicity in all manner of illiberal political projects, from restrictions on immigration to excessive celebration of ethnic minorities to economic protectionism. I have discussed this critically in Calhoun 1999.

5 This hyperTocquevillianism appears famously in Robert Putnam (2000), but has in fact been central to discussions since at least the 1980s, including prominently Robert Bellah *et al.* (1984). The embrace of a notion of civil society as centrally composed of a 'voluntary sector' complementing a capitalist market economy has of course informed public policy from America's first Bush administration with its 'thousand points of light' forward. Among other features, this approach neglects the notion of a political public sphere as an institutional framework of civil society and grants a high level of autonomy to markets and economic actors; it is notable for the absence of political economy from its theoretical bases and analyses. As one result, it introduces a sharp separation among market, government and voluntary association (non-profit) activity that obscures the question of how social movements may challenge economic institutions, and how the public sphere may mobilize government to shape economic practices.

6 See Stephen Toulmin's analysis (1990) of the seventeenth-century roots of the modern liberal rationalist worldview. As Toulmin notes, the rationalism of Descartes and Newton may be tempered with more attention to sixteenth-century forebears. From Erasmus, Montaigne and others we may garner an alternative but still humane and even humanist approach emphasizing wisdom that included a sense of the limits of rationalism and a more positive grasp of human passions and attachments.

7 Amartya Sen (1999) lays out an account of 'capacities' as an alternative to the discourse of rights. This is also adopted by Martha Nussbaum in her most recent cosmopolitan arguments. While this shifts emphases in some useful ways (notably from 'negative' to 'positive' liberties in Isaiah Berlin's terms), it does not offer a substantially 'thicker' conception of the person or the social nature of human life. Some cosmopolitan theorists, notably David Held, also take care to acknowledge that people inhabit social relations as well as rights and obligations.

8 See, for examples, Habermas's surprisingly sharp-toned response to Charles Taylor's 'The Politics of Recognition', both in Amy Gutman (1994); or Janna Thompson's distorting examination of 'communitarian' arguments (1998).

9 One is reminded of Malaysian Prime Minister Mahathir Mohamad's account of human rights as the new Christianity. It makes Europeans feel entitled, he suggested, to invade countries around the world and try to

subvert their traditional values, convert them, and subjugate them. Mahathir was of course defending an often abusive government as well as local culture, but a deeper question is raised.

10   'Jihad and McWorld operate with equal strength in opposite directions, the one driven by parochial hatreds, the other by universalizing markets, the one re-creating ancient subnational and ethnic borders from within, the other making war on national borders from without. Yet Jihad and McWorld have this in common: they both make war on the sovereign nation-state and thus undermine the nation-state's democratic institutions' (Barber 1995: 6). David Held similarly opposes 'traditional' and 'global' in positioning cosmopolitanism between the two ('Opening Remarks' to the Warwick University Conference on 'The Future of Cosmopolitanism').

11   Arguing against Archibugi's account of the nation-state, Brennan rightly notes the intrinsic importance of imperialism, though he ascribes rather more complete causal power to it than history warrants.

12   Richard Sennett cites (and builds on) a French usage of 1738: 'a cosmopolite ... is a man who moves comfortably in diversity; he is comfortable in situations which have no links or parallels to what is familiar to him' (1977: 17).

13   For a sampling, see Archibugi and Held (1995), Archibugi et al. (1998). David Held's argument (1995) is perhaps the best sustained theoretical account of what such a cosmopolitan politics might look like, and how it might differ from an international politics dominated by nation-states.

14   Local communities are of course not unitary; many are culturally heterogeneous and were so before the latest wave of global migrations. Local traditions, thus, are not simply products of homogenous local communities but include local versions of and resources for good relations across lives of difference. Such local versions of 'actually existing cosmopolitanism' are as important and as integral to local communities and especially cities as enmities, rivalries, and suspicions among groups.

References

Adorno, T.W. (1974) *Minima Moralia*, London: Verso.
Archibugi, D. (1998) 'Principles of Cosmopolitan Democracy', in D. Archibugi, D. Held and M. Köhler (eds), *Re-imagining Political Community: Studies in Cosmopolitan Democracy*, Stanford: Stanford University Press, 198–228.
Archibugi, D. (2000) 'Cosmopolitical Democracy', *New Left Review*, 4, 137–50.
Archibugi, D. and D. Held (eds) (1995) *Cosmopolitan Democracy: an Agenda for a New World Order*, Cambridge: Polity.
Archibugi, D., D. Held and M. Köhler (eds) (1998) *Re-Imagining Political Community: Studies in Cosmopolitan Democracy*, Stanford: Stanford University Press.
Barber, B. (1995) *Jihad vs. McWorld*, New York: Times Books.

Bellah, R., W.M. Sullivan, S.M. Tipton, R. Madler and A. Swidler (1984) *Habits of the Heart: Individualism and Commitment in American Life*, Berkeley: University of California Press.

Bellamy, R. and D. Castiglione (1998) 'Between Cosmopolis and Community', in D. Archibugi, D. Held and M. Köhler (eds), *Re-Imagining Political Community: Studies in Cosmopolitan Democracy*, Stanford: Stanford University Press, 152–78.

Bourdieu, P. (1999) *Acts of Resistance: Against the Tyranny of the Market* (trans. by Richard Nice), New York: New Press.

Bourdieu, P. (2001) *Contre-Feux II*, Paris: Raisons d'Agir.

Brennan, T. (2001) 'Cosmopolitanism and Internationalism', *New Left Review*, 7, 75–85.

Calhoun, C. (1983) 'The Radicalism of Tradition: Community Strength or Venerable Disguise and Borrowed Language?', *American Journal of Sociology*, 88 (5), 886–914.

Calhoun, C. (1993) *Social Theory and the Politics of Identity*, Cambridge, Mass.: Blackwell.

Calhoun, C. (1995) *Critical Social Theory: Culture, History, and the Challenge of Diversity*, Cambridge, Mass.: Blackwell.

Calhoun, C. (1997) *Nationalism*, Buckingham: Open University Press.

Calhoun, C. (1999) 'Nationalism, Political Community, and the Representation of Society: or, Why Feeling at Home is not a Substitute for Public Space', *European Journal of Social Theory*, 2 (2), 217–31.

Calhoun, C. (2002) 'Constitutional Patriotism and the Public Sphere: Interests, Identity and Solidarity in the Integration of Europe', in P. de Greiff and C. Cronin (eds), *Global Ethics and Transnational Politics*, Cambridge, Mass.: MIT Press, 275–312.

Coase, R. (1937) 'The Nature of the Firm', *Economica*, 4, 386–405.

Habermas, J. (1994) 'Struggles for Recognition in the Democratic Constitutional State', in Amy Gutman (ed.), *Multiculturalism: Examining the Politics of Recognition*, Princeton: Princeton University Press, 107–48.

Habermas, J. (2000) *The Inclusion of the Other: Studies in Political Theory* (ed. by C. Cronin and P. De Greill), Cambridge, Mass.: MIT Press.

Held, D. (1995) *Democracy and the Global Order: From the Modern State to Cosmopolitan Governance*, Cambridge: Polity.

Kymlicka, W. (1995) *Multicultural Citizenship*, New York: Oxford University Press.

Marx, K. and F. Engels (1976) *Manifesto of the Communist Party*, in *Collected Works*, London: Lawrence and Wishart, 477–519.

Nadel, S. (1957) *A Theory of Social Structure*, London: Cohen and West.

Putnam, R. (2000) *Bowling Alone*, New York: Simon and Schuster.

Rawls, J. (1993) *Political Liberalism*, New York: Columbia University Press.

Rawls, J. (1999) *A Theory of Justice*, Cambridge, Mass.: Harvard University Press.

Rawls, J. (2001) *The Law of Peoples*, Cambridge, Mass.: Harvard University Press.

Robbins, B. (1993) *Secular Vocations: Intellectuals, Professionalism, Culture*, London: Verso.

Robbins, B. (2001) 'The Village of the Liberal Managerial Class', in V. Dharwadker (ed.), *Cosmopolitan Geographies: New Locations in Literature and Culture*, New York: Routledge, 15–32.

Sen, A. (1999) *Development as Freedom*, New York: Oxford University Press.

Sennett, R. (1977) *The Fall of Public Man*, New York: Knopf.

Taylor, C. (1994) 'The Politics of Recognition', in Amy Gutman (ed.), *Multiculturalism: Examining the Politics of Recognition*, Princeton: Princeton University Press, 25–74.

Taylor, C. (2002) 'Modern Social Imaginaries', *Public Culture*, 14 (1).

Thompson, E.P. (1971) 'The Moral Economy of the English Crowd in the Eighteenth Century', *Past and Present*, 50, 76–136.

Thompson, J. (1998) 'Community Identity and World Citizenship', in D. Archibugi, D. Held and M. Köhler (eds), *Re-imagining Political Community: Studies in Cosmopolitan Democracy*, Cambridge: Polity, 179–97.

Toulmin, S. (1990) *Cosmopolis: The Hidden Agenda of Modernity*, New York: Free Press.

Wallerstein, W. (1974) *The Modern World System*, volume 1, New York: Academic Press.

White, H. (1992) *Identity and Control*, Princeton: Princeton University Press.

Williamson, O.E. (1975) *Markets and Hierarchies*, New York: Free Press.

Williamson, O.E. (1991) 'Introduction', in O.E. Williamson and S.G. Winter (eds), *The Nature of the Firm: Origins, Evolution, and Development*, New York: Oxford University Press, 3–17.

7

Belongings: in between the Indigene and the Diasporic

Nira Yuval-Davis

Nationalist ideologies and practices have sought to appropriate and reconstruct notions of belonging. Various historians and theoreticians of nationalism have shown how nationalist discourses have come to replace other forms of belonging, whether local, religious or associated with specific lines of loyalties to specific political hierarchies. Under hegemonic discourses of nationalist politics of belonging the 'nation-state' has come to be the Andersonian (1983) 'imagined community' in which people, territories and states are constructed as immutably connected and the nation is a 'natural' extension of one's family to which one should be prepared, if necessary, to sacrifice oneself. Or is it?

This chapter aims to deconstruct some contemporary notions of belonging as they relate to ethnic and national processes. In particular, it aims to explore alternative narratives to hegemonic discourses of 'national self-determination' and to suggest a model of belonging that encompasses both identity and citizenship. Its main focus will be the contrasting, multi-layered and paradoxical narratives of the 'authentic indigenes' versus those of the 'diasporic strangers'. While doing so it will also engage in a comparative gaze with an earlier attempt of such deconstruction by the Austrian Marxist theorist Otto Bauer.

Several books and collections of papers have recently been published on the question of belonging (for example Fortier 2000; Geddes and Favell 1999; Lovell 1998). In some way, one could claim that one of the prime concerns of sociological theory since its establishment and hence its writings, has been the differential ways people belong to collectivities and states – as well as the social, economic, and political effects of instances of the displacement of such belongings as a result of industrialization and/or migration. Some basic classical examples are Toennies' distinction between *Gemeinschaft* and *Gesellschaft*, Durkheim's division of mechanical and organic solidarity or Marx's notion of alienation. Anthony Giddens (1991) has argued that during modernity, people's sense of belonging becomes reflexive and recently Manuel Castells (1997) claimed that contemporary society has become the 'network society' in which effective belonging has moved from civil societies of nations and states into reconstructed defensive identity communities.

So, what does it mean to belong? Elspeth Probyn (1966) has emphasized what she calls the affective dimension of belonging – not just that of be-ing but of longing, or yearning, as Avtar Brah (1996) relates to a similar emotion among diasporic people.

The Oxford Dictionary gives three interconnected definitions of the term: (1) to be a member of (club, household, grade, society etc.); (2) be resident in or connected with; (3) be rightly placed or classified (in, under, etc.); fit a specified environment. These definitions, although they do not relate to the deep emotional meaning belonging – and displacement from belonging – can evoke, highlight the scope of belonging as encompassing both formal and informal membership, the spatial dimension that belonging usually is wound in, and its multi-layered characteristic – that we all belong to more than one community or collectivity. Primordial definitions of nations would impose a specific hierarchy among the different layers of belonging. Other discourses on the politics of belonging would not necessarily do so.

My own work on belonging has been developing out of my work on multi-layered identity politics on the one hand and citizenship on the other hand. Part of my frustration with identity politics (Yuval-Davis 1994, 1997a and b) has been a result of its usually not differentiating between elements of identification and elements of participation in the construction of belonging, and I shall come back to this point later. Another reason for my decision to focus on the issue of 'belonging' is that recently there have been attempts to develop alternative narratives of belonging to that of the 'nation-state' around the notions of 'diaspora' and 'diasporic space' (Boyarin

1994; Brah 1996; Gilroy 1997; Raz-Karkotzkin 1994). Alternative narratives of a different kind have been developed around narratives of belonging of indigenous people movements, such as the Zapatistas (Marcos 2001; see also Dickanson 1992; Geschiere and Gugler 1998; Reynolds 1996). The growing voices of these narratives have been part of what Charles Taylor (1994) has called 'the politics of recognition', the normative move to recognize previously excluded or marginalized voices and identities, and to legitimize their 'right of self determination'. This 'politics of recognition' can be part of a wider movement to build a global civil society in which, it is often assumed, all these different claims for self-determination can be accommodated, at least in principle. I find this very problematic, especially when these claims are constructed in ethnic, rather than territorial basis, as is most often the case.

I shall, therefore, explore first some general issues of 'the politics of belonging' and 'self determination' and then move to examine some of the specific issues concerning this that emanate from narratives constructed as 'indigenous' and 'diasporic'.

The politics of belonging

Several issues are of central concern when analysing the notion of belonging. Like other hegemonic constructions, belonging tends to become 'naturalized' and thus invisible in hegemonic formations. It is only when one's safe and stable connection to the collectivity, the homeland, the state, becomes threatened, that it becomes articulated and reflexive. It is then that individual, collective and institutional narratives of belonging become politicized.

This politicization tends to focus, as John Crowley argues, on the 'dirty work of boundary maintenance' (1999: 30). Adrian Favell argues that the 'boundary problem' is archetypal to the politics of belonging (1999: 211). Constructing borders and boundaries that differentiate between those who belong and those who do not, determine and colour the meaning of the particular belonging. All too often people talk about Otherness on the one hand and crossing the borders on the other, without paying attention to the ways these borders and boundaries are actually imagined by people who are positioned in differential ways *vis-à-vis* them. At the same time, so many recent theories of identity emphasize – and often celebrate – the ever changing, fluctuating and contested nature of identities. Such theoretical articulations can sometimes disguise the power dimension that often fixes identities and creates what Amrita Chhachhi calls 'forced identities' and Kubena Mercer in somewhat different conditions calls 'the burden of representation'.

However, it is important to relate the notion of belonging to the differential positionings from which belongings are imagined and narrated, in terms of gender, class, stage in the life cycle, and so on, even in relation to the same community and in relation to the same boundaries and borders. These boundaries and borders can be contested not only between those who are in or out of them, but also as a result of the differential social locations and differential social values of people who see themselves and are seen as belonging to the same collectivity, or even by the same people in different times and situations. The contested and shifting nature of these boundaries and borders may reflect not only dynamic power relations between individuals, collectivities and institutions but also subjective and situational processes.

One of the crucial intervening factors in these dynamics is the fact that people tend to belong – in a differential way and in a differential intensity – to more than one collectivity and polity. Local, ethnic, national, inter- and supranational political communities are just some of the 'imagined communities' (Anderson 1983) with which people may identify, in which they are active, at least to a certain degree and to which they may feel a certain sense of attachment. One level of exploration, following Anne-Marie Fortier (2000), is the ways common histories, experiences and places are created, imagined and sustained in what Vicky Bell (1999) calls 'the performativity of belonging'. Another level, however (although interwoven in the first) is the examination of the hierarchy and dynamics of power that are exercised in between these collectivities and their degree of co-operation or conflict. In other words, the relationship between society and polity is crucial to the understanding of the multi-layered and multiplex constructions of belongings of both individuals and groupings.

Following a terminology first used by Michael Walzer (1994), Crowley argues that the idea of 'belonging' is an attempt to give a 'thicker' account of political and social dynamics of integration to that of citizenship, which he relates to as a formal membership in a nation-state (1999: 22).

In my own work on citizenship (1991, 1997a and b; Yuval-Davis and Werbner 1999) I have followed a wider definition of the concept. Using Marshall's definition of citizenship (1950, 1975) as a 'full membership of the community with rights and responsibilities', I have argued that the concept can be used in relation to other polities than that of the 'nation-state', to the extent that membership in other collectivities endows them with rights and obligations in a similar manner. Historically, citizenship emerged as active participation in political communities that evolved in cities (the Greek Polis) and then developed as legal status in empires (such as the Roman Empire). Jean Cohen (1999) argues that in the nation-state these two elements of citizenship have come together. However, as Yasemin Soysal (1994), David Held (1995) and others have argued, new transnational and supranational forms of citizenship are developing, forms what Bryan Turner has called (1998) Post-Fordist citizenship. As I have argued, human rights international legislation can be seen from such a perspective just as another layer of citizenship. At the same time I have pointed out that in terms of affecting personal lives and constructing rights and obligations, sub-national and cross-national communities can also become bearers of significant citizenships, in specific local, religious and ethnic contexts.

The notion of citizenship, however, needs to be differentiated from that of belonging. If citizenship signifies the participatory dimension of belonging, identification relates to the more emotive dimension of association. Feeling that one is part of a collectivity, a community, a social category, or yearning to be so, is not the same as actually taking part in a political community with all the rights and responsibilities involved. This is a differentiation that identity politics has tended not to make, and thus has tended to neglect the complex and contested relationship between individuals, groupings and collectivities. In identity politics, individuals and collectivities are interchangeable emotionally and thus questions of representation, accountability and governability have tended to be ignored (Bourne 1987; Cain and Yuval-Davis 1990; Yuval-Davis 1994, etc.). One of the normative agenda points of this paper is to call for a politics of belonging that encompasses the politics of citizenship as well as that of identification. I shall come back to this point towards the end of the chapter.

'National self-determination' and narratives of belonging

Since the 1789 French Revolution, the principle of 'national self-determination' has gradually grown in legitimacy and become endorsed by both Right and Left. It received full global acceptability in the post-First World War period (around the divisions of the Austro-Hungarian Empire) and was incorporated into the 1945 United Nations charter that states that 'all peoples have the right to national self determination'. In August 2000 the First International Conference on the Right to Self-Determination and the United Nations took place in Geneva and one of its resolutions was to call for a special UN High Commissioner for Self-Determination. The image constructed by this principle is of an ideal world in which all people constitute part of specific homogenous nations, mutually exclusive, who reside in particular non-overlapping territories, enclosed by 'natural' borders, governed by autonomous nation-states. Such an ideal world has never existed, and the cartography of the world is always going through processes of contestation and challenge, especially in periods in which the global social order is unstable as a result of major wars or the disintegration of empires. As Azmi Bishara claims, self-determination has tautologically to do with determining the self. It assumes the existence of a collective national self and presupposes that this self is represented and articulated by its spokespeople. The political projects of such spokespeople can vary greatly, and as a result so would be the national 'imagined community', its collectivity boundaries and the territorial borders claimed as belonging to the nation. Indeed, this might explain why the UN has never defined precisely what is meant by 'national', by 'self' and by 'determination' (Yiftachel 2001). This created a vague context (conveniently open for shifting political circumstances) in which a variety of political struggles in different parts of the globe could find space and legitimation for themselves.

In the 2000 special UN conference very different claims for 'self-determination' have taken place, ranging from claims for regional independence, such as in the case of Kashmir, to claims of justice and reparations, such as in the case of the African Americans. Various discussions in the conference defined the right for self-determination as a 'natural', or at least a legal, result of other principles and struggles, not necessarily compatible, against discrimination, for a desire to expand democracy, for cultural autonomy and reproduction and for governance rights.

And yet, as Delanty (1995) has commented, the principle of national self-determination has first and foremost created a 'moralization of geography'. This morality can gain a fundamentalist flavour when the holy unity of the trinity of people, state and territory is sanctified in religious, as well as nationalist, discourse. This happens often, but not exclusively, in settler societies, when the claim for self-determination is reinforced by narratives about the 'promised land', 'the chosen people' and 'New Jerusalem' (Stuart and Yuval-Davis 1999). 'How...does space become place?', ask Carter *et al.* in the introduction to their edited volume *Space & Place* (1993: xii). 'By being named:...by embodying the symbols and imaginary of a population. Place is space to which meaning has been ascribed.'

The meaning of the 'homeland' can be inscribed via its physical nature. 'What is a man who has no landscape?' asks Athos in *Fugitive Pieces* (Michaels 1996: 86): 'Nothing but mirrors and tides.' The inscription can also be highly ideological: 'Every true republican has drunk in love of country, that is to say love of law and liberty, along with his mother's milk. This love is his whole existence' said Rousseau (1953 [1772]).

Women/mothers are often constructed as the embodiments of the homeland. In peasant societies, the dependence of the people on the fertility of 'Mother Earth' has no doubt contributed to this close association between collective territory, collective identity and womanhood. Women are associated in the collective imagination with children and therefore with the collective, as well as the familial, future. Women represent the homeland, as well as the home.

A figure of a woman, often a mother, symbolizes in many cultures the spirit of the collectivity, whether it is Mother Russia, Mother Ireland or Mother India. In the symbol of the French Revolution, 'La Patrie' was a figure of a woman giving birth to a baby, and in Cyprus a crying woman refugee on roadside posters was the embodiment of the pain and anger of the Greek Cypriot collectivity after the Turkish invasion.

However, it is not the figures of the women/mothers alone that symbolize the homeland but rather the imaginary social relations networks in which they are embedded. As Doreen Massey stated (quoted by Robins 1993: 325): 'Instead, then, of thinking of places as areas with boundaries around, they can be imagined as articulated movements in networks of social relations and understanding.'

I would disagree with Massey concerning the relative (lack) of importance of borders and boundaries of both places and collectivities (as mentioned above and see Yuval-Davis and Stoetzler, forthcoming). However, she does highlight the crucially important trait of the imagining of places in general and homelands in particular, as embedded in social relations and history around which narratives of belonging are woven. As Michael Billig has elaborated in his book *Banal Nationalism* (1995), discourses of nationalism and belonging do not necessarily need to appear in the form of specific and explicit discourses. On the contrary, their power is in their 'naturalness'. Claiming a flower, a fruit, a bird, let alone specific lands as belonging to a particular national collectivity is one of the most affective as well as effective ways in which belonging are claimed. When governments and regimes are rejected, it is memories of smells and landscapes that are remembered and yearned for by political exiles. And in cases of contestations, such as in Israel/Palestine, the 'Jaffa orange', the prickly pear 'Sabres' as well as territorial 'holy sites' and archeological digs become centre spatial and object-related sites for the contesting politics of belonging to take place.

Otto Bauer and the deconstruction of the 'holy trinity'

Attempts to construct alternative narratives of belonging and of citizenship to those in the conventional national self-determination principle have existed for a long time. Notable among those attempts has been the one by Otto Bauer (2001), the Austro-Hungarian Marxist, who attempted to find a solution to the 'national question' in the Austro-Hungarian Empire. Bauer made general important contributions to the theorization of nationalism and nations, especially in his emphasis on the element of 'common destiny' rather than 'common origin' as a primary element around which narratives of national solidarity tend to be constructed. In addition, Bauer rejected the common approach that saw national boundaries as necessarily constructed around what he called 'the territoriality principle'. He argued that any socio-political historical context which would reproduce the boundaries of the nation as a distinct 'community of communication' would do (he thus defined the Jews in pre-capitalist Europe as a distinct nation, although he claimed that under capitalism the historical conditions that have made them into a separate nation do not exist anymore and therefore were no longer a nationality). Bauer called for self-determination according to what he called 'the personality principle'. Given the large waves of migration that accompany the processes of industrialization and urbanization, Bauer saw continued territorial ethnic segregation within the Empire as impossible. Even within the rural areas, patterns of residence were becoming less and less segregated, the larger the demarcated region. And in the cities there were concentrations of migrant workers from a variety of regions and nationalities. The territorial principle of national autonomy, therefore, claimed Bauer, would of necessity bring an end to instances of national oppression, exclusion and conflict.

Unlike the territorial principle, the personality principle relates to self-governance of people, the members of each national grouping. They would govern only their own members and would be responsible for their schools, libraries, theatres, museums and institutions of popular education, and of providing the nation's members with translators and legal assistance when dealing with the state authorities.

Three important characteristics differentiate between Bauer's proposal and contemporary forms of multiculturalism. First, in order, as Bauer argues, to defend the nations from the state, he recommends that the administration of the state would be recruited from the national bodies. Thus, the nations would have autonomy from the state, but the state would not have autonomy from the nations. Second, each national body would have the right to raise taxes from its members. This would give the nations financial autonomy from the territorial state and would also not put members of different nationalities in direct competition for the financial resources of the state. Bauer admitted that this would not solve the differential class positionings of the members of the different nations, but he did not see any real solution for this except a socialist revolution. Third, each nation would have its own formal register (based on the free choice of the mature members of the society) and the representative bodies of each nation would be democratically elected by the members of these registers. This would make the national/ethnic bodies accountable to their members and not just 'represent' them by self-appointed community leaders and activists.

As mentioned above, Bauer developed his model in relation to the Austro-Hungarian empire, and the personality principle of self-determination aimed at maintaining the territorial borders of the state after the demise of the empire. As we know, this did not happen, and US President Woodrow Wilson oversaw the territorial division of the Empire after the First World War into nation-states in which the holy trinity was supposed to uphold. As we know, until today, when a forceful hegemonic central state power is absent, the inadequacies of the 'territorial principle' to solve the national question in the area are glaring and even flared to attempts of 'ethnic cleansing' in many different locations in the globe.

As Michael Walzer commented in his book *On Toleration* (1995), however, multinational states similar to the kind Bauer recommended (e.g. Lebanon, Cyprus and Former Yugoslavia) tend to be most stable and tolerant under the non-democratic control of Empires. When the polity contains only two national groupings, any major political or demographic change can upset the balance. This raises some fundamental questions, which I shall not even try to answer within the boundaries of this chapter, about the relationship between justice, democracy and conflict resolution. After a period of initial euphoria, the failure or only very partial success of the negotiations to resolve the Israel/Palestine and Northern Ireland situations, or even the process of 'truth and reconciliation' in South Africa might attest to it. However, the alternative, which is foreign occupation in places such as Kosovo and Beirut is not really any more palatable.

Going back to the model suggested by Bauer, it seems to be more democratic than either assimilatory nation-state on the one hand, or multicultural state structures in many Western states on the other. However, it does reify and reproduce the separate boundaries of the different national groupings. Bauer's argument, like that of many multiculturalists would be that there are no just 'humans' – everyone has their own ethnicity, language and culture. He felt it was more democratic to try and cultivate these different cultures as important and positive assets, except in cases where he judged that, with all the help in the world, there would not be suitable conditions to reproduce these distinguished 'communities of communication', such as, he considered, in the case of the Jews.

Another issue is that Bauer related only to ethnic and national minorities that 'belonged' to the Empire, rather than also to outside migrations as is the case in contemporary multicultural societies. As in Norman Tebbit's infamous 'Cricket Test' in the UK (in which he suggested that all those who would cheer the opponent team to the British one in international cricket games are not 'truly' British) the challenge to racial and ethnic minorities is often in terms of the informal emotive dimensions of belonging rather than just the formal ones of citizenship rights. On the contrary, given the strength of supranational polities, the autonomy of states in these respects would often be quite limited. In the contemporary global political economy, belonging is being threatened and reinforced at the same time. The global-market McDonalization effect as well as ease of movement of commodities and people also enable the global distribution of cultural artefacts and personal/communal communications that enable the continuous reproduction of cultures, loyalties and most importantly identities in a way Bauer could not have foreseen. As the barrister Majid Tramboo, the Executive Director of the International Council of Human Rights asserted at the UN conference on self-determination in 2000, the popular political concept of self-determination has developed a division into, what is called in the report, 'two limbs': an external one 'confined to populations of fixed territorial entities' and an internal one 'evolving into what seems to be an articulation of the type of rights most often demanded by national minorities' (http://www.hri.ca/racism/Submitted/Author/self-det.htm).

In an attempt to explore some of the characteristics of the politics of belonging that accompany these contemporary forms of the demand for self-determination, I now explore narratives of belonging of both indigenousness and of diasporism. Unlike in multicultural discourses, and also unlike Bauer's suggested solution to ethnic diversity and governance, neither of these approaches is inherently linked to state governance.

Discourses of indigenousness

To be an indigene, means to 'really' belong to a place, and to have the most 'authentic' claim for rights over it. The discourse of 'indigenousness' has been used by hegemonic majorities who have used it as an exclusionary means to limit immigration, prevent citizenship rights, call for repatriation and, in its most extreme forms, for 'ethnic cleansing'. In such a discourse, the immutable links of people, state and territory is formulated in its most racialized forms.

However, the discourse of 'indigenousness' has also played a central role in the politics of inclusion and recognition, of claiming rights. It is used by movements of the largely excluded, dispossessed and marginalized remnants of the societies that existed before or on the margins of settler and other nation-states (Feldman 2001). They are frequently seen, by themselves and by others, as an 'organic' part of the land and the landscape, and all other inhabitants as part of the 'imposing society' (to use an Australian Aboriginal expression) who dispossessed them. A major focus in their struggles is the call to recognize their 'land rights' and claims of ownership of the lands where they used to reside before the European invasion. As they had often been stateless nomadic populations, they had no official land titles registered to their name, as would be the case under bureaucratic state apparatus. As the land they claim is often now privately as well as state owned, their claims have frequently been faced with fierce resistance by settler societies and states, as well as endorsed by human rights discourse.

One of the questions that arise in the attempts to define who are the indigenous inhabitants of a particular territory concerns the temporal dimension. Although in the narratives of indigenous people's movements 'they have occupied a specific territory from time immemorial' (Abu-Saad and Champagne 2001: 158), usually the crucial date of authenticity is fixed as that of occupation at the time of European colonization. This can prove to be Eurocentric. It constructs the past as if history started when the contact with the Europeans was established, and covers up previous population movements and colonizations (as happened in Algeria, for instance, with the Arab settlement and with Amerindians in empires such as the Aztec and the Maya).

Another question, however, even more central to our discussion here concerns the form of ownership to be claimed by those 'land rights' movements. Should it be given to individual members of the 'first nation', in a way that would not limit their freedom to sell the land? Or should it be transferred collectively, as a Trust (and who should have the decision-making power in these Trusts)? And most importantly – should their rights to the land be exclusive, or could other members of the society (as well as the state itself) continue to have rights on the land as well? What are the political, let alone the economic conclusions of indigenous land claims? Indigenous people often claim a spiritual unity with the land:

> We are the land. More than remembered, the Earth is the mind of the people as we are the mind of the earth ... It is not a means of survival ... It is rather part of our being, dynamic, significant, real.
> (A Laguna author, quoted in Tsosie 2001: 184)

Feldman (2001) would argue that such claims are part of critical transformative pedagogy, a 'strategic essentialism', to use a well-known expression of Gayatri Spivak, which can prepare the ground for an exclusive claim on the land, when enough political power would be accumulated, for self-government as an enclave within the nation-state, if there is not enough power/number of people to claim a full 'take-over' of the state (as happened in Algeria, Zimbabwe and in a somewhat different manner, South Africa). However, there are also arguments (e.g. Reynolds 1996) that the aboriginal perception, for instance, that 'they belong to the land', rather than that the land belongs to them, paves the way for an alternative, non-exclusive, mode of ownership and sovereignty, more similar to the Bauerian 'people's principle' than the 'land principle'.

Such a claim, for an alternative nationalist discourse, has also been argued by Gilroy (1997) and others (e.g. Boyarin 1994; Raz-Karkotzkin 1994), as applying to diasporic discourses.

Diasporism as an alternative discourse of belonging

Gilroy attempted to contrast nationalist sentiments based on 'notions of soil, landscape and rootedness' with the idea of diaspora as 'a more refined and more worldly sense of culture' (1997: 328). Avtar Brah (1996) incorporated into her normative notion of 'diaspora space', not just racialized diasporic minorities but also the hegemonic majority in a decentred and non-privileged positioning. Postmodernist discourses on 'travelling cultures' (Clifford 1992), 'nomadism' (Bradiotti 1991), 'hybridity' (Bhabha 1994) and 'living at the border zones' (Anzaldua 1987), both inspired and echoed these constructions of diasporism.

Unfortunately, some of the critiques of such literature (e.g. Anthias 1998; Helmreich 1992; Ifekwunigwe 1999; Yuval-Davis 1997a and b) pointed out the binary, naturalized and essentialist ideas about kinship, nature and territory, so characteristic of more traditional nationalist rhetoric, that often creep up 'through the back door' in these theorizations. Moreover, diasporic politics often tend to have a very different set of values and political dynamics. Unlike the Simmelian (1950) and Schutzian (1976) constructions of 'the stranger', members of diasporic communities are often engaged in narratives of belonging, or of yearning to belong, not only in relation to the country/society where they live, or even a 'cosmopolitan' boundary-less humanity, but also in relation to their country, nation and/or state of origin. As Sara Ahmed (2000) pointed out, the construction of 'the stranger' is a form of fetishism that is produced in the naming and is devoid of any real human characteristics. It is just a reflection of the gaze of the one who has named her/him as such.

As Robin Cohen (1997) has shown, diasporas are much more heterogeneous than the above theories would have us believe. Moreover, as the NGO document of the 2001 World Conference Against Racism in Durban pointed out, Western people living in the third world are often described as 'ex-patriates' while third-world people living in the West are described as migrants or immigrants. The hegemonic Western gaze prevailed in this, as in so many other instances.

Also, what Gilroy and some others do not take into account is the effects 'diaspora yearning and ambivalence' can have on 'the homeland'. Mechanisms of identity regulation which have symbolic meaning of boundary reproduction to the members of the diaspora in the countries where they live, can have serious effects on the continuation of national and ethnic conflicts in 'the homeland' (Anthias 1998; Yuval-Davis 1997a). Contributing funds to various 'causes' and struggles in the homeland can often be the easiest and least threatening ways to members of the diaspora to express their membership and loyalty to the collectivity. Such acts of symbolic identification, which are part of contemporary identity politics (Safran 1999; see also Yuval-Davis 1997b) can, however, have very radical political and other effects in the 'homeland', a fact that might often be only of marginal interest to the people of the diaspora. As Ben Anderson (1995) has pointed out, diasporic politics is often reckless politics without accountability and without due democratic processes. At the same time, as more and more ethnocracies develop, in Central and Eastern Europe as well as in the third world, laws parallel to the Israeli and German 'laws of return' are being developed, and states are constructed that see as their body of citizens all the members of their ethnic collectivity rather than those who are living in their territory.

In states like Lithuania – but also Ireland – the presidents of the state have been living all or most of their lives, until being called to fill in the post, outside the borders of the state.

As I mentioned earlier in the chapter, the development of transport and communicative technologies in the second half of the twentieth century has produced new possibilities of maintaining contact between diasporas and homelands. The spatial/temporal shortcuts in the communication between diasporas and homelands have intensified the level of information as well as the level of interaction between the two. Maybe even more importantly, they have enabled people in the diaspora who were previously isolated from each other new possibilities of getting together and might have changed discourses of belonging.

Conclusion

Today – and probably always – belonging is multiplex and multilayered, continuous and shifting, dynamic and attached. This is true both in terms of the subjective and in terms of the political. The notion of belonging should be examined not as an abstract notion but as one that is embedded in specific discourses of power, in which gender, class, sexual and racialized social divisions are intermeshed. The task is to explore the extent to which, given this, it is possible to develop a political form of participation in which differential belongings as well as positionings are acknowledged in a non-exclusionary way. The extent to which transnational models of citizenship (Delanty 1995; Held 1995; Soysal 1994) could incorporate the 'thicker' politics of belongings and relate different constructions of national belonging – e.g. multiculturalism (Goldberg 1994; Rex 1996), multinationalism (Bauer 2001 [1924]) and hybridity (Bhabha 1990, 1994) to questions of democracy and governability.

Constructions of nationhood were changing in the late twentieth century. Today, territories continue to carry crucial symbolic and emotional meaning to nationalist discourses but in a world in which technologies of transport and communication as well as free market ideologies rule, it is ethnic boundaries that are playing a more and more central role in nationalist ideologies. The Bauerian 'people's principle' can become regressive defensive identities leading to spatial as well as social segregation rather than the democratic principle of pluralist societies Bauer sought.

References

Abu-Saad, I. and D. Champagne (2001) Guest editorial for the Special Issue on Indigenous Peoples, *Hagar*, 2 (2), 157–63.

Ahmed, S. (2000) *Strange Encounters*, London: Routledge.

Anderson, B. (1983) *Imagined Communities: Reflections on the Origins and Spread of Nationalism*, London: Verso.

Anderson, B. (1995) 'Ice Empire & Ice Hockey: Two Fin-de Siècle Dreams', *New Left Review*, 214, 146–50.

Anthias, F. (1998) 'Evaluating Diaspora: Beyond Ethnicity?', *Sociology*, 3, 557–80.

Anzaldua, G. (1987) *Borderlines/La Frontera*, San Francisco: Spinsters/Aunt Lute Books.

Bauer, O. (2001) [1924] *The National Question and Social Democracy*, Minnesota: University of Minnesota Press.

Bell, V. (1999) 'Performativity and Belonging: An Introduction', *Theory, Culture & Society*, special issue on 'Performativity and Belonging', 16 (2), 1–10.

Bhabha, H. (ed.) (1990) *Nation and Narration*, London: Routledge.

Bhabha, H. (1994) *The Location of Culture*, London: Routledge.

Billig, M. (1995) *Banal Nationalism*, London: Sage.

Bourne, J. (1987) 'Homelands of the Mind: Jewish Feminism and Identity Politics', a *Race and Class* Pamphlet, 11.

Boyarin, J. (1994) 'The Other Within and the Other Without', in L.J. Silberman and R.L. Cohn (eds), *The Other in Jewish Thought and History: Constructions of Jewish Thought and Identity*, New York and London: New York University Press, 424–52.

Bradiotti, R. (1991) *Patterns of Dissonance*, Cambridge: Polity.

Brah, A. (1996) *Cartographies of Diaspora*, London: Routledge.

Cain, H. and N. Yuval-Davis (1990) ' "The Equal Opportunities Community" and the Anti-Racist Struggle', *Critical Social Policy*, 29, 5–26.

Carter, E., J. Donald and J. Squires (eds) (1993) *Space and Place: Theories of Identity and Location*, London: Lawrence and Wishart.

Castells, M. (1997) *The Power of Identity*, Oxford: Blackwell.

Chhachhi, A. (1991) 'Forced identities: the state, communalism, fundamentalism and women in India', in D. Kandiyoti (ed.) *Women, Islam and the State*, Basingstoke: Palgrave Macmillan, 144–75.

Clifford, J. (1992) 'Travelling Culture', in L. Grossberg, C. Nelson and P. Treichler (eds), *Cultural Studies*, New York: Routledge.

Cohen, J.L. (1999) 'Changing Paradigms of Citizenship and the Exclusiveness of the Demos', *International Sociology*, 14 (3), 245–68.

Cohen, R. (1997) *Global Diasporas*, London: UCL Press.

Crowley, J. (1999) 'The politics of belonging: some theoretical considerations', in A. Geddes and A. Favell (eds), *The Politics of Belonging: Migrants and Minorities in Contemporary Europe*, Aldershot: Ashgate, 15–41.

Delanty, G. (1995) *Inventing Europe: Idea, Identity, Reality*, Basingstoke: Macmillan (now Palgrave Macmillan).

Dickanson, O.P. (1992) *Canada's First Nations*, Toronto: McClelland & Stanley.

Favell, A. (1999) 'To belong or not to belong: the postnational question', in A. Geddes and A. Favell (eds), *The Politics of Belonging: Migrants and Minorities in Contemporary Europe*, Aldershot: Ashgate, 209–27.

Feldman, A. (2001) 'Transforming People and Subverting States', *Ethnicities*, 1 (2), 147–78.

Fortier, A. (2000) *Migrant Belongings: Memory, Space, Identities*, Oxford: Berg.

Geddes, A. and A. Favell (eds) (1999) *The Politics of Belonging: Migrants and Minorities in Contemporary Europe*, Aldershot: Ashgate.

Geschiere, P. and J. Gugler (eds) (1998) *The Politics of Primary Patriotism*, special issue of *Africa*, 68 (3).

Giddens, A. (1991) *Modernity and Self Identity: Self and Society in the Late Modern Age*, Cambridge: Polity.

Gilroy, P. (1993) *The Black Atlantic*, London: Verso.

Gilroy, P. (1997) 'Diaspora and the Detours of Identity', in K. Woodward (ed.), *Identity and Difference*, London: Sage, 299–343.

Goldberg, D. (ed.) (1994) *Multiculturalism: A Critical Reader*, Oxford: Blackwell.

Held, D. (1995) *Democracy and the Global Order*, Cambridge: Polity.

Helmreich, S. (1992) 'Kinship, Nation and Paul Gilroy's Concept of Diaspora', *Diaspora*, 2 (2), 243–50.

Ifekwunigwe, J.O. (1999) *Scattered Belongings*, London: Routledge.

Lovell, N. (ed.) (1998) *Locality and Belonging*, London: Routledge.

Marcos, Subcomandante Insurgente (2001) *Our Word is Our Weapon*, London: Serpent's Tail.

Marshall, T.H. (1950) *Citizenship and Social Class*, Cambridge: Cambridge University Press.

Marshall, T.H. (1975) *Social Policy in the Twentieth Century*, London: Hutchinson.

Marshall, T.H. (1981) *The Right to Welfare and Other Essays*, London: Heinemann Educational Books.

Mercer, K. (1990) 'Welcome to the jungle: identity and adversity in postmodern politics', in J. Rutherford (ed.) *'Identity, Community, Culture, Difference*, London: Lawrence and Wishart.

Michaels, A. (1996) *Fugitive Pieces*, London: Bloomsbury.

Neuberger, B. (1986) *National Self-Determination in Post-Colonial Africa*, Boulder, CO: Lynne Rienner.

Probyn, E. (1996) *Outside Belongings*, London: Routledge.

Raz-Karkotzkin, A. (1994) 'Diaspora Within Sovereignty: A Critique of the Negation of Diasporism in the Israeli Culture', *Theory & Critique*, 4 and 5.

Reynolds, H. (1996) *Aboriginal Sovereignty, Three Nations, One Australia?*, Sydney: Allen & Unwin.

Rex, J. (1996) *Ethnic Minorities in the Modern Nation-State: Working Papers in the Theory of Multiculturalism & Political Integration*, Basingstoke: Macmillan (now Palgrave Macmillan).

Robins, K. (1993) 'Prisoners of the city: Whatever could a postmodern city be?', in E. Carter *et al.* (eds), *Space and Place: Theories of Identity and Location*, London: Lawrence and Wishart, 303–31.

Rousseau, J.J. (1953) [1772] *Political Writings* (trans. by F. Watkins), New York: Nelson.

Safran, W. (1999) 'Comparing Diasporas: A Review Essay', *Diaspora*, 8 (3), 255–92.

Schutz, A. (1976) [1944] 'The Stranger: An Essay in Social Psychology', in A. Brodersen (ed.), *Alfred Schutz: Studies in Social Theory; Collected Papers II*, The Hague: Martinus Nijhoff, 91–106.

Simmel, G. (1950) 'The Stranger', in H.K. Wolff (ed.), *The Sociology of George Simmel*, New York: The Free Press of Glencoe, 402–9.

Soysal, Y. (1994) *Limits of Citizenship: Migrants and Postnational Membership in Europe*, Chicago, IL: University of Chicago Press.

Stuart, D. and N. Yuval-Davis (1999), 'Homelands, Landscapes and the Construction of Collectivities: Imaginary Geographies and the Internet', paper presented at a departmental seminar, University of Greenwich, May.

Taylor, C. (1994) 'Examining the Politics of Recognition', in Amy Gutmann (ed.), *Multiculturalism and the Politics of Recognition*, Princeton: Princeton University Press, 25–74.

Tsosie, R. (2001) 'Land, Culture and Community: Envisioning Native American Sovereignty and National Identity in the 21st Century', *Hagar*, 2 (2), 183–200.

Turner, B. (1998) 'National Citizenship and Cosmopolitan Virtue – Some Issues With Globalization', paper presented at a departmental seminar, University of Greenwich, November.

Walzer, M. (1994) 'Comment', in A. Gutman (ed.), *Multiculturalism and the Politics of Recognition*, Princeton: Princeton University Press.

Walzer, M. (1997) *On Toleration*, New Haven: Yale University Press.

Yiftachel, O. (2001) 'Homeland and Nationalism', in A.J. Motyl (ed.), *Encyclopedia of Nationalism*, New York: Academic Press.

Yuval-Davis, N. (1991), 'The Citizenship Debate: Women, the State and Ethnic Processes', *Feminist Review*, 39, 58–68.

Yuval-Davis, N. (1994) 'Women, Ethnicity & Empowerment', *Feminism and Psychology*, special issue on 'Shifting Identities, Shifting Racisms' (ed. by K. Bhavnani and A. Phoenix), 4 (1), 179–98.

Yuval-Davis, N. (1997a) *Gender and Nation*, London: Sage.

Yuval-Davis, N. (1997b) 'National Spaces and Collective Identities: Borders, Boundaries, Citizenship and Gender Relations', Inaugural Lecture, University of Greenwich, London, May 22.

Yuval-Davis, N. (1997c), 'Women, Citizenship and Difference', special issue of *Feminist Review* on 'Women and Citizenship: Pushing the Boundaries', 57, 4–27.

Yuval-Davis, N. (1999) 'The Multi-Layered Citizen at the Age of Glocalization', *International Feminist Journal of Politics*, 1 (1), 119–36.

Yuval-Davis, N. and M. Stoetzler (forthcoming), *Borders, Boundaries, and the Situated Imagination*, contract under negotiation.

Yuval-Davis, N. and P. Werbner (eds) (1999) *Women, Citizenship & Difference*, London: Zed Books.

8
Conclusion: the Futures of Nationalism

Will Kymlicka

We can identify four main questions that underlie many recent debates about nationalism, including the chapters in this volume. I will label these as questions about the *nature* of nationalism, the *value* of nationalism, the *alternatives* to nationalism, and the *global diffusion* of nationalism. In this short conclusion, I will say a few words about each, and about why I think they remain so controversial.

The first question, then, concerns the nature of the 'nation' as it is imagined within nationalist movements and ideologies. In particular, how does nationhood relate to, or differ from, other categories, such as race, ethnicity, religion, cultural lifestyles? To what extent does a sense of common nationality depend on the sharing of these other features of social life, and to what extent do nationalist movements and nationalizing states seek to homogenize its members with respect to these features?

A cursory glance around the world would suggest that nationalist movements differ considerably on this issue. Some are more 'thicker' than others, in the sense of requiring or seeking a much higher degree of racial, religious, ethnic or cultural uniformity. Other nationalisms, by contrast, seem 'thinner', in the sense of allowing and tolerating a high level of diversity within the nation, although virtually all nations at least seek to diffuse a common language and common political values throughout their territory.

In the past, this contrast often was discussed in the terminology of 'civic' versus 'ethnic' nationalism, which was then often rephrased as 'tolerant' versus 'intolerant', or 'inclusive' versus 'exclusionary'. It is now increasingly realized that this is not the most helpful way to identify or explain the differing natures of nationalisms. The problem is not only that all nationalist movements contain a mixture of both 'ethnic' and 'civic' elements, but also that there are many different dimensions of thinness and thickness, and nationalist movements can be thick on some while thin on others. There is not a single comprehensive choice to be made about being civic or ethnic, but rather a hundred ongoing decisions to be made about education, immigration, citizenship, language policies, symbols, settlement decisions, legal structures, and so on. All of these decisions, in all nations, reflect ongoing contests about the necessary forms of commonality and homogeneity, and the desirable forms of tolerance and diversity. Nations can be inclusive/tolerant one day on one issue, but not the next day, or on the next issue. Attempts to categorize groups as either 'civic' or 'ethnic' obscure this more contested and fluid reality.[1]

A more salient question, perhaps, concerns the changing nature of nationalism. More specifically, can we say that there has been a general trend, at least in the West, towards a thinner and more tolerant conception of nationhood? In his chapter, John Hall describes the late nineteenth century and early twentieth century as an age of 'homogenization' in Europe. Can we say that the late twentieth and early twenty-first centuries are an age of increasingly benign and tolerant ideas of nationhood, reconceived as multiracial, multicultural and multi-faith societies?

This is one of the central dividing lines in contemporary debates, including the chapters in this book. For some authors, contemporary conditions in the West of economic prosperity, collective security, and democratic consolidation make it possible, perhaps even inevitable, that conceptions of nationhood will become thinner and more inclusive. Others argue, however, with Nira Yuval-Davis, that nationalisms are becoming 'more and more ethnic', or in any event that exclusionary conceptions of the nation are always present at least latently, below the surface, ready to emerge whenever the economy weakens, or whenever the hegemony of a dominant group starts to be challenged.

This debate is related to the second major question, about the value of nationalism, and more specifically whether nationalism can be reconciled with the basic liberal-democratic values of human freedom, equality and democracy. Those observers who see a stable trend towards more inclusive forms of nationalism are inclined to assume that nationalism can be rendered consistent with liberal-democratic values. Indeed, an entire school of thought recently emerged defending the ideal of a 'liberal nationalism'. According to these writers, nationhood is not only consistent with liberal-democratic values, but in fact provides the best home for them. Liberal democracy, it is said, is only viable, or most viable, within national political units. Nationhood provides the trust, solidarity and mutual understanding needed for liberal-democratic institutions to flourish (see, for example, Tamir 1993; Miller 1995; Canovan 1996).

By contrast, those who assume that ethnic and racial exclusion are always latent within nationalist ideologies dispute the compatibility of nationalism and liberalism. These competing normative assessments of the value of nationalism are complicated by the fact that nationalism has both an internal and external dimension. Nationalisms that are tolerant and open internally might nonetheless seek to oppress or conquer other nations. This raises the question, emphasized by Fred Halliday, about the consistency of nationalism. Nations often demand rights of self-determination or of self-defence which they deny to other nations.

According to Halliday, this is the most important challenge to the normative acceptability of nationalism. Nationalism is not inherently inconsistent with universal values of human rights in the way it organizes its own domestic society, but it is prone to hypocrisy and selectivity in its response to the claims of other nations.[2]

Halliday is surely right, although one could question whether this problem does not arise for all political ideals and movements. White males who demanded political rights in the name of 'democracy' were often reluctant to extend the same rights to women and blacks. Yet this hypocrisy does not undermine the legitimacy of the ideal of democracy: rather, the task is to fight for a more consistent application of the ideal. So too, one might argue, with the ideal of national self-determination.

So people disagree about the likelihood that nationalism can be reformed to make it more consistent with liberal-democratic values, either internally or externally. This then leads to the third question, concerning the alternatives to nationalism. Whether we decide to invest our energies in reforming nationalism in a liberal direction will depend, at least in part, on whether we think there is a viable alternative that avoids some of the problems associated historically with nationalism. This alternative is typically labelled as some sort of 'post-nationalist' or 'cosmopolitan' alternative. Indeed, many commentators believe that we already live in such a post-nationalist world, and that the era of nation-states and nation-building states is over.

This is perhaps the deepest disagreement in the contemporary literature. Many of those who defend the liberal nationalist position do so, not because they are unaware of the difficulties it raises, but simply because they believe there is no serious alternative to nationhood as a way of organizing modern political life. By contrast, those who emphasize the intrinsic dangers of nationalism do so, in part, because they think there is a post-nationalist alternative that we should all be working towards.

What would such a post-nationalist alternative look like? A range of phenomena has been identified by scholars as the harbingers or vanguard of a new post-national order. These include the rise of immigrant transnationalisms, substate regionalisms, international human rights law, international advocacy networks, international regulatory bodies, like the World Bank, and transnational political institutions, like the European Union. All of these are said to challenge traditional ideals or models of sovereign, territorially bounded, unitary, homogenous nation-states, and hence to be pushing us towards a post-national or cosmopolitan reordering of political space.

My own view, however, is that most of these allegedly 'post-national' phenomena presuppose the ongoing existence and vitality of territorially-bounded national political units, and that indeed none of them offers any alternative model of how to organize self-governing political communities or to allocate democratic political authority.[3] Moreover, as John Hutchinson emphasizes in his chapter, there are good reasons to think that the globalization of trade and communications is in fact stimulating nationalisms and contributing to nation-formation.

In any event, even if there is a post-nationalist alternative waiting in the wings, we would still need to ask whether this alternative is preferable to an international order based on liberal nation-states. Or, put another way, for whom would it be preferable to move to a post-national order?

This is the central issue discussed in Craig Calhoun's chapter. He emphasizes the danger that cosmopolitan conceptions of politics are likely to be even more elitist than our current nation-based polities. National political units have the important virtue of tying political elites to the masses. They share a common national language, a common national media, often are educated in a common national educational system, as well as sharing national myths, symbols and narratives. This is very different from pre-national political systems in Europe, in which aristocratic or imperial elites were typically separated from the masses linguistically, culturally and institutionally. One could argue that many proposals for a 'post-national' political reordering of political space run the danger of returning to this pre-national condition in which elites govern in a language, culture and institutions that the masses view as foreign to them. Indeed, I think this helps to explain why the general public has shown little enthusiasm for the construction of post-national political institutions, and why nationalism, which originated as an elite project, is now most firmly defended by the working class. As both Calhoun and Partha Chatterjee argue, empowering people means empowering them within their own communities and traditions.[4]

So we have three interconnected debates about the nature and value of, and alternatives to, nationalism. These are all important debates, and are unlikely to be resolved soon. But one could argue that in the West there is actually relatively little at stake in these debates. After all, whatever one views as the likely future evolution of national identities and national political units in the West, it is unlikely to dramatically affect certain basic features of Western life, such as the rule of law, human rights, democratic elections, economic prosperity and individual freedom. No matter how the balance between national and post-national elements evolves in Western Europe, no matter whether we see the European Union as supplementing or supplanting the nation-state, this is unlikely to affect the overwhelming public support for, and institutional consolidation of, liberal-democratic values.

Yet things are very different in most other parts of the world, where peace, freedom and democracy are far from consolidated. And it is here, in the post-communist and developing worlds, where the nature and evolution of nationalism is a matter of critical importance for people's well-being, and perhaps even a matter of life and death. And this raises our final question, about the global diffusion of nationalism.

As Hutchinson notes, many developing countries are in the first stages of nation-formation, and it is far from clear how these nation-building efforts will fare. It is unlikely to simply follow the Western trajectory, since, as Hall notes in his chapter, the nature of the ethnic differences and ethnic conflicts that need to be managed are very different. Many postcolonial states lack a dominant national group, but instead are composed of many different smaller groups, none of which has a clear majority. Moreover, the tools available to the state today for nation-building are more constrained than those available to the West 150 years ago.[5] Under these circumstances, efforts to simply mimic Western models of nationalizing states are unlikely to succeed, and may instead generate greater instability and violence.

And yet here again we confront the question of the alternatives to the nation-state. If nation-states have indeed been 'black man's burden' (Davidson 1992), it is not as if post-national experiments at constructing pan-African (or pan-Arab or pan-Asian) forms of political community have had much success either. It is here, outside the West, where the future of nationalism will be most contested. And whatever we think about the legitimacy of the liberal-democratic nation-states that have arisen over the last two centuries in the West (and that are perhaps now receding), our ultimate moral judgement of the legitimacy and value of nationalism must surely depend on how well or badly it serves the needs of the billions of people for whom the era of the nation-state is just beginning.

Notes

1   For an attempt to distinguish these different dimensions on which nationalist movements can be liberal or illiberal, see Kymlicka and Opalski 2001: 53–60.
2   Michael Walzer calls this the test of 'the next nation'. See Walzer 1990.
3   Or so I argue in Kymlicka 2002.
4   Whereas Calhoun invokes this argument in defence of nationalism against cosmopolitanism, Chatterjee invokes it in defence of subnational ethnic politics against centralized nationalist politics.

5 David Laitin provides a nice example of how our views regarding state coercion have changed over the centuries:

> It is said that in Spain during the Inquisition gypsies who were found guilty of speaking their own language had their tongues cut out. With policies of this sort, it is not difficult to understand why it was possible, a few centuries later, to legislate Castilian as the sole official language. But when Emperor Haile Selassie of Ethiopia pressed for policies promoting Amharic, infinitely more benign than those of the Inquisition, speakers of Tigray, Oromo, and Somali claimed that their groups were being oppressed, and the international community was outraged. Nation-building policies available to monarchs in the early modern period are not available to leaders of new states today. (Laitin 1992: xi)

References

Canovan, M. (1996) *Nationhood and Political Theory*, Cheltenham: Edward Elgar.

Davidson, B. (1992) *The Black Man's Burden: Africa and the Curse of the Nation-State*, New York: Times Books.

Kymlicka, W. (2002) 'New Forms of Citizenship', in T. Courchene and D. Savoie (eds), *The Art of the State: Governance in a World without Frontiers*, Montreal: Institute for Research on Public Policy.

Kymlicka, W. and M. Opalski (2001) *Can Liberal Pluralism be Exported? Western Political Theory and Ethnic Relations in Eastern Europe*, Oxford: Oxford University Press.

Laitin, D. (1992) *Language Repertoires and State Construction in Africa*, Cambridge: Cambridge University Press.

Miller, D. (1995) *On Nationality*, Oxford: Oxford University Press.

Tamir, Y. (1993) *Liberal Nationalism*, Princeton: Princeton University Press.

Walzer, M. (1990) 'Nation and Universe', *Tanner Lectures on Human Values Vol XI: 1990* (edited by G. Peterson), Salt Lake City: University of Utah Press.

Index

153